RNA Silencing

METHODS IN MOLECULAR BIOLOGY™

John M. Walker, Series Editor

METHODS IN MOLECULAR BIOLOGY™

RNA Silencing

Methods and Protocols

Edited by

Gordon G. Carmichael

Department of Genetics and Developmental Biology,
University of Connecticut Health Center, Farmington, CT

HUMANA PRESS ✸ TOTOWA, NEW JERSEY

© 2005 Humana Press Inc.
999 Riverview Drive, Suite 208
Totowa, New Jersey 07512

www.humanapress.com

This publication is printed on acid-free paper. ∞
ANSI Z39.48-1984 (American Standards Institute)

Permanence of Paper for Printed Library Materials.

Cover illustration: Figure 2 from Chapter 12, "siRNA Delivery In Vivo" by M. Sioud.

Cover design by Patricia F. Cleary.

For additional copies, pricing for bulk purchases, and/or information about other Humana titles, contact Humana at the above address or at any of the following numbers: Tel.: 973-256-1699; Fax: 973-256-8341; E-mail: orders@humanapr.com; or visit our Website: www.humanapress.com

Printed in the United States of America. 10 9 8 7 6 5 4 3 2 1

Library of Congress Cataloging in Publication Data
RNA silencing : methods and protocols / edited by Gordon G. Carmichael.
         p. ; cm. — (Methods in molecular biology ; 309)
     Includes bibliographical references and index.
     ISBN 1-58829-436-6 (alk. paper) eISBN 1-59259-935-4
     1. Small interfering RNA—Laboratory manuals. 2. Gene silencing—Laboratory manuals.
     [DNLM: 1. RNA Interference. 2. Gene Expression Regulation—genetics. 3.
     MicroRNAs—physiology. 4. RNA, Small Interfering—physiology. QU 58.7
     R6273 2005] I. Carmichael, Gordon G. II. Series: Methods in molecular biology
     (Clifton, N.J.) ; v. 309.
     QP623.5.S63R634 2005
     572.8'65—dc22   2004023107

Preface

The past decade has witnessed a true revolution in our understanding of how RNA molecules can act as regulators of gene expression. Central to this new understanding is a growing appreciation that short duplex RNAs can trigger potent and highly specific inhibitory effects, most in the cytoplasm but some also in the nucleus. The underlying regulatory pathway is the ancient and conserved RNA interference (RNAi) pathway utilizing small duplex RNAs. RNAi is carried out in two steps. In the first, long dsRNAs are processed into short 21–23 nt-long effector dsRNAs (siRNAs, small interfering RNAs; 21–25 nt in plants) by the action of the enzyme Dicer. In the second step, the siRNAs are assembled into protein-RNA complexes (RNA-induced silencing complexes, RISC) that direct the specific cleavage of target mRNAs. In these complexes, the short dsRNA duplex is unwound, generating active RISC complexes containing single siRNA strands.

In addition to siRNAs, there is another abundant and important class of small ~22 nucleotide noncoding RNA species in cells, the microRNAs (miRNAs). These RNAs are involved in many processes, including regulation of gene expression during development and defense against viruses. In the past several years there has been great progress in the biochemical and computational identification of novel miRNA species, and there are now hundreds of these small RNAs known. Most current models envision miRNAs acting as negative regulators of translation by acting in a still unknown way on translating ribosomes.

Though it is important to understand and appreciate the concepts and mechanisms of RNAi and miRNA action, it is also critical to be able to apply these emerging technologies in a rational and effective manner. *RNA Silencing: Methods and Protocols* is intended to facilitate the translation of gene silencing concepts into practical applications, and includes a broad but useful set of RNA silencing protocols.

The first three chapters deal with biochemical aspects of the silencing machinery. Siomi and Siomi describe a powerful biochemical method for the purification and identification of RNAi components. Pham and Sontheimer then describe electrophoretic methods to separate distinct silencing complexes. Then Lee and Kim describe the purification of the Drosha protein, which is a central player in the generation of miRNAs.

Chapters 4–6 describe detailed methods for RNA silencing in nonmammalian organisms. Dudley and Goldstein detail methods for RNAi in the nematode *C. elegans*. Clayton et al. then describe how to carry out RNAi-mediated gene

v

silencing in *Trypanosoma brucei*. Finally, Mette et al. describe methods for the study of transcriptional gene silencing in the nuclei of plant cells.

Then, there follow a series of chapters detailing ways to design, prepare, and use RNAs to silence gene expression. These include the use of small RNA duplexes (siRNAs). In Chapter 7, Sioud and Walchli describe methods for the design of siRNA libraries. In Chapter 8 Myers and Ferell offer an incredibly valuable and detailed description of how to design, prepare, and use pools of siRNAs generated by in vitro Dicer cleavage. In Chapter 9 Chae and Hla describe a similar approach. Chapters 10 and 11 (Sui and Shi, and Harper and Davidson) describe methods for the design and use of DNA vectors that express short RNA hairpins that can lead to RNAi silencing in vivo.

RNA interference cannot be effective in cells unless the active RNAs are efficiently introduced into them, or expressed in them. Therefore, Chapters 12–14 describe several current methods and strategies for the in vivo delivery of siRNAs and silencing vectors. Sioud (Chapter 12) describes some approaches for the delivery of siRNAs, and Simeoni et al. (Chapter 13) describe a novel peptide-based strategy for this purpose. In Chapter 14, Li and Rossi describe how lentiviral vectors can be used for the very efficient delivery and expression of active siRNAs into cultured and primary hematopoietic cells.

Chapter 15 describes how RNAi technology can be used to fine-tune the regulation of some gene expression by targeting specific isoforms of a given gene.

Chapters 16–18 describe methods for the study and use of microRNAs. Maniataki et al. describe the isolation of the ribonucleoprotein particles that contain miRNAs, and methods for miRNA cloning. Kiriakidou et al. then describe how to detect miRNAs as well as assays for miRNA function. Sioud and Rosok detail a high-throughput analysis of miRNA expression.

Finally, I think that it is important to keep in mind that the RNAi and miRNA pathways are not the only ways that cells can use small RNAs to regulate gene expression using natural and endogenous mechanisms. Small RNAs are important in an number of biological processes, including pre-mRNA splicing. It has been shown recently that one of the key components of the splicing machinery, U1 snRNA, can be redirected to nuclear RNA targets in a way that results in a striking downregulation of gene expression. In contrast to siRNAs and miRNAs, which act primarily in the cytoplasm of mammalian cells, U1 snRNA acts only in the nucleus. Thus, Chapter 19 details a U1-based approach that should be considered as an alternative to RNAi, and which is associated with distinct strengths and weaknesses.

Gordon G. Carmichael

Contents

Contributors

VINCENT P. ALIBU • *ZMBH, Heidelberg, Germany*

WERNER AUFSATZ • *Gregor Mendel Institute of Molecular Plant Biology, Austrian Academy of Sciences, Vienna, Austria*

GORDON G. CARMICHAEL • *Department of Genetics and Developmental Biology, University of Connecticut Health Center, Farmington, CT*

ALICIA M. CELOTTO • *Department of Genetics and Developmental Biology, University of Connecticut Health Center, Farmington, CT*

SUNG-SUK CHAE • *Department of Cell Biology, Center for Vascular Biology, University of Connecticut Health Center, Farmington, CT*

CHRISTINE E. CLAYTON • *ZMBH, Heidelberg, Germany*

BEVERLY L. DAVIDSON • *Departments of Internal Medicine, Neurology, Physiology, and Biophysics, University of Iowa, Iowa City, IA*

LUCIA DAXINGER • *Gregor Mendel Institute of Molecular Plant Biology, Austrian Academy of Sciences, Vienna, Austria*

GILLES DIVITA • *Department of Molecular Biophysics & Therapeutics, Centre de Recherches de Biochimie Macromoléculaire, CNRS, Montpellier, France*

NATHANIEL R. DUDLEY • *Biology Department, University of North Carolina, Chapel Hill, NC*

ANTONIO M. ESTÉVEZ • *Instituto de Parasitologia y Biomedicina "Lopez-Neyra," CSIC, Armilla, Granada*

JAMES E. FERRELL, JR. • *Departments of Molecular Pharmacology and Biochemistry, Stanford University Medical School, Stanford, CA*

MARK FIELD • *Department of Pathology, University of Cambridge, Cambridge, UK*

BOB GOLDSTEIN • *Biology Department, University of North Carolina, Chapel Hill, NC*

BRENTON R. GRAVELEY • *Department of Genetics and Developmental Biology, University of Connecticut Health Center, Farmington, CT*

SCOTT Q. HARPER • *Department of Internal Medicine, University of Iowa, Iowa City, IA*

CLAUDIA HARTMANN • *ZMBH, Heidelberg, Germany*

FREDERIC HEITZ • *Department of Molecular Biophysics & Therapeutics, Centre de Recherches de Biochimie Macromoléculaire, CNRS, Montpellier, France*

TIMOTHY HLA • *Department of Cell Biology, Center for Vascular Biology, University of Connecticut Health Center, Farmington, CT*

DAVID HORN • *London School of Hygiene & Tropical Medicine, London, UK*

TATSUO KANNO • *Gregor Mendel Institute of Molecular Plant Biology, Austrian Academy of Sciences, Vienna, Austria*

V. NARRY KIM • *School of Biological Science, Seoul National University, Seoul, South Korea*

MARIANTHI KIRIAKIDOU • *Division of Neuropathology, Department of Pathology and Laboratory Medicine, University of Pennsylvania School of Medicine, Philadelphia, PA*

STELLA LAMPRINAKI • *Division of Neuropathology, Department of Pathology and Laboratory Medicine, University of Pennsylvania School of Medicine, Philadelphia, PA*

JOO-WON LEE • *Department of Genetics and Developmental Biology, University of Connecticut Health Center, Farmington, CT*

YOONTAE LEE • *School of Biological Science, Seoul National University, Seoul, South Korea*

MINGJIE LI • *Division of Molecular Biology, Beckman Research Institute of the City of Hope, Duarte, CA*

ALEXANDER LICHTLER • *Department of Genetics and Developmental Biology, University of Connecticut Health Center, Farmington, CT*

PENG LIU • *Department of Genetics and Developmental Biology, University of Connecticut Health Center, Farmington, CT*

ELISAVET MANIATAKI • *Division of Neuropathology, Department of Pathology and Laboratory Medicine, University of Pennsylvania School of Medicine, Philadelphia, PA*

ANTONIUS J. M. MATZKE • *Gregor Mendel Institute of Molecular Plant Biology, Austrian Academy of Sciences, Vienna, Austria*

MARJORI MATZKE • *Gregor Mendel Institute of Molecular Plant Biology, Austrian Academy of Sciences, Vienna, Austria*

M. FLORIAN METTE • *Gregor Mendel Institute of Molecular Plant Biology, Austrian Academy of Sciences, Vienna, Austria*

MAY C. MORRIS • *Department of Molecular Biophysics & Therapeutics, Centre de Recherches de Biochimie Macromoléculaire CNRS, Montpellier, France*

ZISSIMOS MOURELATOS • *Division of Neuropathology, Department of Pathology and Laboratory Medicine, University of Pennsylvania School of Medicine, Philadelphia, PA*

JASON W. MYERS • *Department of Molecular Pharmacology, Stanford University Medical School, Stanford, CA*

PETER NELSON • *Division of Neuropathology, Department of Pathology and Laboratory Medicine, University of Pennsylvania School of Medicine, Philadelphia, PA*

JOHN W. PHAM • *Department of Biochemistry, Molecular Biology, and Cell Biology, Northwestern University, Evanston, IL*

ØYSTEIN RØSOK • *Molecular Medicine Group, Institute for Cancer Research, Oslo, Norway*

JOHN J. ROSSI • *Division of Molecular Biology, Beckman Research Institute of the City of Hope, Duarte, CA*

PHILIPP ROVINA • *Gregor Mendel Institute of Molecular Plant Biology, Austrian Academy of Sciences, Vienna, Austria*

DAVID W. ROWE • *Department of Genetics and Developmental Biology, University of Connecticut Health Center, Farmington, CT*

MARIA DELS ANGELS DE PLANELL SAGUER • *Division of Neuropathology, Department of Pathology and Laboratory Medicine, University of Pennsylvania School of Medicine, Philadelphia, PA*

ANUP SHARMA • *Division of Neuropathology, Department of Pathology and Laboratory Medicine, University of Pennsylvania School of Medicine, Philadelphia, PA*

YANG SHI • *Department of Pathology, Harvard Medical School, Boston, MA*

FREDERICA SIMEONI • *Department of Molecular Biophysics & Therapeutics, Centre de Recherches de Biochimie Macromoléculaire, CNRS, Montpellier, France*

MIKIKO C. SIOMI • *Institute for Genome Research, University of Tokushima, Tokushima, Japan*

HARUHIKO SIOMI • *Institute for Genome Research, University of Tokushima, Tokushima, Japan*

MOULDY SIOUD • *Molecular Medicine Group, Institute for Cancer Research, Oslo, Norway*

ERIK J. SONTHEIMER • *Department of Biochemistry, Molecular Biology and Cell Biology, Northwestern University, Evanston, IL*

MARY LOUISE STOVER • *Department of Genetics and Developmental Biology, University of Connecticut Health Center, Farmington, CT*

GUANGCHAO SUI • *Department of Pathology, Harvard Medical School, Boston, MA*

SÉBASTIEN WÄLCHLI • *Molecular Medicine Group, Institute for Cancer Research, Oslo, Norway*

1

Identification of Components of RNAi Pathways Using the Tandem Affinity Purification Method

Mikiko C. Siomi and Haruhiko Siomi

1. Introduction

RNA interference (RNAi) is rapidly becoming a standard laboratory technique for understanding and regulating the function of specific genes in evolutionarily diverse organisms, including plants, *Caenorhabditis elegans*, *Drosophila*, and mammalian cells *(1–10)*. RNAi is initiated by the conversion of double-stranded RNA (dsRNA) into 21- to 23-nucleotide (nt) fragments of dsRNA by Dicer enzymes. These short, interfering RNAs, or siRNAs as they are known, are incorporated into an RNAi effector complex, the RNA-induced silencing complex (RISC), which uses them as guides to target and destroy complementary messenger RNA (mRNA). Recent findings point to a tight connection between microRNA (miRNA) and RNAi molecular machineries. Recent study also has led to the unmasking of a widespread biological regulatory mechanism involving miRNAs *(11–17)*, and there is a wide agreement that the core RNAi machinery carries out numerous cellular functions by an endogenous pathway important for normal development in many organisms, including gene regulation, virus resistance, and chromatin remodeling *(1–17)*. Although some components of the RNAi cellular machinery have been identified, the overall picture is far from clear. We describe the tandem affinity purification (TAP) method *(18–20)* to isolate protein components of RISC in cultured *Drosophila* Schneider-2 (S2) cells. This purification method has allowed us to identify several RISC components, including Argonaute 2 (AGO2), the *Drosophila* homolog of fragile X mental retardation protein (dFMR1), and a DEAD-box RNA helicase Dmp68 *(21,22)*. Identification of components of RNAi/miRNA pathways by the TAP method eliminates the need for large-scale sample

From: *Methods in Molecular Biology, vol. 309: RNA Silencing: Methods and Protocols*
Edited by: G. G. Carmichael © Humana Press Inc., Totowa, NJ

preparation and overcomes, for example, the potential hazards associated with the use of a radioisotope, thereby placing the method within the scope of the average laboratory.

The expression of TAP-tagged AGO2, which is an essential component for a siRNA-directed RNAi response *(22,23)*, will be used to illustrate the method utilized for isolation of components of RNAi. This method will be useful for studying many aspects of RNAi/miRNA machineries.

2. Materials

1. *Drosophila* Schneider-2 (S2) cells.
2. pRmHa-C-FLAG-His expression vector.
3. pRmHa-TAP expression vector.
4. Schneider's *Drosophila* medium.
5. Spinner bottles.
6. CELLFECTIN.
7. pCoBlast.
8. Blastcidin S Hydrochloride.
9. TAP buffer: 10 mM Tris-HCl (pH 8.0); 150 mM NaCl; 0.5% Triton X-100 (can be replaced by 0.05–0.1% NP-40); protease inhibitors.
10. IgG Sepharose.
11. TEV protease.
12. TEV buffer: 10X Stock buffer and ditheithreitol (DTT) solution are supplied with TEV enzyme when purchased from Invitrogen. Prepare 1X TEV buffer from the stock solution according to the manufacturer's instruction.
13. Calmodulin affinity resin.
14. Calmodulin-binding buffer: 10 mM Tris-HCl (pH 8.0), 150 mM NaCl, 0.5% Triton X-100; 10 mM β-Mercaptoethanol, 1 mM MgOAc, 1 mM Imidazole, 2 mM CaCl$_2$.
15. Calmodulin elution buffer: 10 mM Tris-HCl (pH 8.0), 150 mM NaCl, 0.5% Triton X-100; 10 mM β-Mercaptoethanol; 1 mM MgOAc, 1 mM Imidazole, 2 mM EGTA (the concentration of EGTA can be higher (up to 10 mM) if the elution efficiency is lower than desired).
16. Sodium dodecyl sulfate–polyacrylamide gel electrophoresis (SDS-PAGE) equipment.
17. Coomassie brilliant blue staining solution: 0.22% Coomassie blue R; 50% Methanol; 10% Acetic acid.
18. SYPRO Ruby (Molecular Probes, Inc.). SYPRO Ruby staining is done according to the manufacture's manual (*see* **Subheading 3.4.**).
19. Silver staining kit (Wako, Inc.). Silver staining is done according to the manufacturer's manual (*see* **Subheading 3.4.**).

3. Methods

The methods described below outline (1) the construction of the TAP-tagged expression plasmid, (2) the transfection of the plasmid DNA and the induction of protein expression, (3) the purification of the TAP-tagged protein and its associated proteins from S2 cells, and (4) the identification of associated proteins by mass spectrometric peptide sequencing.

3.1. Expression Plasmid

The pRmHa-C expression system was a kind gift of Dr. Frank Lafont *(24)* and is based on the pRmHa-3 *(25)*. pRmHa-3 contains the *Drosophila* metallothionein promoter, which is activated by Cu^{2+} and a polyadenylation signal. pRmHa-3 was modified by introducing a multiple cloning site (*Eco*RI-*Sac*I-*Nhe*I-*Kpn*I-*Sma*I-*Bam*HI-Flag tag-*Eco*RV-10 × His-*Sal*I), FLAG-tag, and His × 10 tag downstream of the metallothionein promoter to generate pRmHa-C *(25)*. The pBS1479 plasmid was a kind gift of Dr. Bertrand Seraphin *(18,19)* and is a yeast expression vector that contains a TAP tag consisting of two IgG-binding domains of *Staphylococcus aureus* protein A and a calmodulin-binding peptide (CBP) separated by a TEV protease cleavage site. A *Bam*HI-*Hin*dIII fragment containing the TAP tag of pBS1479 was subcloned into pBluScript SK (Stratagene, Inc.), which had been cut with *Bam*HI and *Hin*dIII. A recombinant plasmid was isolated and designated pBlue-TAP. A *Bam*HI-*Sal*I fragment of pBlue-TAP was subcloned into pRmHa-C that had been digested with *Bam*HI and *Sal*I, and the desired plasmid was designated pRmHa-TAP. A cDNA encoding a protein of interest (in this case, AGO2) was subcloned into a multiple cloning site (*Eco*RI-*Sac*I-*Nhe*I-*Kpn*I-*Sma*I-*Bam*HI) of the plasmid (*see* **Note 1**), which gave rise to a C-terminal TAP tag fusion gene (*see* **Fig. 1**).

3.2. Protein Induction

The next steps in this process involve the transfection of the Schneider line 2 (S2) of *Drosophila* with the expression plasmid for TAP-tagged AGO2 followed by induction with $CuSO_4$ to achieve the production of the fusion protein.

3.2.1. S2 Cells

The S2 line is a highly versatile expression system, useful for both the analysis of gene function and generating substantial quantities of expressed protein *(25,26)*. It was originally derived from primary cultures of late-stage *Drosophila melanogaster* male embryos. S2 cells have also been demonstrated to be amenable to RNAi, thereby permitting selective silencing of gene activity.

The S2 cells were grown at room temperature (25°C) under normal atmosphere in Schneider's *Drosophila* medium (Gibco, Inc.) supplemented with 10% fetal calf serum (FCS) at densities between 5×10^5 and 5×10^6 cells/mL. Mass cultures were grown in 1-L spinner bottles in 400 mL of medium.

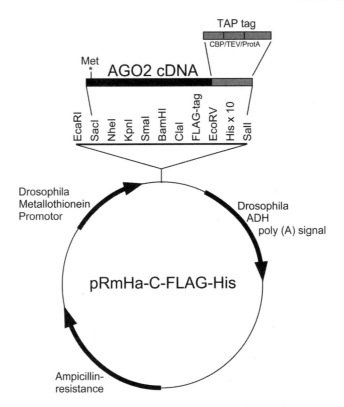

Fig. 1. Schematic drawing of pRmHa-AGO2-TAP. The TAP is tagged at the C-terminus of AGO2. CBP: calmodulin-binding peptide; TEV: TEV protease cleavage site; ProtA: two IgG binding domains of *Staphylococcus aureus* protein A.

3.2.2. Transfection

Transfection was performed using CELLFECTIN (Life Technologies, Inc.) as follows:

1. Put 250 μL of FCS-free medium in each of two 1.7-mL Eppendorf tubes.

    Tube 1: Add 2–6 μg of plasmid DNA into the tube and mix well by pipetting. If wishing to make a stable line, add 0.1–0.3 μg of a plasmid containing cDNA of a blasticidin-resistant gene (pCoBlast; Invitrogen, Inc.) additionally.

    Tube 2: Add 8 μL of CELLFECTIN into the tube and mix well by pipetting.

2. Add the well-mixed DNA solution into the CELLFECTIN solution tube and mix well by pipetting.

3. Incubate at room temperature (RT) for 20 min.

4. During the incubation time, wash S2 cells (about 1×10^7 cells) with FCS-free medium one or two times.

5. Resuspend the cells with 0.8 mL of FCS-free medium in a 15-mL conical tube.

6. Add the DNA/CELLFECTIN solution into the cell suspension and mix by gentle tapping.
7. Incubate the mixture for 3 h at 25°C.
8. After 3 h incubation, spin down the cells at 400–700g for 5 min. Discard the supernatant and resuspend the cells with 8 mL of S2 medium containing 10% FCS and incubate at 25°C.
9. For heterologous protein expression, on d 2, one-half of the transfected cells are incubated with $CuSO_4$ at 1 mM, and cells are harvested after 12 h.
10. To generate stably transfected S2 cell lines, on the same day, add blasticidin (Blastcidin S Hydrochloride; Waken, Inc.) at 25 µg/mL to the other half of the transfected cells. After 2–3 wk of selection, resistant clones have grown. Cells are replated in fresh medium every 4 d over the selection period. These clones are kept together and are used as polyclonal cell lines. Note that it is difficult to obtain monoclonal S2 cell lines mainly because S2 cells grow poorly at low densities. Frozen stocks are kept in 80% FCS, 10% dimethyl sulfoxide at –80°C.

3.3. TAP Purification

Protein purification by the TAP method is performed as follows.

1. Harvest S2 cells expressing the TAP-tagged protein of interest.
2. Wash the cells twice with cold PBS.
3. Resuspend the cells in cold TAP buffer. One milliliter of TAP buffer is needed per 10^7–10^8 cells.
4. Pass the cell suspension through a 25-gauge needle attached to a syringe five times. During this step, the sample should be kept on ice. If cells are not disrupted enough at this point, pass the sample several times through a 30-gauge needle after the 25-gauge needle passage to ensure cell lysis.
5. Transfer the sample to 1.7-mL microtubes and spin at 16,000g for 1–2 min at 4°C.
6. Take the supernatant and transfer it to new microtubes. This is the cytoplasmic lysate used for further purification steps. If you wish to obtain the whole-cell lysate at this point, the cell suspension at **step 4** should be sonicated on ice and **steps 5** and **6** then followed.
7. The supernatant is now mixed with IgG Sepharose™ 6 Fast Flow beads (Amersham Biosciences, Inc.) in microtubes and rocked for 1–2 h at 4°C. Prior to this step, an aliquot of the IgG beads (approx 30–40 µL) is placed into a 1.7-mL microtube, washed with TAP buffer several times, and kept on ice.
8. Wash the IgG beads with TAP buffer five times (*see* **Note 2**). After the last wash, spin the tubes briefly to collect residual buffers to the bottom, suck them up, and discard.
9. Wash with TEV buffer once. Drain the beads well, as in **step 8**. Add TEV buffer to the beads. The buffer volume used here is about 500 µL per tube.
10. Add TEV (Invitrogen) to the beads. The enzyme volume added is 1/100 vol of the buffer. Rock at 16°C for 2 h.
11. Spin at 16,000g for 1–2 min at 4°C. Transfer the supernatant into new microtubes.

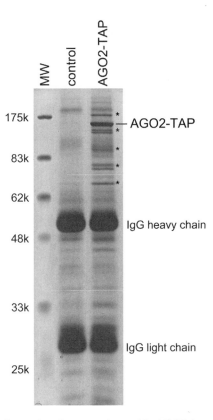

Fig. 2. Purification of proteins that associate with AGO2 by using a C-terminal TAP tag. The protein components in the "IgG bound" obtained from S2 cells expressing the C-terminal TAP-tagged AGO2 (AGO2-TAP) and the parental cells (control) were resolved on SDS-PAGE and visualized by Coomassie blue staining. Several distinct bands (indicated with asterisks) are observed only in the AGO2–TAP lane. The purification was done from 150 mL of S2 cell culture. The left lane represents molecular markers.

12. Add 100 µL of TEV buffer to the beads, and mix well by tapping. Take the supernatant after the spinning at 16,000g for 1–2 min at 4°C and combine it with the sample obtained at **step 11**.
13. Spin again at 16,000g for 1–2 min at 4°C to make sure all of the beads sediment at the bottom of the tubes. Take the supernatant and transfer it to new tubes.
14. Add 3 mL of calmodulin-binding buffer and 2 µL of 1 M CaCl$_2$ per 1 mL of the sample.
15. Mix with calmodulin affinity resin (Stratagene, Inc.) and rock for 1 h at 4°C. An aliquot of the calmodulin beads (approx 30–40 µL) is placed in each microtube and washed with calmodulin-binding buffer several times prior to use.
16. Wash the beads with calmodulin-binding buffer five times. After the last wash, spin briefly to collect the residual buffer to the bottom and drain well.

17. Add EGTA-containing elution buffer to the beads; rock at RT for 15 min or longer to elute the TAP complexes. If the volume is higher than desired at this point, perform a trichloroacetic acid (TCA) precipitation to concentrate the protein sample.
18. Run the sample on a protein gel and stain to visualize the protein bands.

3.4. Identification of TAP Affinity Purified Proteins

The final extracts were visualized on SDS-PAGE gels by Coomassie brilliant blue (ICN Biomedicals, Inc.) staining, SYPRO Ruby staining using a SYPRO Ruby Protein gel stain kit (Molecular Probes, Inc.), or silver staining using SilverQuest Silver Staining kit (Invitrogen) or Silver Stain II kit (Wako, Inc). Stained bands (*see* **Fig. 2**) were excised from the gels, digested with trypsin, and processed for mass spectrometric fingerprinting as described *(21)*. Using mass spectrometry and the complete *Drosophila* genome (the Celera/Berkeley *Drosophila* protein database), it is relatively easy to identify each of the specific bands on the gels.

4. Notes

1. Usually, standard DNA cloning procedures can be used to introduce the C-terminal TAP tag in frame with the coding region of the protein of interest. The upstream sequence of the initiation ATG is often very important for getting sufficient expression levels of the TAP-tagged protein of interest. If the Kozak sequence of the gene of interest is not close to the consensus Kozak sequence of *Drosophila melanogaster*, you might need to modify the sequence upstream (and downstream) of the initiation ATG of the gene.
2. If the background is high when you visualize proteins recovered following TAP purification, you might use a higher salt concentration in the washing buffer. However, of course, a high salt level would disrupt the interaction of TAP-tagged proteins with proteins. Therefore, you might have to selectively adjust TAP purification conditions (salt and/or detergent concentrations, incubation time for binding, and so on).

Acknowledgments

The authors thank Dr. B. Seraphin for providing the TAP plasmids. We also thank Dr. F. Lafont for the pRmHa plasmid. This work was supported by grants from the Ministry of Education, Culture, Sports, Science, and Technology of Japan (MEXT) and the Japan Society for the Promotion of Science (JSPS).

References

1. Fire, A., Xu, S., Montgomery, M. K., Kostas, S. A., Driver, S. E., and Mello, C. C. (1998) Potent and specific genetic interference by double-stranded RNA in *Caenorhabditis elegans*. *Nature* **391**, 806–811.
2. Novina, C. D. and Sharp, P. A. (2004) The RNAi revolution. *Nature* **430**, 161–164.

3. Dykxhoorn, D. M., Novina, C. D., and Sharp, P. A. (2003) Killing the messenger: short RNAs that silence gene expression. *Nat. Rev. Mol. Cell. Biol.* **4,** 457–467.

4. Pederson, T. (2004) RNA interference and mRNA silencing, 2004: how far will they reach? *Mol. Biol. Cell* **15,** 407–410.

5. Denli, A. M. and Hannon, G. J. (2003) RNAi: an ever-growing puzzle. *Trends Biochem. Sci.* **28,** 196–201.

6. Zamore, P. D. (2004) Plant RNAi: how a viral silencing suppressor inactivates siRNA. *Curr. Biol.* **14,** R198–R200.

7. Hutvagner, G. and Zamore, P. D. (2002) RNAi: nature abhors a double-strand. *Curr. Opin. Genet. Dev.* **12,** 225–232.

8. Matzke, A. M. and Matzke, A. J. M. (2004) Planting the seeds of a new paradigm. *PLoS Biol.* **2,** 582–586.

9. Matzke, M. and Matzke, A. J. M. (2003) RNAi extends its reach. *Science* **301,** 1060–1061.

10. Siomi, H., Ishizuka, A. and Siomi, M. C. (2004) RNA interference: a new mechanism by which FMRP acts in the normal brain? What can *Drosophila* teach us? *Ment. Retard. Dev. Disabil. Res. Rev.* **10,** 68–74.

11. He, L. and Hannon, G. J. (2004) MicroRNAs: small RNAs with a big role in gene regulation. *Nat. Rev. Genet.* **5,** 522–531.

12. Bartel, D. P. and Chen, C. Z. (2004) Micromanagers of gene expression: the potentially widespread influence of metazoan microRNAs. *Nat. Rev. Genet.* **5,** 396–400.

13. Bartel, D. P. (2004) MicroRNAs: genomics, biogenesis, mechanism, and function. *Cell* **116,** 281–297.

14. Bartel, B. and Bartel, D. P. (2003) MicroRNAs: at the root of plant development? *Plant Physiol.* **132,** 709–717.

15. Lai, E. C. (2003) microRNAs: runts of the genome assert themselves. *Curr. Biol.* **13,** R925–R936.

16. Aravin, A. A., Lagos-Quintana, M., Yalcin, A., et al. (2003) The small RNA profile during *Drosophila melanogaster* development. *Dev. Cell* **5,** 337–350.

17. Stark, A., Brennecke, J., Russell, R. B., and Cohen, S. M. (2003) Identification of Drosophila MicroRNA Targets. *PLoS Biol.* **1,** E60.

18. Rigaut, G., Shevchenko, A., Rutz, B., Wilm, M., Mann, M., and Seraphin, B. (1999) A generic protein purification method for protein complex characterization and proteome exploration. *Nat. Biotechnol.* **17,** 1030–1032.

19. Puig, O., Caspary, F., Rigaut, G., et al. (2001) The tandem affinity purification (TAP) method: a general procedure of protein complex purification. *Methods* **24,** 218–229.

20. Forler, D., Kocher, T., Rode, M., Gentzel, M., Izaurralde, E., and Wilm, M. (2003) An efficient protein complex purification method for functional proteomics in higher eukaryotes. *Nat. Biotechnol.* **21,** 89–92.

21. Ishizuka, A., Siomi, M. C., and Siomi, H. (2002) A *Drosophila* fragile X protein interacts with components of RNAi and ribosomal proteins. *Genes Dev.* **16,** 2497–2508.

22. Okamura, K., Ishizuka, A., Siomi, H., and Siomi, M. C. (2004) Distinct roles for Argonaute proteins in small RNA-directed RNA cleavage pathways. *Genes Dev.* **18,** 1655–1666.

23. Hammond, S. M., Boettcher, S., Caudy, A. A., Kobayashi, R., and Hannon, G. J. (2001) Argonaute2, a link between genetic and biochemical analysis of RNAi. *Science* **293,** 1146–1150.

24. Lafont, F., Lecat, S., Verkade, P., and Simons, K. (1998) Annexin XIIIb associates with lipid microdomains to function in apical delivery. *J. Cell. Biol.* **142,** 1413–1427.

25. Benting, J., Lecat, S., Zacchetti, D., and Simons, K. (2000) Protein expression in *Drosophila* Schneider cells. *Anal. Biochem.* **278,** 59–68.

26. Towers, P. R. and Sattelle, D. B. (2002) A *Drosophila melanogaster* cell line (S2) facilitates post-genome functional analysis of receptors and ion channels. *Bioessays* **24,** 1066–1073.

2

Separation of *Drosophila* RNA Silencing Complexes by Native Gel Electrophoresis

John W. Pham and Erik J. Sontheimer

1. Introduction

   Large, multicomponent complexes mediate many stages of eukaryotic gene expression, from transcription to translation. Despite their size, these complexes and their precursors can often be resolved and analyzed by native gel electrophoresis. Although other techniques exist for large-scale separations, native gels require very little sample, making them superior for analytical applications. We have developed a native gel system for rapidly and reliably separating complexes that form on short interfering RNAs (siRNAs) *(1)*. This approach has facilitated our analysis of RNA silencing complexes and holds promise for additional mechanistic studies, particularly those that are limited by sample size [such as experiments with mutant lysates *(2)*]. In this chapter, we describe the procedures involved in reproducibly detecting *Drosophila* siRNA/protein complexes and discuss significant issues as well as techniques for optimization.

2. Materials

   1. Oligonucleotide siRNAs (Dharmacon or IDT).
   2. T4 Polynucleotide kinase (New England Biolabs).
   3. [γ-^{32}P]ATP (MP Biomedicals).
   4. 1X Annealing buffer: 30 mM HEPES, pH 7.5, 100 mM potassium acetate, 2 mM magnesium acetate.
   5. 1X Lysis buffer: 1X annealing buffer with 5 mM dithiothreitol (DTT) and 1 mg/mL Pefabloc SC (Boehringer Mannheim).
   6. 20% 19 : 1 Acrylamide : bisacrylamide (Bio-Rad).
   7. Elution buffer: 0.5 M sodium acetate, 50 mM Tris-HCl, pH 7.5, 1 mM EDTA, pH 8.0, 0.1% sodium dodecyl sulfate (SDS).
   8. 40% Acrylamide (Bio-Rad).
   9. 2% Bisacrylamide (Bio-Rad).

From: *Methods in Molecular Biology, vol. 309: RNA Silencing: Methods and Protocols*
Edited by: G. G. Carmichael © Humana Press Inc., Totowa, NJ

10. Ammonium persulfate (Fisher).
11. TEMED (Bio-Rad).
12. 2X TBE: 178 mM Tris base, 178 mM boric acid, 4 mM EDTA, (final pH 8.3).
13. Glycerol (Fisher).
14. Creatine phosphate (Sigma).
15. Creatine kinase (Calbiochem).
16. 100 mM ATP (Amersham).
17. Dithiothreitol (Sigma).
18. *Drosophila* embryo extract.
19. Native gel dye: 2 mM Tris-HCl, pH 7.4, 30 mg/mL Ficoll-400, 0.04% bromophenol blue.
20. Heparin mix: 60 mM potassium phosphate, 3 mM magnesium chloride, 3% PEG$_{8000}$, 8% glycerol, 4 mg/mL heparin.
21. BioMax XAR film (Kodak).
22. RNA guard ribonuclease inhibitor (Amersham).
23. 5X RNA; mix: 125 mM creative phosphate, 5 mM ATP, 25 mM dithiotheritol.

3. Methods

3.1. siRNA Preparation

3.1.1. siRNA Labeling

Synthetic single-stranded siRNAs can be labeled either at their 3′ or 5′ ends, using poly(A) polymerase (Amersham) with [γ-^{32}P]cordycepin 5′-triphosphate (New England Nuclear) or T4 polynucleotide kinase (PNK) with [γ-^{32}P]ATP, respectively, using the manufacturer's instructions (*see* **Note 1**). Isolate the labeled siRNAs in a 15% (19:1 acrylamide:bisacrylamide) denaturing gel run in 1X TBE and recover them from the gel by gentle agitation at room temperature in 400 µL of elution buffer.

Approximately 75–85% of the siRNA will elute from the gel in 2 h. After transferring the aqueous phase to a fresh tube, precipitate the siRNAs by adding 1.1 mL of absolute ethanol, chilling the mixture for at least 20 min at −20°C, and spinning in a centrifuge at 4°C for 15 min at 16,000g. Wash the pelleted siRNAs with 70% ethanol and quantify by Cerenkov counting.

3.1.2. siRNA Annealing

Anneal the siRNAs by combining an equimolar amount of each strand in 1X annealing buffer, heating to 95°C for 2 min and cooling at 37°C for 1 h (*see* **Note 2**). A typical annealing mixture contains at least 0.5 µM RNA in a 10-µL volume. The siRNAs should be stored at −20°C at this concentration and diluted in 1X annealing buffer to approx 7500 cpm/µL just prior to use (*see* **Note 3**).

3.2. Native Gel Preparation

1. To prepare the 4% (40:1 acrylamide:bisacrylamide) native gel, combine 13.8 mL distilled water (dH$_2$O), 20 mL 2X TBE, 4 mL 40% acrylamide, and 2 mL 2% bisacrylamide, and mix well.

2. Add 180 µL 10% ammonium persulfate and 43.3 µL TEMED, and mix. This recipe produces a solution adequate for a 0.8-mm × 17-cm × 23-cm gel. The combs we use produce wells that are approx 0.6 cm wide and are inserted into the gel to a depth of roughly 1 cm.
3. Pour an excess amount of gel solution on top of the comb after inserting it into the gel to facilitate polymerization of the wells. It might be necessary to raise the top end of the gel (approx 25°) to prevent the solution from leaking out of the top. The gel will polymerize in approx 20 min.
4. After the gel polymerizes, assemble the gel apparatus and fill the chamber with 1X TBE.
5. Remove the comb and, using a razor blade, cut off any excess polymerized material at the top of the gel. The gel should be chilled at 4°C until the temperature of the buffer drops to 6°C (*see* **Note 4**).
6. Pre-run the gel for at least 30 min at constant power (10 W, which corresponds to approx 360 V using a gel of the dimensions noted above).

3.3. RNAi Reactions

3.3.1. Lysate Preparation

1. Prepare *Drosophila* embryo lysates as previously described *(3)*. Briefly, collect 0–2 h embryos in a mesh-lined basket and treat them with 50% bleach for 4 min.
2. After rinsing away the bleach with H_2O, blot the embryos on Kimwipes until blotting no longer leaves water on the Kimwipes.
3. Carefully transfer the embryos to a preweighed microcentrifuge tube and weigh.
4. Add 1 µL of 1X lysis buffer per milligram of embryos and homogenize (on ice) using a pestle.
5. Spin the lysate at 16,000g at 4°C for 25 min to pellet the cellular debris.
6. Collect the lysate, avoiding the insoluble material, and spin for 5 min. Repeat as necessary until the lysate is free of insoluble material.
7. Flash-freeze the lysate in liquid N_2 and store at −80°C.

3.3.2. Sample Preparation and Analysis

1. A typical RNAi reaction mixture supplemented with glycerol can be used for native gel analysis. However, the conditions can be modified to optimize complex formation (*see* **Note 5**). We observe consistent, robust complex formation using the following mixture: 0.75 µL of *Drosophila* embryo extract, 1.75 µL of 1X lysis buffer, 1 µL of 5X RNAi mix, 0.375 µL of 0.2 µg/mL creatine kinase, 0.125 µL of RNAguard, 0.5 µL of glycerol, and 0.5 µL of siRNA duplex (*see* **Subheading 3.1.2.**). This mixture can be prepared in larger volumes and separated into 5 µL aliquots. Incubate the 5 µL reaction mixtures at 25°C for 30–60 min (*see* **Note 5**) and transfer the tubes to ice.
2. Add 1 µL of heparin mixture, mix well by pipetting up and down, and spin quickly at 4°C.
3. Immediately load onto a prechilled, prerun native gel (*see* **Section 3.2**).
4. Load the native gel dye into an unused well.
5. Run the gel at constant power (10 W) until the bromophenol blue dye migrates 11 cm from the bottom of the well. The total run time should be approx 1.5 h.

Fig. 1. Native polyacrylamide gel analysis of complexes formed on a radiolabeled siRNA duplex in *Drosophila* embryo lysate. The name of each complex *(1)* is given on the right, (*see* **Note 6**).

6. Transfer the gel to Whatman paper, dry in a gel-drying apparatus, and expose to film (with an intensifying screen) at −80°C or to a phosphorimager screen overnight. A typical autoradiograph of a native gel lane is shown in **Fig. 1**. We initially named the three complexes R1, R2, and R3, in order of decreasing mobility *(1)*. The R2 complex has a somewhat variable mobility, but it always migrates between R1 and R3. R1 and R2 are assembly intermediates, whereas R3 is a large (approx 80S) functional form of RISC *(1)* (*see* **Note 6**).

4. Notes

1. We have used both methods and prefer 5′-end labeling because the reagents are comparatively inexpensive. Furthermore, T4 PNK can be used to label siRNAs that already have a 5′-phosphate—prior treatment with a phosphatase is not required. Although an initial report indicated that siRNA 5′-phosphate groups are rapidly removed in *Drosophila* embryo lysates *(4)*, later work *(5)* indicated that this is not a consistent phenomenon.
2. For most applications, only one strand of the siRNA duplex needs to be labeled. One should be mindful, however, of the asymmetry rules that determine which RNA strand is incorporated into RISC *(5,6)*.

3. It may be necessary in some circumstances to control the exact molar amount of siRNA added to a reaction. In these cases, adjust your dilution protocol accordingly. When this level of control is not required, the dilution protocol described will provide a robust signal in an overnight exposure to film (using an intensifying screen), provided that the label has not decayed much beyond two half-lives.

4. It is critical that the gel be prechilled in order to see all of the complexes (particularly R2, which is less stable than the other complexes). It takes about 2 h to chill 1 L of buffer from room temperature to approx 6°C in a 4°C cold room. This amount of time can be significantly reduced by adding frozen 1X TBE to the buffer.

5. A number of factors contribute to consistent and robust complex formation. These factors include incubation time, total protein concentration in the lysate, gel buffer conditions (temperature), and sample volume. Because the complexes appear to be linked in a pathway *(1)*, longer incubation times will favor formation of higher-order complexes (R3, holo-RISC). Likewise, higher protein concentrations will favor holo-RISC formation (presumably by providing more of a limiting factor, although too much protein disrupts complex formation for unknown reasons). We recommend performing an extract titration and time-course to determine the conditions most suitable for analysis. Once the optimal protein concentrations are determined, it would be wise to normalize all extract preparations to match the one determined to be optimal. As noted above, R2 is more stable at lower temperatures. It is also more consistently observed in 5-µL reaction mixtures compared to 10-µL mixtures. This may be due to a surface tension effect, as the mixture spreads to cover the bottom of the microcentrifuge tube at lower volumes.

6. Zamore and co-workers independently developed an agarose native gel system for similar purposes *(7)* and observed two complexes, A and R, which may resemble R2 and R3, respectively. They also detected a complex (B) that was not apparent in our polyacrylamide gels; conversely, they did not observe a complex that was clearly analogous to R1 *(7)*.

Acknowledgments

We are grateful to Richard Carthew and members of his laboratory for help with *Drosophila* manipulations and for advice and discussions. This work was supported by an NIH Biophysics Training Grant (J. W. Pham), a Burroughs Wellcome Fund New Investigator Award in the Basic Pharmacological Sciences (E. J. Sontheimer), and NIH grant GM068743-01 (R. W. Carthew and E. J. Sontheimer).

References

1. Pham, J. W., Pellino, J. L., Lee, Y. S., Carthew, R. W., and Sontheimer, E. J. (2004) A Dicer-2-dependent 80S complex cleaves targeted mRNAs during RNAi in *Drosophila*. *Cell* **117,** 83–94.

2. Lee, Y. S., Nakahara, K., Pham, J. W., et al. (2004) Distinct roles for *Drosophila* Dicer-1 and Dicer-2 in the siRNA/miRNA silencing pathways. *Cell* **117,** 69–81.

3. Tuschl, T., Zamore, P. D., Lehmann, R., Bartel, D. P., and Sharp, P. A. (1999) Targeted mRNA degradation by double-stranded RNA *in vitro*. *Genes Dev.* **13,** 3191–3197.

4. Nykanen, A., Haley, B., and Zamore, P. D. (2001) ATP requirements and small interfering RNA structure in the RNA interference pathway. *Cell* **107,** 309–321.

5. Schwarz, D. S., Hutvagner, G., Du, T., Xu, Z., Aronin, N., and Zamore, P. D. (2003) Asymmetry in the assembly of the RNAi enzyme complex. *Cell* **115,** 199–208.

6. Khvorova, A., Reynolds, A., and Jayasena, S. D. (2003) Functional siRNAs and miRNAs exhibit strand bias. *Cell* **115,** 209–216.

7. Tomari, Y., Du, T., Haley, B., et al. (2004) RISC assembly defects in the *Drosophila* RNAi mutant *armitage*. *Cell* **116,** 831–841.

3

Preparation and Analysis of Drosha

Yoontae Lee and V. Narry Kim

1. Introduction
1.1. Drosha, the Class II RNase III Protein

Drosha is a member of the ribonuclease (RNase) III family. Like other RNase III proteins, Drosha is a double-stranded RNA (dsRNA)–specific endonuclease that introduces staggered cuts on each strand of the RNA helix *(1,2)*. RNase III proteins are classified based on domain organization. Drosha and its homologs belong to class II, in which each member contains tandem RNase III catalytic motifs and one C-terminal dsRNA-binding domain (dsRBD) *(3,4)*. Drosha also possesses an extended N-terminal domain whose function is currently unknown. Class I proteins are simpler and possess only one RNase III catalytic motif and a dsRNA-binding domain (dsRBD). The Dicer homologs of class III proteins contain a putative helicase domain, a PAZ domain, and a DUF283 domain apart from tandem nuclease domains and a dsRBD. The human genome encodes only two RNase III proteins: Drosha and Dicer. Whereas Dicer homologs are found in a broad range of eukaryotic organisms, Drosha homologs are present only in animals but not in yeast or plants.

Drosha was initially identified in *Drosophila* as an open reading frame adjacent to the *RNase H* gene *(3)*. The human homolog was first described as a protein that interacts with the transcription factor Sp1 in a yeast two-hybrid screening, although the significance of this interaction is still unclear *(5)*. The first functional study was carried out by Stanley Crooke and colleagues, where human Drosha was referred to as "human RNase III" and was shown to be a nuclear protein *(6)*. The role of Drosha was implicated in pre-rRNA processing in this study because downregulation of Drosha using antisense oligonucleotides resulted in moderate accumulation of pre-rRNA processing

From: *Methods in Molecular Biology, vol. 309: RNA Silencing: Methods and Protocols*
Edited by: G. G. Carmichael © Humana Press Inc., Totowa, NJ

intermediates. It remains to be determined, however, whether Drosha is directly involved in pre-rRNA processing.

1.2. Drosha and MicroRNA Processing

A more recent analysis of Drosha revealed that it functions in the initiation step of microRNA processing *(7,8)*. MicroRNAs (miRNAs) are approx 22-nt RNAs that base-pair with mRNAs, leading to the cleavage of the mRNA or to the inhibition of protein synthesis *(9)*. The target genes of miRNA are known to function in cellular differentiation, development, apoptosis, and organogenesis.

In animals, miRNA genes are transcribed to yield long primary transcripts (pri-miRNAs), often as long as several kilobases *(10)*. Pri-miRNAs contain local hairpin structures, and the miRNA sequences are located in the stem of the hairpin. From these long pri-miRNA transcripts, mature miRNAs are generated through two sequential processing events. First, Drosha recognizes the substrate and makes cleavage in the nucleus *(8)*. The resulting hairpin RNA (termed pre-miRNA) is approx 70 nt in length and contains an approx 2-nt 3′ overhang. Pre-miRNAs are then exported out of the nucleus by exportin 5, a member of the Ran-dependent nuclear transport receptor family *(11–13)*. In the cytoplasm, another RNase III protein, Dicer, cleaves the pre-miRNA at approx 22 nt away from the bottom of the stem *(14–18)*.

MicroRNA maturation appears to be a well-coordinated procedure. Drosha generates short 3′ overhangs on its products, which are believed to be the defining feature of RNAs involved in a small RNA pathway. In fact, this short 3′overhang is effectively recognized by exportin 5 and Dicer, thereby facilitating miRNA biogenesis *(8,12)*. Apart from enhancing the overall yield of miRNA biogenesis, Drosha contributes to the accuracy of miRNA maturation. Drosha creates one end of miRNA by precisely cutting at the lower part of the hairpin.

1.3. How to Prepare the Drosha Protein

Partially because of its large size (160 kDa in humans) and low solubility, purification of the full-length recombinant Drosha in *Escherichia coli* has been unsuccessful. As an alternative, the Drosha protein can be immunopurified from mammalian cells that overexpress Drosha (*see* **Subheading 3.1.**). To accomplish this, FLAG epitope-tagged Drosha expression plasmid is transfected into HEK293T cells for the transient expression of Drosha. Total-cell lysate or nuclear extract is incubated with anti-FLAG antibody conjugated to agarose beads for immunoprecipitation. The immunoprecipitated Drosha protein can be characterized in enzyme assay (*see* **Subheading 3.2.**).

It should be noted that the immunoprecipitation procedure described here does not aim for the purification of the Drosha protein to homogeneity. In fact,

Drosha forms a large complex together with essential co-factor(s) (Lee and Kim, unpublished results). Therefore, complete purification of Drosha is not desirable in the preparation of a catalytically active protein.

1.4. How to Design the Substrates for Drosha

The RNA segments to be cleaved in a cropping reaction can be transcribed in vitro. The DNA segment containing miRNA sequences should be amplified from genomic DNA by polymerase chain reaction (PCR) and placed under a T7 promoter (or SP6 promoter). The PCR product can be inserted into a cloning vector containing a T7 promoter (or SP6 promoter). The resulting plasmid should be linearized at a restriction enzyme site located downstream of the pri-miRNA sequences before transcription. Alternatively, the PCR product can be used directly for in vitro transcription. In this case, the forward primer should contain the promoter sequences to drive transcription directly from the PCR product.

The complicated part is in deciding how much of the region around the miRNA should be included to achieve efficient processing. The full-length pri-miRNAs are usually over several kilobases long, which is too long to handle in this kind of experiment. By the rule of thumb, we amplify a genomic segment containing a given miRNA stem loop plus the surrounding sequences approx 100 bp from each side of the stem loop (*see* **Fig. 1**). This should include all of the *cis*-acting element(s) recognized by the Drosha complex.

1.5. Depletion of Drosha

The function of Drosha can be studied in vivo by depleting Drosha. This can be achieved by RNA interference (RNAi), which involves transfection of siRNAS targeting Drosha mRNA. The effect on miRNA biogenesis can be examined by reverse transcription (RT)-PCR, RNase protection assay (RPA), or Northern blot analysis. RT-PCR is the easiest means of determining the pri-miRNA level. The primers are usually designed to bind about 50 nt away from the stem loop. RNase protection assay can be used to detect pri-miRNA, pre-miRNA, and mature miRNA at the same time. However, it is often difficult to optimize the probe and condition. Northern blot analysis is to detect mature miRNA as well as pre-miRNA. Both mature and pre-miRNA diminishes after depletion of Drosha. However, it might take more than 6 d to obtain a dramatic effect on mature miRNA because mature miRNA has a long half-life. For prolonged incubation, it is desirable to transfect twice; split the cells on the third day and repeat transfection on the fourth day.

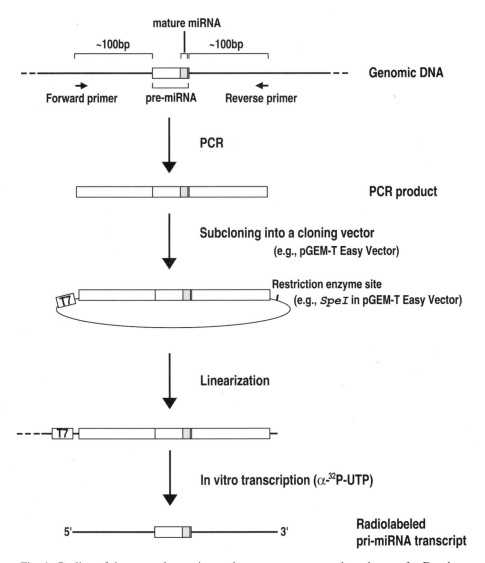

Fig. 1. Outline of the general experimental strategy to prepare the substrate for Drosha.

2. Materials

2.1. Preparation of the Drosha Protein

1. Mammalian expression plasmid containing human Drosha cDNA and a FLAG epitope fused in frame.
2. HEK293T cells, culture medium (Dulbecco's modified Eagle's medium [DMEM] supplemented with 10% fetal bovine serum [FBS]), culture equipments, transfection reagent (calcium phosphate method is the most economic and efficient choice for this experiment).

3. Phosphate-buffered saline (PBS): Dissolve 8 g NaCl, 0.2 g KCl, 1.44 g $Na_2HPO_4 \cdot 2H_2O$, 0.2 g KH_2PO_4 in water, check pH (should be 7.2), set volume to 1 L, and autoclave (*see* **Note 1**).

4. Lysis buffer: 20 mM Tris-HCl (pH 7.8), 100 mM KCl, 0.2 mM EDTA, 20% (v/v) glycerol, and 1 mM PMSF (*see* **Note 2**).

5. Anti-FLAG antibody conjugated agarose (Anti-FLAG M2 Affinity Gel Freezer-Safe; Sigma).

6. Sonicator (Sonics, VC130).

7. Microcentrifuge with cooling.

2.2. Detection o f Pri-miRNA Processing Activity

2.2.1. In Vitro Transcription

1. Template plasmid (1 mg/mL): This plasmid contains the T7 or SP6 promoter followed by pri-miRNA sequences. This plasmid must be linearized at a restriction site located downstream of the pri-miRNA sequences (*see* **Note 3**).

2. 100 mM Stock solutions of ATP, CTP, GTP, and UTP.

3. α-^{32}P-UTP (20 μCi/μL, 800 mCi/mmol).

4. T7 or SP6 RNA polymerase.

5. 10X transcription buffer (use buffer provided with the polymerase).

6. RNase inhibiter (RNasin or equivalent, 40U/μL).

7. RNase-free water.

2.2.2. Purification of Radiolabeled Pri-miRNA

1. Hoeffer gel apparatus (SE600, 18×16 cm) or equivalent, plates, combs (0.75 mm, 10 wells), spacers (0.75 mm), and a power supply.

2. 5X TBE stock solution: 54 g Tris base, 27.5 g boric acid, and 20 mL of 0.5 M EDTA, pH 8.0, dissolved in distilled water to make 1 L.

3. 6% urea–polyacrylamide stock solution: 57 g acrylamide, 3 g *bis*-acrylamide, 100 mL of 5X TBE stock solution (final, 0.5X), and 420.42 g urea (final, 7M) in distilled water to make 1 L (*see* **Note 4**).

4. 20% Ammonium persulfate (APS) solution.

5. N,N,N',N'-Tetramethylethylenediamine (TEMED).

6. Phenol : chloroform : isoamyl alcohol (IAA) (25 : 24 : 1), pH 6.6.

7. 5 M ammonium acetate.

8. 3 M sodium acetate, pH 5.5.

9. Glycogen (5 mg/mL; Ambion).

10. TE buffer: 10 mM Tris-HCl, pH 7.5, and 1 mM EDTA.

11. RNA elution buffer: 0.3 M sodium acetate, pH 5.5, and 2% sodium dodecyl sulfate (SDS).

12. RNA loading buffer: 95% deionized formamide, 0.025% bromophenol blue, 0.025% xylene cyanol, 5 mM EDTA, and 0.025% SDS.

13. G-50 prepacked column (NICK^{TM} column; Amersham).

14. Ethanol: 100% and 75%.

15. Microcentrifuge with cooling.

16. Thermoblock at 37°C/95°C.

17. Kodak X-AR5 autoradiography films.

2.2.3. In Vitro Processing Assay

1. Radiolabeled pri-miRNA transcript.
2. HEK293T whole-cell extract or FLAG–Drosha immunoprecipitate.
3. 10X Reaction buffer (64 mM MgCl$_2$) (*see* **Note 5**).
4. RNase inhibitor (RNasin or equivalent, 40 U/µL).
5. Hoeffer gel apparatus (18 × 16 cm) or equivalent, combs (0.75 mm, 15 wells), spacers (0.75 mm), and a power supply.
6. Urea–polyacrylamide stock solution (12.5%): 118.75 g acrylamide, 6.25 g *bis*-acrylamide, 100 mL of 5X TBE (final, 0.5X), and 420.42 g urea (final, 7 M) in water to make 1 L.
7. APS solution: 20% dissolved in water.
8. TEMED.
9. RNA elution buffer: 0.3 M sodium acetate, pH 5.5, and 2% SDS.
10. Phenol : chloroform : IAA (25 : 24 : 1), pH 6.6.
11. 3 M sodium acetate solution, pH 5.5.
12. RNA loading buffer: 95% deionized formamide, 0.025% bromophenol blue, 0.025% xylene cyanol, 5 mM EDTA, and 0.025% SDS.
13. Thermoblock at 37°C/95°C.
14. Kodak X-AR5 autoradiography films.
15. Autoradiography cassettes with intensifying screens.

2.3. RNAi Against Drosha

1. Drosha-targeting siRNA duplex (20 µM stock) (Samchully).

   Targeted mRNA sequence: 5′ AACGAGUAGGCUUCGUGACUU 3′
   Sense siRNA: 5′ CGAGUAGGCUUCGUGACUUdTdT 3′
   Antisense siRNA: 5′ AAGUCACGAAGCCUACUCGdTdT 3′

2. OLIGOFECTAMINE (Invitrogen) (*see* **Note 6**).
3. Opti-MEM (Gibco-BRL).
4. HeLa cells.

3. Methods

3.1. Preparation of the Drosha Protein (See Note 6)

3.1.1. Preparation of HEK293T Whole-Cell Extract

1. Grow HEK293T cells on 100-mm dish to 95% confluency.
2. Remove the media and rinse the cells with 5 mL of ice-cold PBS.
3. Add 1 mL of ice-cold PBS, collect the cells by pipetting, and transfer to an Eppendorf tube.
4. Centrifuge at 3300 g at 4°C for 5 min.
5. Decant the PBS and resuspend the cell pellet in 500 µL of lysis buffer.
6. Sonicate 10 times (for 5 s each) with short intervals at 30% amplitude.
7. Centrifuge at 13,400 g at 4°C for 15 min.
8. Transfer the supernatant to a fresh tube. This is the HEK293T whole-cell extract.

3.1.2. Immunoprecipitation of the Drosha Protein

1. Prepare HEK293T whole cell extract as described above using HEK293T cells transfected with the Drosha–FLAG expression plasmid.
2. Wash 15 μL of anti-FLAG antibody conjugated beads twice with 1mL of lysis buffer.
3. Add 500 μL of HEK293T whole-cell extract to the washed beads.
4. Rotate the tube at 4°C for 1 h.
5. Centrifuge at 5900 g at 4°C for 2 min.
6. Remove the supernatant, add 1 mL of lysis buffer, and wash the beads by inverting the tube six to seven times.
7. Centrifuge at 5900 g at 4°C for 2 min.
8. Repeat **steps 6** and **7** four times.
9. Drain the beads (15 μL) that hold the immobilized FLAG–Drosha protein. Keep the beads wet and cold on ice until they are used for enzyme assay (*see* **Note 6**).

3.2. Detection of Drosha Processing Activity

3.2.1. In Vitro Transcription (See **Note 6**)

1. Mix the following at room temperature (RT): 1 μL of template DNA (linearized, 1 mg/mL), 2 μL of 10X transcription buffer, 2 μL of NTP mixture (10 m*M* ATP, 10 m*M* GTP, 10 m*M* CTP, 1 m*M* UTP), 0.5 μL of RNase inhibiter (40 U/μL), 1.5 μL of α-^{32}P-UTP (20 μCi/μL, 800 mCi/mmol), 1 μL of RNA polymerase (T7 or SP6), and 12 μL of RNase-free water. Adjust the total volume to 20 μL.
2. Incubate at 37°C for 3 h.

3.2.2. Purification of Radiolabeled Pri-miRNAs

1. Add 220 μL of TE buffer to the above reaction mixture.
2. Add 240 μL of phenol : chloroform : IAA (25 : 24 : 1) and vortex for 30 s. Centrifuge for 5 min at RT and take the upper layer.
3. Add 160 μL of 5 *M* ammonium acetate, 1 μL of glycogen, 1 mL of 100% ethanol. Mix and leave the tube at −80°C for at least 20 min (or 5 min in dry ice–methanol mix, or overnight at −20°C).
4. Centrifuge at full speed (at least 13,400 g) at 4°C for 15 min.
5. Remove the supernatant carefully so as not to disturb the RNA pellet.
6. Wash the pellet with 500 μL of 75% ethanol.
7. Air-dry and resuspend the pellet in 20 μL of RNA loading buffer.
8. Assemble a gel cast.
9. Mix 20 mL of 6% urea–polyacrylamide stock solution, 100 μL of 20% APS, and 20 μL of TEMED. Pour this mixture into the gel cast immediately and insert a comb as quickly as possible because the gel solidifies in a few minutes.
10. Prerun at 350 V for at least 90 min. Running buffer is 0.5X TBE.
11. Load the RNA sample (**step 7**) on 6% urea–polyacrylamide gel and run at 350 V until bromophenol blue reaches the bottom of the gel.
12. Dissemble the gel cast and remove one of the glass plates from the gel. Wrap the gel in Saran film and place an X-ray film on the gel for 30 s to 1 min. Make sure to

mark the position and orientation of the gel on the film so that the gel can be aligned with the film once the film is developed. The radiolabled transcript will appear as a strong band on the developed film.

13. Align the film with the gel and cut out the gel slice containing RNA. Put the gel slice in 350 μL of RNA elution buffer.
14. Incubate overnight at 42°C.
15. Transfer the supernatant (about 300 μL) to a fresh tube.
16. Add 100 μL of RNA elution buffer to the gel slice and vortex. Remove the supernatant and add it to the previous supernatant.
17. Prepare G-50 column by prewashing it with 3 mL of TE buffer.
18. Load the supernatant containing RNA onto the column.
19. Place a fresh tube under the column and elute by adding 400 μL of TE buffer onto the column. This is the first fraction.
20. Repeat **step 19** to collect the second fraction.
21. To each 400-μL fraction, add 1 μL of glycogen, 40 μL of sodium acetate (pH 5.5), and 1 mL of 100% ethanol.
22. Mix and place the tube at −80°C for at least 20 min (or 5 min in dry ice–methanol mix, or overnight at −20°C).
23. Spin at full speed at 4°C for 15 min.
24. Wash the pellet with 500 μL of 75% ethanol.
25. Dry the pellet. Be careful not to overdry the pellet.
26. Count cpm and resuspend in RNase-free water.

3.2.3. In Vitro Processing Assay

1. Mix the following: 3 μL of 10X reaction buffer (final, 6.4 m*M*), 3 μL of radiolabeled pri-microRNA (1×10^4 to 10^5 cpm), 0.75 μL of RNase inhibitor (final, 1U/μL), 8.25 μL of RNase-free water, and 15 μL of HEK293T whole extract or Drosha immunoprecipitate.
2. Incubate at 37°C for 90 min.
3. Add 170 μL of RNA elution buffer to the reaction mixture.
4. Add 200 μL of phenol : chloroform : IAA (25 : 24 : 1) and vortex for 30 s.
5. Centrifuge for 5 min at RT and take the upper layer.
6. Add 20 μL of 3 *M* sodium acetate, 1 μL of glycogen, and 800 μL of 100% ethanol. Mix and leave at −80°C for at least 20 min (or 5 min in dry ice–methanol mix, or overnight at −20°C)
7. Centrifuge at maximum speed at 4°C for 15 min.
8. Remove the supernatant carefully so as not to disturb the RNA pellet. Wash the pellet with 500 μL of 75% ethanol.
9. Air-dry the pellet.
10. Resuspend the pellet in 15 μL of RNA loading buffer.
11. Prepare 12.5% urea–polyacrylamide gel as described above.
12. Heat the RNA sample at 95°C for 5 min. Load 7.5 μL of RNA samples on the gel and run at 350 V until bromophenol blue reaches the bottom of the gel.
13. Expose the gel on an X-ray film overnight at −80°C and develop this film (*see* **Note 7**).

3.3. RNAi Against Drosha (See Note 8)

1. One day before transfection, split HeLa cells in 100-mm plates using 10 mL of medium so that the cells become 30–50% confluent on the day of transfection. DMEM tissue culture medium is supplemented with 10% FBS; no antibiotics should be used.
2. Mix 60 µL of 20 µ*M* siRNA duplex with 1 mL of Opti-MEM. In another tube, mix 60 µL of OLIGOFECTAMINE reagent with 240 µL of Opti-MEM.
3. Incubate at RT for 7–10 min.
4. Combine the solutions and mix gently by pipetting. Do not vortex.
5. Incubate another 20–25 min at RT. The solution turns turbid.
6. Add 760 µL of fresh Opti-MEM to obtain a final volume of 2120 µL.
7. Add the final solution to HeLa cells.
8. After 3 d, extract total RNA or proteins from HeLa cells and analyze the level of Drosha depletion by Northern blotting or Western blotting.

4. Notes

1. It is critical that all of the reagents used are RNase-free and at least of molecular biology grade.
2. Instead of Tris buffer, HEPES and PIPES buffer can also be used as the lysis buffer, as long as the pH ranges between 7.5 and 8.0.
3. To prepare the templates for in vitro transcription, PCR primers are designed to amplify the region covering about 100 bp upstream and downstream from stem-loop pre-miRNA sequences. The PCR products are subcloned into a general cloning vector containing T7 or SP6 promoter (e.g., pGEM-T-easy; Promega). It is also possible to use the PCR product directly as the template in transcription reaction if the T7 promoter sequence is included in the forward PCR primer. For efficient transcription, the promoter sequences should be followed by two consecutive Gs.
4. When preparing urea–polyacrylamide stock solution, dissolve urea in water at 60°C because this reaction is an endothermic reaction.
5. RNase III family proteins require magnesium ion for catalysis. The optimal concentration of magnesium chloride for Drosha is 6.4 m*M* in vitro. The processing efficiency is not significantly reduced down to 3.2 m*M*, but below 3.2 m*M*, the processing efficiency decreases gradually with reduced amount of magnesium in the reaction. The original reaction buffer contained ATP, creatine phosphate, and dithiothreitol (DTT) *(10)*. We recently found that these ingredients are dispensable for pri-miRNA processing.
6. If the Drosha protein is to be used for in vitro processing assay, the pri-miRNA transcript should be prepared (*see* **Subheading 3.2.1.**) before the beginning of immunoprecipitation (*see* **Subheading 3.1.2.**). This is because the Drosha protein immobilized on beads should be prepared freshly for the assay for maximal activity.
7. The product of processing will appear as a band of 60–75 nt depending on the pre-miRNA substrate used in the assay (*see* **Fig. 2**). For instance, pre-miR-30a is

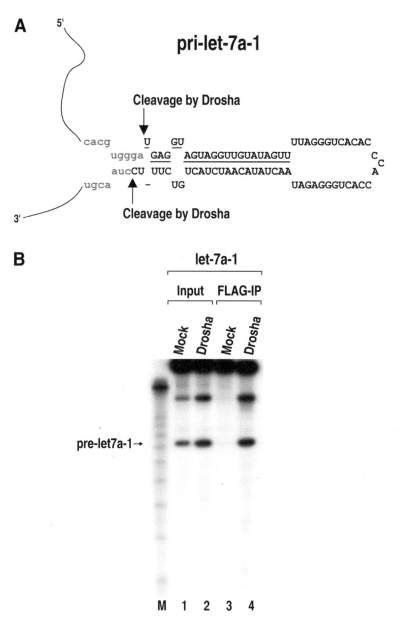

Fig. 2. Determination of the enzymatic activity of Drosha. (**A**) Sequences and structure of a substrate, pri-let-7a-1. The cleavage sites are indicated with arrows. Mature let-7a-1 sequences are underlined. (**B**) A typical result from in vitro processing assay. Pri-let-7a-1 was incubated with HEK293T whole-cell extract (input) or immunoprecitates (FLAG-IP). HEK293T cells had been transfected with either an empty vector, pCK (Mock), or Drosha-FLAG expression plasmid, pCK-Drosha-FLAG (Drosha). Pre-let-7a-1 is produced when pri-miRNA is incubated with Drosha immunoprecipitate or whole-cell extract.

63 nt in length, whereas pre-let-7a- is 72 nt in length. Additional bands may appear, which correspond to the flanking sequences around the stem. Contamination by other nuclease(s) or nonspecific chemical cleavage reaction can also result in unexpected cleavage products.

8. The transfection method described here is essentially identical to that developed by Elbashir et al. *(20)*. The efficiency of siRNA transfection may depend on the cell type but also on the passage number and the confluence of the cells. The time and the manner of formation of siRNA–liposome complexes are also critical. Other transfection reagents are also available. We have successfully used RNAiFect transfection reagent (Qiagen), which involves slightly simpler procedures.

Acknowledgments

This work was supported by a grant (R02-2004-000-10173-0) from the Basic Research Program of the Korea Science & Engineering Foundation and by the BK21 Research Fellowship from the Ministry of Education and Human Resources Development.

References

1. Carmell, M. A. and Hannon, G. J. (2004) RNase III enzymes and the initiation of gene silencing. *Nat. Struct. Mol. Biol.* **11,** 214–218.
2. Zamore, P. D. (2001) Thirty-three years later, a glimpse at the ribonuclease III active site. *Mol. Cell* **8,** 1158–1160.
3. Filippov, V., Solovyev, V., Filippova, M., and Gill, S. S. (2000) A novel type of RNase III family proteins in eukaryotes. *Gene* **245,** 213–221.
4. Fortin, K. R., Nicholson, R. H., and Nicholson, A. W. (2002) Mouse ribonuclease III. cDNA structure, expression analysis, and chromosomal location. *BMC Genomics* **3,** 26.
5. Gunther, M., Laithier, M., and Brison, O. (2000) A set of proteins interacting with transcription factor Sp1 identified in a two-hybrid screening. *Mol. Cell. Biochem.* **210,** 131–142.
6. Wu, H., Xu, H., Miraglia, L. J., and Crooke, S. T. (2000) Human RNase III is a 160-kDa protein involved in preribosomal RNA processing. *J. Biol. Chem.* **275,** 36,957–36,965.
7. Kim, V. N. (2004) MicroRNA precursors in motion: exportin-5 mediates their nuclear export. *Trends Cell. Biol.* **14,** 156–159.
8. Lee, Y., Ahn, C., Han, J., et al. (2003) The nuclear RNase III Drosha initiates microRNA processing. *Nature* **425,** 415–419.
9. Bartel, D. P. (2004) MicroRNAs: genomics, biogenesis, mechanism, and function. *Cell* **116,** 281–297.
10. Lee, Y., Jeon, K., Lee, J. T., Kim, S., and Kim, V. N. (2002) MicroRNA maturation: stepwise processing and subcellular localization. *EMBO J.* **21,** 4663–4670.
11. Bohnsack, M. T., Czaplinski, K., and Gorlich, D. (2004) Exportin 5 is a RanGTP-dependent dsRNA-binding protein that mediates nuclear export of pre-miRNAs. *RNA* **10,** 185–191.

12. Lund, E., Guttinger, S., Calado, A., Dahlberg, J. E., and Kutay, U. (2004) Nuclear export of microRNA precursors. *Science* **303,** 95–98.

13. Yi, R., Qin, Y., Macara, I. G., and Cullen, B. R. (2003) Exportin-5 mediates the nuclear export of pre-microRNAs and short hairpin RNAs. *Genes Dev.* **17,** 3011–3016.

14. Bernstein, E., Caudy, A. A., Hammond, S. M., and Hannon, G. J. (2001) Role for a bidentate ribonuclease in the initiation step of RNA interference. *Nature* **409,** 363–366.

15. Grishok, A., Pasquinelli, A. E., Conte, D., et al. (2001) Genes and mechanisms related to RNA interference regulate expression of the small temporal RNAs that control *C. elegans* developmental timing. *Cell* **106,** 23–34.

16. Hutvagner, G., McLachlan, J., Pasquinelli, A. E., Balint, E., Tuschl, T., and Zamore, P. D. (2001) A cellular function for the RNA-interference enzyme Dicer in the maturation of the let-7 small temporal RNA. *Science* **293,** 834–838.

17. Ketting, R. F., Fischer, S. E., Bernstein, E., Sijen, T., Hannon, G. J., and Plasterk, R. H. (2001) Dicer functions in RNA interference and in synthesis of small RNA involved in developmental timing in *C. elegans. Genes Dev.* **15,** 2654–2659.

18. Knight, S. W. and Bass, B. L. (2001) A role for the RNase III enzyme DCR-1 in RNA interference and germ line development in *Caenorhabditis elegans. Science* **293,** 2269–2271.

19. Chen, C. Z., Li, L., Lodish, H. F., and Bartel, D. P. (2004) MicroRNAs modulate hematopoietic lineage differentiation. *Science* **303,** 83–86.

20. Elbashir, S. M., Harborth, J., Lendeckel, W., Yalcin, A., Weber, K., and Tuschl, T. (2001) Duplexes of 21-nucleotide RNAs mediate RNA interference in cultured mammalian cells. *Nature* **411,** 494–498.

4

RNA Interference in *Caenorhabditis elegans*

Nathaniel R. Dudley and Bob Goldstein

1. Introduction

The introduction of double-stranded RNA (dsRNA) into *Caenorhabditis elegans* hermaphrodites results in the rapid and sequence-specific degradation of endogenous mRNAs *(1,2)*. This RNA-mediated interference, or RNAi, effectively shuts down expression of the target gene and can phenocopy loss-of-function mutations. RNAi is also remarkably potent, requiring only substoichiometric amounts of dsRNA to elicit a response *(1)*. Another notable aspect of RNAi in *C. elegans* is that it is systemic in that the silencing can spread between tissues throughout the adult as well as its progeny *(1,3,4)*. Many neurons, however, are refractory to the spreading effect in wild-type backgrounds *(5)*.

Although discovered in *C. elegans,* the use of dsRNA to silence gene expression has quickly become a widely used tool to study gene function in a number of organisms, including mammals *(6,7)*. RNAi may have evolved from an ancient phenomenon used to regulate gene expression and combat transposable elements and viruses *(8)*.

The ability to use RNAi to target single or multiple transcripts and the ability to apply RNAi to large-scale genomic screens further highlight the power of this technique *(9–12)*.

2. Materials

1. Wild-type *C. elegans*.
2. NGM agar plates: 2.5 g peptone, 17 g agar, 3 g NaCl, 975 mL distilled water (dH$_2$O). Autoclave (30 min). Let cool to 55°C and add the following sterile solutions: 1 mL of 5 mg/mL cholesterol (stock in EtOH), 1 mL of 1 *M* CaCl$_2$, 1 mL of 1 *M* MgSO$_4$, 25 mL of 1 *M* KH$_2$PO$_4$.

From: *Methods in Molecular Biology, vol. 309: RNA Silencing: Methods and Protocols*
Edited by: G. G. Carmichael © Humana Press Inc., Totowa, NJ

3. OP50 *Escherichia coli.*
4. Agarose, low electroendosmosis (EEO) electrophoresis grade (Fisher Scientific, Suwanee, GA).
5. Large plastic Petri dishes, 100×15 mm (Fisher Scientific).
6. Sterile M9 buffer: 3 g KH_2PO_4, 6 g Na_2HPO_4, 5 g NaCl, 1 mL of 1 M $MgSO_4$; fill to 1 L with dH_2O.
7. 15-mL polypropylene tube.
8. 50–100–20 buffer: 50 mM Tris-HCl, 100 mM NaCl, 20 mM EDTA, pH 7.5.
9. 50–100–20 + Sodium dodecyl sulfate (SDS) + proteinase K: 1% SDS, 200 µg/mL proteinase K in 50–100–20 buffer.
10. Phenol : chloroform (1 : 1).
11. 3 M sodium acetate, pH 5.2.
12. 100% ethanol.
13. Worm pick: 36-gauge platinum wire attached to glass Pasteur pipet (Fisher Scientific).
14. TE buffer: 10 mM Tris-HCl, 1 mM EDTA, pH 7.6.
15. DNase-free RNase A.
16. Oligonucleotide primers specific to target gene.
17. Oligo $(dT)_{12-18}$ (500 µg/mL).
18. Sterile diethylpyrocarbonate (DEPC) H_2O.
19. 5X first strand buffer, supplied with enzyme (Stratagene, La Jolla, CA).
20. 0.1 M dithiothreitol (DTT), supplied with enzyme (Stratagene).
21. 10 mM dNTP stock in DEPC–dH_2O (Stratagene).
22. SuperScript II enzyme (200 U/µL) (Stratagene).
23. 0.5 M EDTA.
24. 2 mM dNTPs.
25. *Taq* polymerase.
26. Polymerase chain reaction (PCR) thermocycler and supplies.
27. PCR purification kit (Qiagen, Valencia, CA).
28. Phenol : chloroform : isoamylalcohol (25 : 24 : 1).
29. Chloroform.
30. 10 M NH_4Ac.
31. 70% EtOH.
32. Agarose gel electrophoresis apparatus.
33. Ultraviolet (UV) transilluminator.
34. Ethidium bromide (EtBr) (0.3 µg/mL).
35. Ultraviolet spectrophotometer.
36. Gel extraction kit (Qiagen).
37. T7 transcription kit (Ambion Megascript T7, Austin, TX or Promega, T7 Ribomax, Madison, WI).
38. Speed vacuum apparatus.
39. Borosilicate capillaries model 1B100F (World Precision Instruments, Sarasota, FL).
40. Needle puller (Sutter Instruments, Novato, CA).
41. Halocarbon oil, series 700 (Sigma, St. Louis, MO).

42. Injection pads (24 × 50 mm #1 cover slip with 2% agarose).
43. RNAi soaking buffer: 0.05% gelatin, 5.5 mM KH$_2$PO$_4$, 2.1 mM NaCl, 4.7 mM NH$_4$Cl, 3 mM spermidine. Add spermidine fresh before each use.
44. NGM plates + 1 M Isopropyl-β-D-thiogalactopyranoside (IPTG) + ampicillin poured fresh (1–3 d before use).
45. Ampicillin 100 mg/mL.
46. Luria–Bertani (LB) ampicillin agar plates (50 µg/mL).
47. LB tetracycline agar plates (12.5 µg/mL).
48. Sterile, RNase free dH$_2$O.
49. Siliconized 0.6 mL eppendorf tubes.
50. Siliconized 200 µL pipet tips.
51. *C. elegans* feeding library (MRC Geneservice).
52. IPTG.

3. Methods

The methods below outline (1) the preparation of genomic DNA from *C. elegans*, (2) dsRNA synthesis, and (3) RNA interference methods. Methods for handling *C. elegans* are described in **ref. *13*** and detailed methods for injection are described in **ref. *14***.

3.1. Preparation of Genomic DNA from C. elegans

1. Grow up one to two large plates of worms on NGM plates, using agarose in place of the agar.
2. Wash worms off of a recently starved plate (to avoid excess bacteria) using sterile M9 buffer, into a 15-mL polypropylene tube.
3. Pellet worms by centrifugation and carefully discard supernatant.
4. Wash worms with 10 mL of 50–100–20 buffer.
5. Pellet worms by centrifugation and carefully discard supernatant.
6. Add 2 mL of 50–100–20 + 1% SDS + 200 µg/mL proteinase K and incubate at 65°C.
7. Periodically agitate until suspension is viscous (use gentle agitation to avoid shearing DNA). This takes about 30 min.
8. Add 2 mL of phenol : chloroform and invert gently for 5 min.
9. Centrifuge to separate phenol : chloroform and aqueous phases.
10. Transfer top (aqueous) phase to a clean 15-mL tube.
11. Add 270 µL of 3 M sodium acetate. This results in a final concentration of 0.4 M sodium acetate.
12. Add 2.5 vol of 100% ethanol.
13. Wind out DNA precipitate onto the tip of a sealed glass Pasteur pipet (DNA visible as thin threads).
14. Dissolve DNA in 400 µL of TE (*see* **Note 1**).
15. Add RNase to 20 µg/mL final concentration.
16. Incubate at 37°C for 20 min.
17. DNA is sufficiently clean for use as template in polymerase chain reaction (PCR). Store at 4°C (*see* **Note 2**).

3.2. dsRNA Synthesis

3.2.1. Primer Design

Design primers to amplify approx 1 kb of mostly coding sequence (*see* **Note 3**), although 200–500 bp of target sequence often works well and may also reduce the chance of targeting other homologous transcripts. Be sure to target mostly exonic sequences and ensure that you are not targeting other homologous transcripts (*see* **Note 4**).

3.2.2. First Strand cDNA Synthesis (If Unable to Use Genomic DNA)

In those cases where only small exons separated by large introns are available, cDNA must be used to ensure you have sufficient exonic sequence to target. Before starting, make sure that all of the materials and reagents are sterilized and/or nuclease-free.

1. Mix 1 μL of oligo (dT)$_{12-18}$ (500 μg/mL), 1–5 μg of mRNA, sterile DEPC H$_2$O to 12 μL.
2. Heat the mixture to 70°C for 10 min, and then quickly chill on ice.
3. Centrifuge briefly to collect mixture at bottom of tube.
4. Add the following: 4 μL of 5X first strand buffer, 2 μL of 0.1 M DTT, 1 μL of mixed dNTP stock.
5. Mix the contents and centrifuge briefly to collect. Place tube at 42°C for 2 min to equilibrate temperature.
6. Add 1 μL of SuperScript II enzyme (200 U/μL) for each microgram of mRNA used in **step 1** to the mix.
7. Incubate at 42°C for 1 h.
8. Add 4 μL of 0.5 M EDTA or place tube at 70°C for 15 min in order to terminate the reaction.
9. Store cDNA at –20°C until use.

3.2.3. PCR

Amplification is performed in two steps. Designing an in vitro transcription template from a two-step PCR makes synthesizing multiple primers less expensive and may increase the yield of PCR products bearing complete T7 sites. The first step uses two 35-base pair primers, and each primer contains 15 bases of T7 sequence plus 20 bases of sequence from the gene of interest (*see* **Note 5**).

Example:

partial T7   <u>gene of interest</u>

Forward primer: 5′ CGACTCACTATAGGG<u>CGATGAGGGCCTATTTATTC</u> 3′

Reverse primer: 5′ CGACTCACTATAGGG<u>GAGAAAGTACACGATATAGC</u> 3′

The first PCR product is cleaned using the Qiagen PCR Purification kit (as per manufacturer's instructions) and eluted in 33 μL final volume. One-tenth of

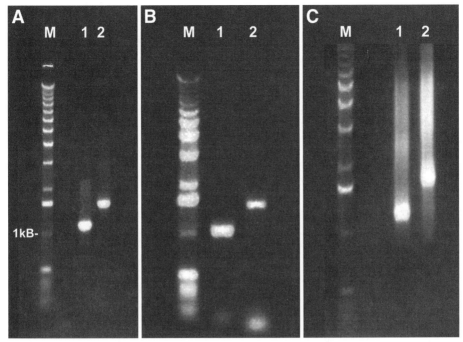

Fig. 1. (**A**) *mlc-4* (1) and *nmy-2* (2) PCR product amplified from genomic DNA with primers having partial T7 sequences. (**B**) *mlc-4* and *nmy-2* second-round PCR products that have been gel-purified using complete T7 primers and restriction sites. (**C**) Amplified and purified *mlc-4* and *nmy-2* dsRNAs. M is the DNA marker in each panel.

the recovered product is analyzed on an agarose gel (*see* **Fig. 1A**) and visualized using EtBr and UV illumination.

The second PCR uses the same primer on each end—a primer containing the full T7 polymerase promoter plus restriction sites for use with cloning: *Eco*RI, *Xba*I, and *Hind*III (*see* **Note 6**):

EXHT7:   5′ ATAGAATTCTCTAGAAGCTTAATACGACTCACTATAGGG 3′

One-tenth of the recovered product is analyzed on an agarose gel (*see* **Fig. 1B**) and visualized using EtBr and UV illumination.

3.2.4. In Vitro Transcription

The second PCR product is then gel purified (Qiagen Gel Extraction kit, as per manufacturer's instructions) and 1–2 µg of PCR product are used as the template for transcription. Commercially available in vitro transcription kits such as Ambion's Megascript-T7 kit or Promega's T7 Ribomax kit can be used to obtain high yields of dsRNA (*see* **Note 7**).

3.2.5. RNA Recovery

Pass the RNA through a Qiagen PCR purification column (treating the RNA as if it were a PCR product) to purify the RNA. RNA-specific columns can be used instead, although these are more expensive. Alternatively, dsRNA can be purified using phenol–chloroform extraction followed by ethanol precipitation (*see* **Subheading 3.2.7.**). Assess dsRNA integrity on an agarose gel (*see* **Fig. 1C**) and determine concentration using a UV spectrophotometer.

3.2.6. Storage

Mix the resulting solution with 2 vol 100% EtOH and store at –80°C. Before use, the EtOH should be evaporated in a speed-vaccum apparatus and the dsRNA pellet can be resuspended in sterile distilled water at your desired concentration.

3.2.7. Phenol–Chloroform Extraction and Ethanol Precipitation

Instead of using expensive columns (*see* **Subheading 3.2.5.**) you can purify your dsRNA using phenol–chloroform extraction followed by ethanol precipitation.

1. Add an equal volume of phenol : chloroform : isoamylalcohol (25 : 24 : 1) to the in vitro transcription reaction.
2. Mix (vortex) the contents of the tube until an emulsion forms.
3. Centrifuge the mixture at 11,750g for 15 s in a microcentrifuge at room temperature.
4. Transfer the upper (aqueous) phase containing dsRNA to a fresh tube. Discard the interface and the lower (organic) phase.
5. Add an equal volume of chloroform and repeat extraction steps (**steps 2–4**).
6. Add 1/3 vol of 10 M NH$_4$Ac.
7. Add 2.5 vol of 100% EtOH and put at –20°C for 30 min. The dsRNA can be stored at this step at –20°C if desired.
8. Spin down for 10 min at 11,750g in a microcentrifuge at 4°C to pellet dsRNA.
9. Wash pellet with 70% EtOH.
10. Air-dry pellet.
11. Resuspend in 100 µL of TE, pH 7.6.
12. dsRNA should be stored at –20°C in 1–2 vol of 100% EtOH.

3.3. Microinjection

The dsRNA for microinjection can be injected anywhere in the body cavity or gonad of the *C. elegans* hermaphrodite (*1*). Additional details can be found in **ref. 14.**

3.3.1. Preparing and Storing Needles

1. Pull capillaries to a finely tapered tip using a needle puller.
2. Use a Petri dish containing a strip of modeling clay to hold the needles. If kept dust-free, needles can be stored indefinitely.

3.3.2. Injection Pads

1. A drop of 2% agarose (in dH$_2$O) is flattened between two 24 × 50-mm #1 cover slips.
2. After hardening, remove the top cover slip and let the pad dry in open air (Alternatively, pads can be baked at 60°C for 1 h). If letting the pads air-dry, they should be made at least 1 d prior to use. Pads made either way can be stored for several weeks at room temperature.
3. Inject dsRNA solution into hermaphrodite gut or gonad.

3.3.3. Loading and Breaking the Needle

1. Place a 0.5 µL drop of the dsRNA solution on the back (blunt) end of the needle; the needle contains an inner glass filament that will wick the dsRNA solution to the tapered end.
2. Load needle onto an injection microscope. Use the edges of the agarose injection pad to gently break the needle. This step here takes practice to ensure that the needle does not have too large an opening, which would tend to damage the worms during injection (*see* **Note 8**).

3.3.4. Mounting and Injecting Worms

1. Place a drop of Halocarbon oil on the pad.
2. Using a worm pick, transfer young adult hermaphrodites to the oil drop on the pad (*see* **Note 9**).
3. Pat the animal down on the agarose pad with your pick until it is firmly stuck to the pad. Minimize the time that the worms spend drying out on the pad during injection to increase the viability of the worms.

3.3.5. Recovery

Put the pad under the dissecting scope and place a drop of M9 buffer on the oil drop containing your worms. Your worms will rehydrate and begin to swim. Transfer worms to a fresh NGM plate.

3.4. Feeding

RNA interference can be performed in *C. elegans* by feeding RNAse III-resistant bacteria expressing dsRNA to *C. elegans* (**5**). A genomic feeding library has been constructed (**15**) and is available commercially. The library consists of genomic fragments cloned into an IPTG-inducible T7 polymerase vector (the PCR product made in **Subheading 3.2.3.** can also be cloned into this feeding vector). The vectors are transformed into the RNAse III-resistant bacterial strain.

1. Streak out –80°C stock feeding strain that contains the desired construct onto LB tetracycline plates (tetracycline will select for feeding competent bacteria).
2. Pick a bacterial colony from streaked plates with a sterile toothpick (using sterile technique) and place it in approx 2 mL of LB/amp (ampicillin will select for the cloned gene of interest).

3. Shake overnight at 37°C.

4. Inoculate 2 mL fresh LB plus ampicillin with 20 µL of the overnight culture. Grow at 37°C until the optical density is between 0.35 and 0.40 (this usually takes about 3 h).

5. Seed NGM + IPTG + ampicillin plates with actively growing culture and allow to dry at room temperature (this will happen anytime between 3 h to overnight) (*see* **Note 10**).

6. Add worms to dsRNA-producing bacteria and allow worms to feed. Transfer worms every 24 h to new feeding plates (*see* **Note 11**).

3.5. Soaking

RNA interference can be performed in *C. elegans* by soaking worms in dsRNA *(16)*; this method has been used in large-scale screens *(12)*.

1. Pick four young adults onto a fresh unseeded NGM plate to clean off bacteria (*see* **Note 12**).

2. Wash worms with a drop of M9 and allow them to crawl away.

3. Add the four worms to 2 µL of RNAi soaking buffer in 0.6-mL siliconized Eppendorf tubes (final dsRNA concentration should be at 1 mg/mL).

4. Place at 20°C for 24 h.

5. Transfer worms to a fresh, seeded plate using siliconized pipet tips.

6. Wash Eppendorf tube with 100 µL of M9 to ensure transfer of all of the soaked worms.

7. Transfer to new plates daily and score progeny.

4. Notes

1. Swirl the tip of the glass Pasteur pipet containing the DNA in a microfuge tube containing 400 µL of TE.

2. Do not freeze genomic sample, as thaw cycles can shear genomic DNA. Multiple genomic DNA preparations can be used as templates when a product does not amplify from a single preparation.

3. Visit www.wormbase.org for gene structure information.

4. Visit http://www.sanger.ac.uk/cgi-bin/blast/submitblast/c_elegans for BLAST tools needed to search for and avoid dsRNAs that target other homologous transcripts.

5. When designing primers, be sure to avoid creating primers that contain primer dimers. Also, primers should not have a high GC content (>50%) and be sure that your primers are also free from strong secondary structure.

6. These DNA templates can be easily cloned into the feeding construct (L4440) discussed in **Subheading 3.4.**

7. Purchasing rNTPs, T7 polymerase, DNase, and RNase inhibitors separately and then purifying the dsRNA using phenol : chloroform followed by EtOH precipitation also works well (*see* **Subheading 3.2.7.**).

8. You can also dip needles in hydrofluoric acid to open needle tips.

9. Be sure to limit the amount of bacteria transferred, as excess bacteria can make it difficult for the worms to adhere to the pad.

10. Be sure that all media for plates are cooled to 50°C before adding IPTG, ampicillin, or tetracycline.
11. You can also place young larvae on plates or have worms lay eggs onto the feeding plates for a couple of hours, followed by removal of the mothers.
12. This protocol can be scaled up as necessary.

Acknowledgments

Our work using RNAi in *C. elegans* has been supported by NIH R01 GM68966 and NSF MCB-0235654. B.G. is a Pew Scholar in the Biomedical Sciences. We thank members of the Goldstein Lab for careful reading of the manuscript.

References

1. Fire, A., Xu, S., Montgomery, M. K., Kostas, S. A., Driver, S. E., and Mello, C. C. (1998) Potent and specific genetic interference by double-stranded RNA in *Caenorhabditis elegans. Nature* **391,** 806–811.
2. Montgomery, M. K. and Fire, A. (1998) Double-stranded RNA as a mediator in sequence-specific genetic silencing and co-suppression. *Trends Genet.* **14,** 255–258.
3. Grishok, A., Tabara, H., and Mello, C. C. (2000) Genetic requirements for inheritance of RNAi in *C. elegans. Science* **287,** 2494–2497.
4. Winston, W. M., Molodowitch, C., and Hunter, C. P. (2002) Systemic RNAi in *C. elegans* requires the putative transmembrane protein SID-1. *Science* **295,** 2456–2459.
5. Timmons, L., Court, D. L., and Fire, A. (2001) Ingestion of bacterially expressed dsRNAs can produce specific and potent genetic interference in *Caenorhabditis elegans. Gene* **263,** 103–112.
6. Tijsterman, M., Ketting, R. F., and Plasterk, R. H. (2002) The genetics of RNA silencing. *Annu. Rev. Genet.* **36,** 489–519.
7. Maine, E. M. (2000) A conserved mechanism for post-transcriptional gene silencing? *Genome Biol.* **1,** REVIEWS1018.
8. Plasterk, R. H. (2002) RNA silencing: the genome's immune system. *Science* **296,** 1263–1265.
9. Dudley, N. R., Labbe, J. C., and Goldstein, B. (2002) Using RNA interference to identify genes required for RNA interference. *Proc. Natl. Acad. Sci. USA* **99,** 4191–4196.
10. Fraser, A. G., Kamath, R. S., Zipperlen, P., Martinez-Campos, M., Sohrmann, M., and Ahringer, J. (2000) Functional genomic analysis of *C. elegans* chromosome I by systematic RNA interference. *Nature* **408,** 325–330.
11. Gonczy, P., Echeverri, C., Oegema, K., et al. (2000) Functional genomic analysis of cell division in *C. elegans* using RNAi of genes on chromosome III. *Nature* **408,** 331–336.
12. Maeda, I., Kohara, Y., Yamamoto, M., and Sugimoto, A. (2001) Large-scale analysis of gene function in *Caenorhabditis elegans* by high-throughput RNAi. *Curr. Biol.* **11,** 171–176.

13. Sulston, J. and Hodgkin, J. (1998) *The nematode* Caenorhabditis elegans. Cold Spring Harbor Laboratory Press, Cold Spring Harbor, NY.

14. Mello, C. C. and Fire, A. (1995) DNA Transformation, in *Caenorhabditis elegans: Modern Biological Analysis of an Organism*, Epstein, H. F. and Shakes, D. C. (ed.). Academic Press, San Diego, CA, **Vol. 48,** pp. 452–480.

15. Kamath, R. S. and Ahringer, J. (2003) Genome-wide RNAi screening in *Caenorhabditis elegans. Methods* **30,** 313–321.

16. Tabara, H., Grishok, A., and Mello, C. C. (1998) RNAi in *C. elegans*: soaking in the genome sequence. *Science* **282,** 430–431.

5

Down-Regulating Gene Expression by RNA Interference in *Trypanosoma brucei*

Christine E. Clayton, Antonio M. Estévez, Claudia Hartmann, Vincent P. Alibu, Mark Field, and David Horn

1. Introduction

RNA interference (RNAi) in *Trypanosoma brucei* was first reported in 1998 *(1)*. As in other eukaryotes, interference involves digestion of the interfering double-stranded RNA into short fragments *(2)*, a polysome-associated complex *(3)*, and Argonaute protein *(4,5)*. *T. brucei* is an ideal organism for testing gene function by RNAi. A complete genome sequence is available, and liquid suspension culture is unproblematic. Various combinations of RNAi vector and host trypanosomes, with different advantages, are available; if procedures are working optimally, it should be possible to obtain a stable cell line with inducible RNAi within 2 wk. The RNAi process itself is apparently not essential for parasite survival *(4)*, although some adverse effects have been reported *(5)*. To analyze the process itself, it is possible to delete candidate genes completely, as the efficiency of homologous recombination is essentially 100%; *in situ* epitope tagging can also be effected through homologous recombination *(6)*. For general reviews of the peculiarities of trypanosome gene expression and RNA processing, see **refs.** *7* and *8*, and for pre-RNAi methods, such as inducible gene expression and knockouts by homologous recombination, see **ref.** *9*. A few recent applications of RNAi in trypanosomes are found in **refs.** *10–16*. Importantly, in trypanosomes, RNAi can also be used to deplete nuclear RNAs *(8,17)*. In this chapter, we will describe two different options for the construction of RNAi vectors, followed by methods for trypanosome culture and selection of transfectants. At the end, we will highlight various options for testing of phenotypes, as well as various pitfalls with working with RNAi in trypanosomes.

From: *Methods in Molecular Biology, vol. 309: RNA Silencing: Methods and Protocols*
Edited by: G. G. Carmichael © Humana Press Inc., Totowa, NJ

1.1. Choosing the Life-Cycle Stage

T. brucei infects mammals (but not humans), where it replicates in the blood and tissue fluids (bloodstream form), and tsetse flies (procyclic form), where it multiplies in the midgut. Both replicative stages can be cultured in vitro in liquid or on soft agar plates; here, only the liquid culture will be described. The choice of life-cycle stage depends on the gene to be analyzed. Evidently, genes that are expressed exclusively in one stage should be analyzed in that stage. Therefore, it is important to check expression of your target gene, at least at the messenger RNA (mRNA) level, before initiating the experiments. Procyclic trypanosomes are metabolically more complex than bloodstream forms and, in particular, have a much more active mitochondrial metabolism. They grow slightly slower than bloodstream forms. When a gene is essential in both stages, the limited experience so far suggests that experiments on essential genes with procyclic forms are often more informative than with bloodstream forms. Whereas bloodstream forms simply die rapidly, procyclics may show alterations in metabolism or ultrastructure before dying more slowly (see, for example **ref. *14***). Possible reasons are the more diverse metabolism of procyclics or, simply, their slower growth; the advantage is that there is often a greater opportunity to observe the mutant phenotype.

1.2. Inducible RNAi Using Opposing T7 Promoters

The simplest method to achieve RNAi in *T. brucei* is to use vectors with opposing bacteriophage T7 promoters, controlled by binding of the *tet* repressor to *tet* operators *(18–20)*. The host trypanosomes described below already contain stably integrated plasmids constitutively expressing T7 polymerase and the *Tn10 tet* repressor. The main advantage of this method is that cloning of polymerase chain reaction (PCR) products into the RNAi vector is very rapid and efficient. The disadvantage is that the control of T7 transcription by the *tet* repressor can be somewhat leaky: If the "background" transcription of double-stranded RNA (dsRNA) is sufficient to cause growth inhibition, no trypanosomes containing the plasmid will be obtained or the phenotype of the RNAi-expressing trypanosomes may be very unstable (*see* **Notes 1** and **2**). Despite this, a considerable number of essential genes have been studied using this approach and it is definitely the method of first choice.

A particularly impressive use of this method was the construction of a genomic library for RNAi expression. The library is transfected into trypanosomes, the recombinants are grown, and then RNAi is induced. Subsequently, one can select for any loss-of-function mutation that is compatible with survival *(10)*. The major limitation, however, is the low frequency of transfection of trypanosomes: hence, at present, such an approach is only applicable when an extremely good selection method is available.

1.3. Inducible RNAi Using Stem-Loop Contructs

In this strategy, transcription of an RNA stem loop is driven by an RNA polymerase I promoter under tetracycline control *(21,22)*. Construction of the stem-loop-expressing vector involves three cloning steps. Two opposing copies of the target sequence are inserted on either side of a "stuffer" DNA, which is required for stability of the plasmid in *Escherichia coli,* and then the stem loop is cloned into a vector containing an inducible RNA polymerase I promoter *(23,24)*. Disadvantages of this method are the more complicated plasmid construction and the fact that sometimes the chosen sequences resist cloning in opposite orientations. The advantages are that the background RNAi expression is generally lower than with opposing T7 promoters and that the lines are more stable upon cultivation or storage. Examples of the use of this methodology include those described in **refs.** *13* and *14*. For more tips on the choice of strategy, see **Notes 3–7**.

2. Materials
2.1. Material Specific for T7-Polymerase-Driven RNAi

1. Plasmid DNA (*see* **Fig. 1A**).
2. Bloodstream trypanosomes expressing T7 polymerase and the *tet* repressor, which are not resistant to hygromycin, such as the "single-marker" line containing pHD 328 and pLew114hyg5' *(25)*. (Note that this line can also be used for RNAi vectors bearing a phleomycin resistance marker.)

An alternative is procyclic trypanosomes expressing T7 polymerase and the *tet* repressor, which are not resistant to hygromycin, such as the pHD 1313–1333 line, in which T7 polymerase synthesis is tetetracycline inducible, or the pHD1313–514 line, which constitutively expressed the polymerase *(26)*. Results so far suggest that the pHD 1313–1333 line may give the highest transformation efficiencies when the RNAi is very toxic.

2.2. Material Specific to Stem-Loop RNAi

1. "Stuffer" plasmid DNA (*see* **Fig. 1B**).
2. A vector containing a tetracycline-inducible trypanosome RNA polymerase I promoter (*see* **Fig. 1C**) or an inducible vector containing the stuffer (*see* **Fig. 1D**).
3. Bloodstream or procyclic trypanosomes expressing the *tet* repressor, which are not resistant to hygromycin, such as those containing integrated pHD 1313 *(26)* or pHD 449 *(23)*.

A RNAi vectors with opposing T7 promoters

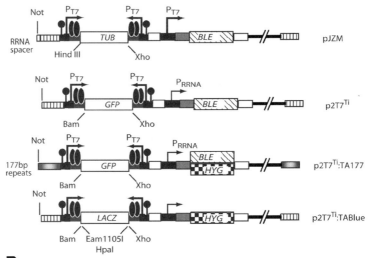

B Vector for stem-loop cloning

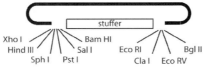

C Inducible polymerase I vectors for insertion of ready-made stem-loops

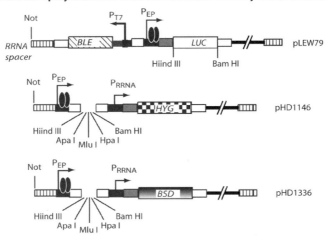

D Inducible polymerase I vector for direct cloning of stem-loops

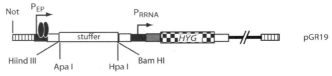

2.3. Bloodstream-Form Culture and Transfection

1. Medium HMI-9:

Component	100X Stock	Amount	Final concentration
IMDM		176.6 g	
Sodium bicarbonate		30.24 g	
Hypoxanthine	13.6 mg/mL	1.36 g	1 mM
Thymidine	3.9 mg/mL	390 mg	0.16 mM
Bathocuproine sulfonic acid	2.82 mg/mL	282 mg	0.05 mM

Dissolve to a final volume of 10 L, filter-sterilize, and aliquot 450 mL into bottles. Store at –20°C.

Component	100X Stock	Final concentration	
Cysteine	18.2 mg/mL	1.5 mM	Filter-sterilize, store no more than 1 mo, –20°C
β-Mercaptoethanol	14 μL /10 mL		Filter-sterilize, make fresh

Before use, add the following to 450 mL medium: 5 mL each cysteine and β-mercaptoethanol solutions and 50 mL fetal calf serum (FCS) (heat to 56°C for 30 min; if consumption is not very rapid, make 50-mL aliquots in sterile tubes and freeze.)

Fig. 1. (*Opposite page*) Cloning vectors. Only useful sites are shown. Sequences and DNA are available from the authors. (**A**) Vectors with opposing T7 promoters. Each of these vectors is designed to be linearized with *Not*I within a "targeting sequence." After transfection, the plasmid integrates by homologous recombination via the sequences either side of the *Not*I site. All integrate into a transcriptionally silent spacer between rRNA genes except p2T7:177, which integrates into silent minichromosomal repeats: In some cases, the minichromosomal site might give a lower background transcription of the RNAi fragment *(25)*. References for the other vectors are as follows: pJZM: **ref.** *18*; p2T7 derivatives: **ref.** *26*. (**B**) Plasmid for cloning of stem loops. Any other similar plasmid could be used. (**C**) Plasmids designed for tetracycline-inducible gene expression in *T. brucei*. Several other similar vectors are available. These vectors can be used for expression of any open reading frame or of stem-loop transcripts. The pHD vectors are derivatives of those described in **ref.** *23* and their use is described in **ref.** *13*. PLEW79 is described in **ref.** *24*. (**D**) Inducible plasmid for direct cloning of stem loops (Estévez, unpublished).

Note that powdered HMI-9 is available from laboratories working on try-panosomes in the United Kingdom: contact the authors for details.

References for trypanosome cultivation, including plating on soft agar, are **refs. 27–29**. There is a detailed discussion of this topic at http://tryps.rockefeller.edu/DocumentsGlobal/tryp_culture_commentary.html.

2. Selective drug concentrations:

Drug	Stock	Amount for 10 mL	Final concentration
Blasticidin	5 mg/mL	10 μL	5 μg/mL
G418	50 mg/mL	1 μL	5 μg/mL
Hygromycin	50 mg/mL	1–3 μL	5–15 μg/mL[a]
Phleomycin/ bleomycin	5 mg/mL	0.4 μL	0.2 μg/mL
Puromycin	1 mg/mL	2 μL	0.2 μg/mL
Tetracycline	5 mg/mL in EtOH	0.2 μL	100 ng/mL

[a]Ideal amount seems to depend on vector and cell line. Try with the lower amount first.

Stocks are stored in small aliquots at $-20°C$. Tetracyline should be kept frozen in the dark for no more than 1 mo. The half-life of tetracycline in tissue culture medium is about 24 h, so induction should be maintained by adding to 75 ng/mL every 2 d if no additional tetracycline-containing medium is added. An alternative is to ini-tially add 1 μg/mL tetracycline; then the cells can be left for 4 d before readding the drug.

3. Transfection medium (modified Cytomix):

Component	Stock	How much	Final concentration
EGTA	0.5 *M*	0.4 mL	2 m*M*
KCl		0.9 g	120 m*M*
CaCl$_2$		0.0017g	0.15 m*M*
Phosphate buffer, pH 7.6[a]	0.1 *M*	10 mL	10 m*M*
HEPES		0.596 g	25 m*M*
MgCl$_2$.6H$_2$O		0.102 g	5 m*M*
Glucose		0.5 g	0.5%
BSA (defatted)		10 mg	100 μg/mL
Hypoxanthine[b]	13.6 mg/mL	1 mL	1 m*M*

[a]Mix 8.66 mL of 1*M* K$_2$HPO$_4$ with 1.34 mL KH$_2$PO$_4$ in 90 mL H$_2$O.
[b]Hypoxanthine: dissolve in 0.1 *M* NaOH.

Dissolve in 90 mL, adjust to pH 7.6 with 1 *M* KOH, make up to 100 mL, and filter-sterilize. Store at 4°C.

2.4. Procyclic-Form Culture and Transfection

1. Medium MEM-Pros:

Component	g/10 L
Salts	
$CaCl_2 \cdot 2H_2O$	2.65
KCl	4.00
$MgSO_4 \cdot 7H_2O$	2.00
NaCl	68.00
$NaH_2PO_4 \cdot H_2O$	1.40
HEPES	71.40
Other	
L-Arg·HCl	1.26
L-Cys·Cys	0.24
L-Gln	2.92
L-His·HCl·H_2O	0.42
L-Ile	0.52
L-Leu	0.52
L-Lys	0.73
L-Met	0.15
L-Phe	1.00
L-Thr	0.48
L-Try	0.10
L-Tyr	1.00
L-Val	0.46
L-Pro	6.00
Adenosine	0.12
Ornithine·HCl	0.10

Many labs order this or a similar medium as a ready-made powder: contact any of the authors for details.

MEM nonessential amino acids	100 mL
MEM vitamins	100 mL
Phenol red (1000X)	10 mL (or 0.1 g in 10 L)

Make salt solution including HEPES in a volume of about 4 L and adjust pH to 7.4 using NaOH. Add all of the other solids and wait until dissolved; dilute up to about 9 L as necessary. Add the MEM solutions, make up to final volume and sterile-filter. Leave out overnight to check for contamination. Store in aliquots of 450 mL at 4°C.

Hemin solution: Dissolve 250 mg of Hemin in 100 mL of 0.1 M NaOH. Autoclave. Aliquot into 10-mL aliquots in sterile tubes and store at 4°C.

Fetal calf serum (FCS): Heat to 56°C for 30 min. Aliquot in 50-mL aliquots in sterile tubes.

Final medium: Add 50 mL heat-inactivated FCS and 1.5 mL hemin per 450 mL medium.

A detailed discussion of trypanosome culture, including copies of older publications, can be found at http://tryps.rockefeller.edu/DocumentsGlobal/tryp_culture_commentary.html (*see* **Note 8**). Some references for MEMPros are **ref. 30** (the glycine should be glycerol) and **ref. 31**.

An alternative medium called SDM79 *(32)* is equally good for continuous cultivation of procyclic trypanosomes. If starting trypanosome culture for the first time, it is usually best to use the medium in which the parasites were previously grown, using the protocols of the laboratory from which the parasites were obtained.

2. Antibiotic concentrations:

Drug	Stock	Amount for 10 mL	Final concentration
Blasticidin	5 mg/mL	20 µL	10 µg/mL
G418	50 mg/mL	3 µL	15 µg/mL
Hygromycin	50 mg/mL	10 µL	50 µg/mL
Phleomycin/			
Bleomycin	5 mg/mL	1 µL	0.5 µg/mL
Puromycin	1 mg/mL	10 µL	1 µg/mL
Tetracycline	5 mg/mL in EtOH	0.2 µL	100 ng/mL

For details, see **Subheading 2.3.**, item 2.

3. Transfection medium ZPFM:

	Amount	Final concentration
NaCl	7.714 g	132 mM
KCl	0.596 g	8 mM
Na_2HPO_4 (anhyd)	1.136 g	8 mM
KH_2PO_4 (anhyd)	0.204 g	1.5 mM
$MgOAc \cdot 4H_2O$	0.107 g	1.5 mM
{$CaOAc_2$	0.016 g	90 µM
or		
$CaCl_2$	0.014 g	

Dissolve in 800 mL, adjust pH to 7.00 ± 0.05 with NaOH or HAc and make up to 1 L. It is also possible to make a 5X stock for storage at 4°C.

2.5. Other Materials

1. Two PCR primers designed to amplify the target gene.
2. Trypanosome genomic DNA.
3. *Taq* polymerase, enzyme buffers, and nucleoside triphosphates.
4. Tissue culture flasks (25 mL, 250 mL) and 24-well plates.

5. Electroporation machine: must be capable of delivering a pulse at 1.5 kV at physiological salt without annulling the guarantee. We use BTX machines.
6. Electroporation cuvets.
7. Standard materials for molecular biology including restriction enzymes and competent *E. coli.*

3. Methods

3.1. Construction of the RNAi Plasmid

3.1.1. Primer Design

RNA interference in *T. brucei* works well with dsRNAs of about 500 bp, and dsRNA will target any sequence with 85% identity or more. In order to target one sequence specifically, you should therefore choose a fragment of about 500 bp that is within the predicted mRNA and has less than 85% identity with other genes *(33)*. It is best to be near the initiator ATG but not to include it (risk of making a truncated protein). The trypanosome genome can be searched from http://www. genedb.org/genedb/seqSearch.jsp?organism=tryp. Primers of 18–20 nt, with compatible melting points, are designed to amplify the fragment from genomic DNA. For unique genes, specific RNAi fragments and appropriate primers with compatible melting points are best designed using the RNAit program, which is a web-based resource *(34)* (http://www.trypanofan.org/software/RNAit.html). If the target gene is present in multiple copies, it will be necessary to design primers manually; if the coding regions are identical, then try using the 3′-untranslated regions (3′-UTRs).

If you intend to build a stem-loop construct, it is necessary to include boundary restriction sites at the 5′ end of the oligonucleotides to allow directional cloning (see below); these sites should be absent from the amplified fragment. Add at least three additional nucleotides 5′ of the restriction sites in order to ensure efficient restriction digestion of the amplified fragments. The choice of enzymes must be made carefully, as some will not cleave close to a double-stranded break.

3.1.2. Expression Vector

3.1.2.1. PCR

1. Mix the following in a PCR tube: 1 µL forward primer (stock = 100 mM); 1 µL reverse primer (stock = 100 mM); 1 µL *T. brucei* genomic DNA template (about 0.1 µg); 1 µL dNTP's (stock = 10 mM); 5 µL 10X buffer [MBI Fermentas buffer + $(NH_4)_2SO_4$]; 3 µL Mg·Cl_2 (stock = 25 mM); 0.5 µL *Taq* DNA polymerase (5 U/mL); 37.5 µL H_2O.

   You can also use your favorite *Taq* polymerase with appropriate buffer, but be sure to use a classical (non-proof-reading) polymerase, which leaves "A" overhangs. Very high fidelity is not an issue here, as low-frequency PCR errors will not affect the specificity of the RNAi.

2. Set the thermal cycler to the following sequence: 30 s at 94°C; 30 s at 55°C; 40 s at 72°C; 30 Cycles: followed by 10 min final elongation at 72°C. This sequence is for oligonucleotides designed by RNAit, with 60°C melting temperature.
3. Run 5 μL of the PCR product on an agarose gel to check quantity and specificity.
4. Clean the remainder using your favorite method, or you can attempt cloning without purification. One suitable kit is the MinElute PCR purification kit (Qiagen cat. no. 28106); elute in 30 μL of elution buffer.
   The PCR products can be stored at –20°C until use, but efficiency may be better if the ligation is started immediately.

3.1.2.2. VECTOR PREPARATION AND FRAGMENT CLONING

The parent vector, p2T7 TAblue, is shown in Fig. 1A. *Eam*1105I sites that flank the *Lac*Z "stuffer" are engineered to leave T-overhanging ends.

1. For approx 20 ligations, prepare the following in a 1.5-mL Eppendorf (scale up if necessary): 4 μL vector (stock = 1 mg/mL); 5 μL 10X reaction buffer (Fermentas); 39 μL dH$_2$O; 2 μL *Eam*1105I (Fermentas).
2. Incubate at 37°C for 1–2 h. (*Note:* Longer incubations may lead to damaged DNA ends and higher background.)
3. Check a small sample on a 1% agarose gel with an uncut vector control. If the digest appears complete, it is not necessary to purify the cut vector.
4. Add the following to 10 μL of PCR product: 1 μL *Eam*1105I digested p2T7; 1.3 μL 10X ligase buffer (NEB); 0.5 μL T4 ligase (NEB: #M0202S).
5. Incubate overnight at 4°C or 16°C.
6. Transfect 5 μL of each ligation into your favorite competent *E. coli.*
7. Plate onto Lurie–Bertani medium (LB) containing ampicillin (100 ug/mL), X-gal (1 mg/plate), and IPTG (2 mg/plate) and grow overnight.
8. Select and label two to four white colonies from each plate for screening.
9. Make minipreps using your favorite method.
10. Digest 2–5 μL of the miniprep with *Xho*I and *Bam*HI. This digest should liberate the RNAi target insert, 400–600 bp. *Xba*I digestion can be used to liberate insert+ approx 1 kbp. *Hpa*I can be used instead of *Bam*HI/*Xho*I.
11. Run digests on a 1% agarose gel and select apparent positive clones.
12. Use the remaining DNA for sequencing. You can use one of your original PCR primers or use a standard primer (5′ CCGCTCTAGAACTAGTGGA 3′).
13. Select correct clones for midi-prep. Check the DNA concentration and bring to 1 mg/mL. Digest a sample of each with *Xho*I/*Bam*HI to confirm that no rearrangement has occurred. *Hpa*I can be used instead of *Bam*HI/*Xho*I.

3.1.3. Stem-Loop Expression Vector

In this procedure, the fragment is cloned twice, in opposite orientations, on either side of a "stuffer fragment," which facilitates maintenance of the inverted repeat in *E. coli.* When transcribed, the "stuffer" forms the loop of the stem loop.

There are two options:

i. Clone the RNAi fragments around the stuffer, then transfer the stem loop to an inducible vector.
ii. Clone the RNAi fragments around the stuffer, which has already been cloned into the inducible vector. This is clearly simpler if the fragments are readily cloned in inverted orientation in *E. coli* and the *HYG* marker is to be used.

In either case, the genomic fragment is first amplified using oligonucleotides bearing nested restriction sites. For both cloning procedures, the structures of initial DNAs involved are as follows:

Amplified fragment:

site 1–site 2–fragment–site 3–site 4

Initial recipient plasmid vector (Fig. 1B,D):

Vector–site 2–site 3–"stuffer"–site 4–site 1–vector

The first cloning intermediate (plasmid A):

Vector– site 2–PCR fragment–site 3–"stuffer"–site 4–site 1–vector

Final construct for insertion into the expression vector (three-step procedure) or final construct in the two-step procedure:

Vector–site 2–PCR fragment–site 3–"stuffer"– site 4–site 3–PCR fragment in inverse orientation–site 2–site 1–vector

3.1.3.1. THREE-STEP PROCEDURE

The following protocol is for cloning either side of a portion of the spliced leader repeat as described in **ref. 22**. Alternative stuffers of similar size can also be used; the nature of the sequence might influence either cloning efficiency or the effectiveness of RNAi, but there has been no systematic analysis of this.

1. The fragment should not contain *Bgl*II, *Sph*I, *Eco*RI, *Sal*I, *Hind*III, or *Not*I sites; however, if it does and you cannot avoid it, the following strategy can be modified by picking other suitable restriction sites.
2. Design two oligos to PCR-amplify that region. The sense oligo should contain a *Bgl*II-*Sph*I linker (5′-GAGA**AGATCT**GCATGC ...) and the antisense oligo, a *Eco*RI-*Sal*I one (5′-CG**GAATTC**GTCGAC ...). Please note that the *Sph*I site contains an ATG codon that can act as a initiator codon (shown in italics), so design the oligo so that ATG is placed out of frame. If you cannot avoid including these restrictions sites within the amplified fragment, use alternatives in the polylinkers.
3. Carry out a PCR reaction using standard conditions. Run a gel and purify the PCR product. Digest half with *Sph*I and *Sal*I and the other half with *Eco*RI and *Bgl*II. Store the latter for **step 7**.

4. Digest the stuffer vector with *Sph*I and *Sal*I. Gel-purify the linearized vector.
5. Ligate the linearized stuffer vector with the PCR product that was digested with *Sph*I and *Sal*I. The resulting plasmid is hereafter referred to as plasmid A. Check minipreps by digesting with *Xho*I + *Bgl*II (pHD1144 yields 2.4 + 0.8 kb; plasmid A yields 2.4 + another band of 0.8 + size of the PCR product).
6. Digest plasmid A with *Eco*RI and *Bgl*II; gel-purify the linearized vector.
7. Ligate the linearized vector with the PCR product digested with *Eco*RI and *Bgl*II (stored in **step 3**). The resulting plasmid is plasmid B. Check minipreps by digestion with *Hin*dIII and *Bgl*II (it should yield two bands: one of 2.4 kb corresponding to the vector backbone and another one of approx 2 kb, depending of the size of the original PCR product). Plasmid A can be included as a control; it yields 2.4 kb + another band whose size is very similar to the one obtained in **step 5** (around 1.4 kb).

    At this point, you may discover that it is impossible to clone your desired fragment in the inverted orientation because of recombination in *E. coli*. If this is so, you can try changing either the fragment you are cloning (if the target gene is long enough to offer alternatives) or the stuffer.
8. Purify the *Hin*dIII- *Bgl*II 2-kb insert band from plasmid B (bearing the RNAi fragment–stuffer–tnemgarf iANR) to be used in **step 10**.
9. Digest pHD1146, pHD 1336, or pLEW79 with *Hin*dIII and *Bam*HI. These plasmids differ mainly in the selectable marker. Gel-purify the linearized plasmid.
10. Ligate both fragments to obtain the final plasmid, plasmid C (*Bgl*II and *Bam*HI leave ends that are compatible). Check by restriction digestion.

3.1.3.2. Two-Step Procedure

This is similar to **steps 1–7** of **Subheading 3.1.3.1.**, but with the following modifications:

1. The fragment should not contain *Bgl*II, *Hpa*I, *Apa*LI, *Hin*dIII, or *Not*I sites, but if it does and you cannot avoid it, the following strategy can be modified by using a single blunt end on the insert fragment. (If you use two blunt ends, it might be very difficult to obtain the inserts in opposite orientations.)
2. Design two oligos to PCR-amplify that region. The sense oligo should contain a *Bgl*II-*Hin*dIII linker (5′-GAGA**AGATCT**AAGCTT ...), and the antisense oligo, a *Hpa*I-*Apa*I one (5′-CG**GTTAAC**GGGCCC ...). (*Hpa*I gives a blunt end so any other blunt-cutting enzyme will also do; *Bam*HI can be used instead of *Bgl*II.)
3. Carry out a PCR reaction using standard conditions. Run a gel and purify the PCR product. Digest half with *Hin*dIII and *Apa*I and the other half with *Hpa*I and *Bgl*II. Store the latter for **step 7**.
4. Digest the stuffer vector pGR19 with *Hin*dIII and *Apa*I. Gel-purify the linearized vector.
5. Ligate the linearized vector with the PCR product that was digested with *Hin*dIII and *Apa*I. Check minipreps.
6. Digest plasmid from **step 5** with *Hpa*I and *Bam*HI; gel-purify the linearized vector.
7. Ligate the linearized vector with the PCR product digested with *Hpa*I and *Bgl*II (stored in **step 3**).

3.2. Bloodstream Trypanosome RNAi Lines

3.2.1. Bloodstream Trypanosome Culture

The bloodstream trypanosomes used for RNAi studies can be cloned as single cells and grow to a maximum cell density of about 7×10^6/mL. It is very important not to exceed this, as the cells will go into an irreversible stationary phase and die. Ideally, do not exceed 2×10^6/mL; then the cells will remain in log phase. The cell density usually increases 10-to 20-fold in 24 h. The cells are aerobic and are grown in a 5% CO_2 incubator at 37°C in flasks with loosened or vented caps. Large volumes can be grown in sealed glass or plastic roller bottles in a warm room, but, in that case, it is important that at least 80% of the volume should be air. This is easiest achieved by equilibrating the bottles for 1 h in the incubator before tightening the caps. An alternative is to use multilayered flasks.

To freeze the cells, take a growing culture ($[0.5–2] \times 10^6$ cells/mL) and add an equal volume of sterile medium containing 20% glycerol. Aliquot 0.5–1 mL of cells per tube and cap tight. Make at least three tubes. Wrap the cells in paper tissue or wadding, with a thickness of about 5 cm. Place in the –70°C freezer. After 1–7 d, place in liquid nitrogen. After a further day, remove a tube and thaw rapidly. Place into medium and incubate overnight to check viability and the ability to grow. Although parasite motility is one indicator of viability, it does not always guarantee that the parasite can grow. Viability of RNAi lines after freezing is very variable (*see* **Notes 1** and **2**); do not throw away a growing culture until you are certain that the frozen stock is satisfactory. To recover cells, it can help to try different densities and to omit selective drugs immediately after thawing. (Glycerol can inhibit growth.)

3.2.2. Electroporation and Selection

3.2.2.1. PREPARATION

1. Cut vector (10 µg/cuvet) with *Not*I. Check 0.2 µg of the digest on a gel. Heat to denature the *Not*I; then ethanol-precipitate.
2. Wash the precipitate with 70% ethanol and then with 100% ethanol. Fill the tube with the ethanol to sterilize it. Remove the 100% ethanol, spin again, and remove the visible liquid. Then, leave the tube(s) open in the laminar flow hood to dry.
3. In the hood, resuspend the DNA in 25 µL (per 10 µg) of sterile water or Tris-EDTA buffer. Make sure that the DNA is well dissolved, heating if necessary. Store at –20°C.
4. The day before transfection, inoculate fresh medium with trypanosomes at a final density of 10^5/mL. You will need $(1–2) \times 10^7$ trypanosomes per transfection, and it is best to do two transfections for each DNA. Thus, for each transfection, you need 10–20 mL of culture at 10^6/mL. Be sure also to allow for no DNA and empty (*Not*-linearized) plasmid controls!

3.2.2.2. TRANSFECTION

1. On the day of transfection, check cell numbers and wait, if necessary, until minimum numbers are attained.
2. Centrifuge the trypanosomes at about 1000g in a standard cell culture centrifuge and resuspend in ice-cold cytomix (1 mL per transfection; keep in the refrigerator). Set out the electroporation cuvets on ice, thaw the DNA, and label recipient flasks. If the electroporation cuvets come with a small Pasteur pipet, you should leave it, sterile, in the wrapper. Array the lids open side down on the sterile work surface in the laminar flow hood.
3. Recentrifuge the cells, and resuspend in 0.5 mL cytomix per cuvet. Keep on ice. During the centrifugation, aliquot 5 mL of growth medium per flask. We also use 36 mL of medium (in a 25-cm flask—kept upright).
4. Aliquot the trypanosomes into the cuvets, on ice.
5. For each transfection, take two cuvets from the ice and place them on the work surface in the hood. Dry them carefully with tissue. Add 10 µg of cut DNA. Mix up and down with a Pasteur pipet or using a Gilson pipetman. It is critical that the DNA be well mixed. Put on the cap.
6. Electroporate both cuvets using settings of 1.5 kV and resistance R2. The time constant should be about 0.3 and there will be a little foaming. If there is a flash and bang, use the cells anyway; the transfection may still have worked.
7. Transfer the contents of each cuvet to a separate flask with medium.
8. Proceed, two cuvets at a time, until all transfections are complete.
9. Take two flasks at random and count the cell density. It should be no more than 5×10^5/mL and there should clearly be dead cells and debris. Dilute if necessary.
10. Place in the incubator overnight.

3.2.2.3. SELECTION AND CLONING

1. Count the cells in each flask.
2. Dilute to 2×10^5/mL and add selective drug. In fact, in most cases there is no risk of overgrowth and the cells do not need counting before adding the drug and putting in multiwell plates, but counting is safer.
3. Plate the cells out into two 24-well plates with 1 mL per well. Any remaining culture can be kept in a larger flask. The no-DNA control can also be left in a flask.
4. Incubate for 5 d. At this time, any well containing live trypanosomes should be recognizable under the inverted microscope. Upon seeing the parasites, count to determine the cell density. It is also easy to see when the parasites start to grow because the medium turns yellower. The controls are usually dead after 4 d, often earlier.
5. When the density reaches 10^6/mL, transfer to a larger volume.
6. If there is no visible growth after 8 d, throw the cultures away and start again.
7. If this happens three times and a positive control experiment with empty vector gives clones, the RNAi is probably leaky and lethal. Under these circumstances, you may want to use an alternative region of the gene or switch to the stem-loop procedure if not already doing so.

3.3. Procyclic Trypanosome RNAi Lines

3.3.1. Procyclic Trypanosome Culture

Procyclic trypanosomes usually die if they are diluted below about 1×10^5 cells/mL, although the presence of other cells can be mimicked by using conditioned medium. The component of conditioned medium that promotes parasite growth is not known. The parasites grow to a maximum cell density of $(1–2) \times 10^7$/mL and they can be left for about a day at this density without dying, although such cultures may take a day or two to start growing after dilution. The cell density usually increases fourfold in 24 h, although behavior varies between lines and laboratories. The cells are aerobic and are classically grown at 27°C, but growth at 30°C is also possible. If the ambient temperature rises above about 22°C, it will be necessary to buy a refrigerated incubator, as most incubators can only operate at temperatures 5°C above the ambient temperature. Note that the ability to grow at 30°C allows the parasites to share an incubator or warm room with *Saccharomyces cerevisiae*; if you do this, contamination can be avoided simply by wrapping the caps of flasks with parafilm.

Small volumes are easiest to grow in plastic, and large volumes can be grown in slowly shaking glass conical flasks.

The parasites are aerobic, but the flasks cannot be left open in refrigerated incubators or warm rooms because the cultures will dry out. The solution to this is to make sure that at least 80% of the volume is air and—for flat flasks—to lay the flask down to ensure a maximum air–liquid interface. Caps should then be screwed down tightly to avoid evaporation. When using 24- or 96-well plates, seal them with parafilm.

Freezing and thawing protocols are exactly as for bloodstream forms but using the appropriate medium.

3.3.2. Electroporation and Selection

3.3.2.1. PREPARATION

Follow **subheading 3.2.2.1., steps 1–3.**

4. The day before transfection, inoculate fresh medium with trypanosomes at a final density of 5×10^5/mL. You can transfect at densities up to, but not exceeding, 5×10^6/mL.

3.3.2.2. TRANSFECTION

Follow **Subheading 3.2.2.2., item 1.**

2a. After centrifuging the trypanosomes, transfer the supernatant (used medium) to a new sterile tube or bottle. Store it in the refrigerator. If you are worried about sterility, you can refilter the supernatant.
2b. The electroporation medium is ZPFM.

3.3.2.3. Selection and Cloning

1. Count the cells in each flask.
2. Dilute the cells to 2×10^5/mL using the conditioned medium that you saved from the previous day. Add the selective drug.
3. Plate on a 24-well plate. You can either put 1 mL of the culture per well or make four serial twofold dilutions. Usually the transfection efficiency is sufficiently low that plating the original culture is sufficient to give clonal populations.
4. Incubate for 7 d. At this time, any well containing live trypanosomes should be recognizable under the inverted microscope. Upon seeing the parasites, count to determine the cell density. It is imperative to count the cells, as the appearance of the deep cultures in wells is deceptive and excessive dilution at this stage will kill the parasites. If there are no cells growing well after 10 d, throw away the plates and start again.
5. When the density reaches 2×10^6/mL, transfer to a larger volume, but do not under any circumstances dilute to less than 5×10^5/ml.
6. If the majority of wells is positive, you may not have clones. This can, if necessary, be corrected later by recloning using conditioned medium or using normal (non-drug-resistant) cells as "feeders."

3.4. Checking for RNAi

The effectiveness of RNAi varies between clones, probably due at least partially to variation between integration sites: There are multiple copies of the rRNA spacer region on several different chromosomes. Therefore, it is nearly always necessary to analyze several independent clones in order to find the clone with the most effective reduction in gene expression.

3.4.1. Cell Growth

The easiest initial test of the clones is to plate each clone into two neighboring wells on a 24-well plate; use 2×10^5/mL for procyclics and 5×10^4/mL for bloodstream forms. Add tetracycline (final concentration, 100 ng/mL) and follow growth over the next 2–3 d. This is only useful if growth effects are seen in some clones, which can then be picked for further analysis.

3.4.2. Northern and Western Blots

Pick up to six clones for each transfection and grow for 2–10 d with and without tetracycline. A Northern blot using any probe for the target gene (including your RNAi fragment) can be done with total RNA from 3×10^7 trypanosomes, Note, however, that effective RNAi in trypanosomes sometimes merely alters the mobility of an RNA rather than eliminating the transcript (*35*). Messenger RNA levels are strongly affected by parasite density (*36*), so be sure to make the RNA from cells at similar densities: We usually use $(0.5–1.0) \times 10^6$/mL for bloodstream forms and $(1–2) \times 10^6$/mL for procyclics.

The ideal initial screen is to do a Western blot, if antibodies are available. For each lane on a polyacrylamide minigel, spin down $(1–2) \times 10^6$ trypanosomes in the microfuge, resuspend in 10 μL of sample buffer, boil, and load.

3.5. More Detailed Analysis

3.5.1. Growth Kinetics

Growth kinetics after RNAi are highly variable and depend on the stability of the target RNA/protein and its role in the cell. Sometimes, growth inhibition is seen the day following tetracycline addition, but in some cases, only after 10 d. It can also take several days to see a reduction in a protein even if the RNA appears to be depleted. Experience so far indicates that many trypanosome proteins are present in 5- to10-fold excess over that required for maintenance of the wild-type phenotype, so at least 80% depletion is required before any change in phenotype can be seen (see, e.g., **refs. *37* and *38***). Phenotypes may also be more severe if the cells are grown at low density on the day before RNAi induction; for example, diluting bloodstream forms to 1×10^4/mL or procyclics to 1×10^5/mL and repeating this the next day just before tetracycline addition. The reasons for this are unknown. Cell death might be preceded by the appearance of cells with abnormal morphology, including elongated or multinucleate cells, especially in procyclics. In bloodstream forms, it is more usual to simply see a reduction or stabilization of cell numbers, or a decrease in the apparent growth rate. In our experience, lysis rapidly follows death so that dead cells are not normally seen at high frequency within the cultures.

The RNAi experiments in trypanosomes are further complicated by the ability of trypanosomes to "escape" from deleterious manipulations (*see* **Note 1**). Usually, if a defect in growth is seen, the parasites will start to escape within a few days. This is easily seen as a regrowth of normal cells. If the RNAi is very deleterious to the cells, the only phenotype that can be observed will be a reduction in numbers, possibly followed by regrowth of normal cells that no longer show the RNAi effect.

4. Notes

1. It is very common to find that the RNAi phenotype of a trypanosome line is lost upon prolonged cultivation in the absence of tetracycline. This is presumed to be caused by "leakage" of dsRNA transcription in the absence of inducer, and consequent negative selection acting on parasites able to express the dsRNA. The escape mechanism could be, for example, elimination of part of the RNAi plasmid or the T7 polymerase (*41*), some epigenetic mechanism, or an alternative adaptation (such as alterations in metabolism) enabling the parasite to bypass the defect. It is, therefore, important to freeze several aliquots of useful lines as soon as possible to minimize passage.

2. Although RNAi lines are always frozen without tetracycline, "leakage" might prevent their recovery after thawing. Alternatively, lines might be recovered but have

lost the RNAi phenotype. This seems to be more frequent with cells made using the p2T7 strategy than with stem loops. For this reason, it is very important to do as many experiments as possible, as rapidly as possible after obtaining the line, and not to rely on the presence of a stabilate.

3. The easiest strategy for RNAi is the p2T7TA Blue approach, so this might be tried first.

4. If RNAi does not give any phenotypic effects—or no clones are obtained—in one life-cycle stage, it is worth trying the other life-cycle stage, assuming that the protein is expressed in that stage.

5. If no clones are obtained that show a reduction in the target RNA, or no clones are obtained at all even thought the empty vector control yields transformants, try using the stem-loop approach.

6. If this too does not work, it will be necessary to use more laborious strategies—either a straight knockout or a knockout that is compensated by the presence of an inducible copy of the wild-type gene *(37–40)*.

7. If it is anticipated that the line will be stored and used for further experiments, and especially if it is planned to retransfect the line with additional plasmids, use a stem-loop strategy.

8. Useful Internet links are as follows:

Cultivation of trypanosomes:	http://tryps.rockefeller.edu/
Molecular Parasitology website:	http://164.67.60.203/par/molpar/index.html
Trypanofan Chromosome I RNA interference project, including protocols and p2T7^{TABlu} sequence:	http://www.trypanofan.org http://homepages.lshtm.ac.uk/~ipmbdhor/ Protocols.htm
Trypanosome genome at the Sanger Centre:	http://www.genedb.org/genedb/tryp/ index.jsp
Trypanosome genome at TIGR:	http://www.tigr.org/tdb/e2k1/tba1/

References

1. Ngo, H., Tschudi, C., Gull, K., and Ullu, E. (2002) Double-stranded RNA induces mRNA degradation in *Trypanosoma brucei*. *Proc. Natl. Acad. Sci. USA* **95,** 14,687–14,692.

2. Djikeng, A., Shi, H., Tschudi, C., and Ullu, E. (2001) RNA interference in *Trypanosoma brucei*: cloning of small interfering RNAs provides evidence for retroposon-derived 24–26-nucleotide RNAs. *RNA* **7,** 1522–1530.

3. Djikeng, A., Shi, H., Tschudi, C., Shen, S., and Ullu, E. (2003) An siRNA ribonucleoprotein is found associated with polyribosomes in *Trypanosoma brucei*. *RNA* **9,** 802–808.

4. Shi, H., Djikeng, A., Tschudi, C., and Ullu, E. (2004) Argonaute protein in the early divergent eukaryote *Trypanosoma brucei:* control of small interfering RNA accumulation and retroposon transcript abundance. *Mol. Cell Biol.* **24,** 420–427.

5. Durand-Dubief, M. and Bastin, P. (2003) TbAGO1, an Argonaute protein required for RNA interference, is involved in mitosis and chromosome segregation in *Trypanosoma brucei. BMC Biol.* **1,** 2.
6. Shen, S., Arhin, G. K., Ullu, E., and Tschudi, C. (2001) In vivo epitope tagging of *Trypanosoma brucei* genes using a one step PCR-based strategy. *Mol. Biochem. Parasitol.* **113,** 171–173.
7. Clayton, C. E. (2002) Developmental regulation without transcriptional control? From fly to man and back again. *EMBO J.* **21,** 1881–1888.
8. Liang, X., Haritan, A., Uliel, S., and Michaeli, S. (2003) *Trans* and *cis* splicing in trypanosomatids: mechanism, factors, and regulation. Euk. *Cell* **2,** 830–840.
9. Clayton, C. E. (1999) Genetic manipulation of Kinetoplastida. *Parasitol. Today* **15,** 372–378.
10. Morris, J. C., Wang, Z., Drew, M. E., and Englund, P. T. (2002) Glycolysis modulates trypanosome glycoprotein expression as revealed by an RNAi library. *EMBO J.* **21,** 4429–4438.
11. McKean, P. G., Baines, A., Vaugha, S., and Gull, K. (2003) Gamma-tubulin functions in the nucleation of a discrete subset of microtubules in the eukaryotic flagellum. *Curr. Biol.* **13,** 598–602.
12. Drew, M. E., Morris, J. C., Wang, Z., et al. (2003) The adenosine analog tubercidin inhibits glycolysis in *Trypanosoma brucei* as revealed by an RNA interference library. *J. Biol. Chem.* **278,** 46,596–46,600.
13. Estévez, A. M., Lehner, B., Sanderson, C. M., Ruppert, T., and Clayton, C. (2003) The roles of inter-subunit interactions in exosome stability. *J. Biol. Chem.* **278,** 34,943–34,951.
14. Guerra-Giraldez, C., Quijada, L., and Clayton, C. E. (2002) Compartmentation of enzymes in a microbody, the glycosome, is essential in *Trypanosoma brucei. J. Cell Sci.* **115,** 2651.
15. Allen, C. L., Goulding, D., and Field, M. C. (2003) Clathrin-mediated endocytosis is essential in *Trypanosoma brucei. EMBO J.* **22,** 4991–5002.
16. Hammarton, T. C., Engstler, M., and Mottram, J. C. (2004) The *Trypanosoma brucei cyclin*, CYC2, is required for cell cycle progression through G1 phase and for maintenance of procyclic form cell morphology. *J. Biol. Chem.* **279,** 24,757–24,764.
17. Mandelboim, M., Barth, S., Biton, M., Liang, X., and Michaeli, S. (2003) Silencing of Sm proteins in *Trypanosoma brucei* by RNA interference captured a novel cytoplasmic intermediate in spliced leader RNA biogenesis. *J. Biol. Chem.* **278,** 51,469–51,478.
18. Wang, Z., Morris, J. C., Drew, M. E., and Englund, P. T. (2000) Inhibition of *Trypanosoma brucei* gene expression by RNA interference using an integratable vector with opposing T7 promoters. *J. Biol. Chem.* **275,** 40,174–40,179.
19. LaCount, D. J., Bruse, S., Hill, K. L., and Donelson, J. E. (2000) Double-stranded RNA interference in *Trypanosoma brucei* using head-to-head promoters. *Mol. Biochem. Parasitol.* **111,** 67–76.
20. Morris, J. C., Wang, Z., Drew, M. E., Paul, K. S., and Englund, P. T. (2001) Inhibition of bloodstream form *Trypanosoma brucei* gene expression by RNA interference using the pZJM dual T7 vector. *Mol. Biochem. Parasitol.* **117,** 111–113.

21. Bastin, P., Ellis, K., Kohl, L., and Gull, K. (2000) Flagellum ontogeny in try-panosomes studies via an inherited and regulated RNA interference system. *J. Cell. Sci.* **113,** 3321–3328.
22. Shi, H., Djikeng, A., Mark, T., Wirtz, E., Tschudi, C., and Ullu, E. (2000) Genetic interference in *Trypanosoma brucei* by heritable and inducible double-stranded RNA. RNA **6,** 1069–1076.
23. Biebinger, S., Wirtz, L. E., and Clayton C. E. (1997) Vectors for inducible over-expression of potentially toxic gene products in bloodstream and procyclic *Trypanosoma brucei. Mol. Biochem. Parasitol.* **85,** 99–112.
24. Wirtz, E., Leal, S., Ochatt, C., and Cross, G. A. M. (1999) A tightly regulated inducible expression system for conditional gene knock-outs and dominant-negative genetics in *Trypanosoma brucei. Mol. Biochem. Parasitol.* **99,** 89–102.
25. Wickstead, B., Ersfeld, K., and Gull, K. (2002) Targeting of a tetracycline-inducible expression system to the transcriptionally silent minichromosomes of *Trypanosoma brucei. Mol. Biochem. Parasitol.* **125,** 211–216.
26. Alibu, P., Storm, L., Haile, S., Horn, D., and Clayton, C. (2005) A doubly inducible system for RNA interference and rapid RNAi plasmid construction in *Trypanosoma brucei. Mol. Biochem. Parasitol.* **139,** 75–82.
27. Duszenko, M., Ferguson, M. A. J., Lamont, G., Rifkin, M. R., and Cross, G. A. M. (1986) Cysteine eliminates the feeder cell requirement for cultivation of *Trypanosoma brucei* bloodstream forms in vitro. *J. Exp. Med.* **162,** 1256–1263.
28. Hirumi, H. and Hirumi, K. (1989) Continuous cultivation of *Trypanosoma brucei* bloodstream forms in a medium containing a low concentration of serum protein without feeder cell layers. *J. Parasitol.* **75,** 985–989.
29. Carruthers, V. B. and Cross, G. A. M. (1992) High efficiency clonal growth of bloodstream- and insect-form *Trypanosoma brucei* on agarose plates. *Proc. Natl. Acad. Sci. USA* **89,** 8818–8821.
30. Ziegelbauer, K., Quinten, M., Schwarz, H., Pearson, T. W., and Overath, P. (1990) Synchronous differentiation of *Trypanosoma brucei* bloodstream to procyclic forms in vitro. *Eur. J. Biochem.* **192,** 373–378.
31. Vassella, E. and Boshart, M. (1996) High molecular mass agarose matrix supports growth of bloodstream forms of pleomorphic *Trypanosoma brucei* strains in axenic culture. *Mol. Biochem. Parasitol.* **82,** 91–105.
32. Brun, R. and Schönenberger, M. (1979) Cultivation and *in vitro* cloning of pro-cyclic culture forms of *Trypanosoma brucei* in a semi-defined medium. *Acta Trop.* **36,** 289–292.
33. Durand-Dubief, M., Kohl, L., and Bastin, P. (2003) Efficiency and specificity of RNA interference generated by intra- and intermolecular double-stranded RNA in *Trypanosoma brucei. Mol. Biochem. Parasitol.* **129,** 11–21.
34. Redmond, S., Vadivelu, J., and Field, M. C. (2003) RNAit: an automated web-based tool for the selection of RNAi targets in *Trypanosoma brucei. Mol. Biochem. Parasitol.* **128,** 115–118.
35. Hendriks, E. F., Abdul-Razak, A., and Matthews, K. R. (2003) *Tb*CPSF30 deple-tion by RNA interference disrupts polycistronic RNA processing in *Trypanosoma brucei. J. Biol. Chem.* **278,** 26,870–26,878.

36. Häusler, T. and Clayton, C. E. (1996) Post-transcriptional control of hsp 70 mRNA in *Trypanosoma brucei. Mol. Biochem. Parasitol.* **76,** 57–72.
37. Krieger, S., Schwarz, W., Ariyanagam, M. R., Fairlamb, A., Krauth-Siegel, L., and Clayton, C. E. (2000) Trypanosomes lacking trypanothione reductase are avirulent and show increased sensitivity to oxidative stress. *Mol. Microbiol.* **35,** 542–552.
38. Helfert, S., Estévez, A., Bakker, B., Michels, P., and Clayton, C. E. (2001) The roles of triosephosphate isomerase and aerobic metabolism in *Trypanosoma brucei. Biochem J.* **357,** 55–61.
39. Maier, A., Lorenz, P., Voncken, F., and Clayton, C. E. (2001) An essential dimeric membrane protein of trypanosome glycosomes. *Mol. Microbiol.* **39,** 1443–1451.
40. Lorenz, P., Meier, A., Erdmann, R., Baumgart, E., and Clayton, C. (1998) Elongation and clustering of glycosomes in *Trypanosoma brucei* overexpressing the glycosomal Pex11p. *EMBO J.* **17,** 3542–3555.
41. Chen, Y., Hung, C.-H., Burderer, T., and Lee, G.-S. M. (2003) Development of RNA interference revertants in *Trypanosoma brucei* cell lines generated with a double stranded RNA expression construct driven by two opposing promoters. *Mol. Biochem. Parasitol.* **126,** 275–279.

6

Analysis of Double-Stranded RNA and Small RNAs Involved in RNA-Mediated Transcriptional Gene Silencing

M. Florian Mette, Werner Aufsatz, Tatsuo Kanno, Lucia Daxinger, Philipp Rovina, Marjori Matzke, and Antonius J. M. Matzke

1. Introduction

RNA-mediated gene silencing is triggered by double-stranded RNAs (dsRNAs) that are processed to short RNAs approx 21–25 nucleotides in length. The short RNAs act as guides for enzyme complexes that degrade, modify, or inhibit the function of homologous target nucleic acids. There are three general types of RNA-mediated silencing: (1) RNA interference (RNAi) and the related processes of posttranscriptional gene silencing in plants and quelling in *Neurospora crassa*, which involve sequence-specific mRNA degradation; (2) microRNA-guided repression of translation; and (3) RNA-directed epigenetic modifications of the genome, such as DNA cytosine (C) methylation and histone methylation *(1)*. This chapter focuses on a specific aspect of the third type of silencing: RNA-directed promoter methylation and transcriptional gene silencing (TGS).

The RNAi-dependent heterochromatin assembly involving histone H3 lysine-9 methylation has been observed in diverse organisms *(2–5)*. By contrast, RNA-directed DNA methylation (RdDM) has been documented so far only in plants *(1,6)*. RdDM is an unusual process that modifies all Cs, regardless of their sequence context, within the region of RNA–DNA sequence homology *(6,7)*. Unlike RNAi-based heterochromatin, which spreads kilobases from the RNA-targeted nucleation site, RdDM does not infiltrate substantially into adjacent sequences *(6,7)*. This suggests that direct RNA–DNA base-pairing creates a substrate that is recognized by DNA methyltransferases. Double-stranded

From: *Methods in Molecular Biology, vol. 309: RNA Silencing: Methods and Protocols*
Edited by: G. Carmichael © Humana Press Inc., Totowa, NJ

Fig. 1. Transgene strategy for inducing RNA-mediated TGS in plants. The silencer locus contains an inverted repeat of target promoter sequences. Transcription of the inverted repeat from an unrelated promoter (shaded box) produces a hairpin RNA, which is processed by a Dicer-Like (DCL) activity into short dsRNAs (21–25 bp) that have two-nucleotide 3′ overhangs. Single-stranded short RNAs are thought to target homologous DNA regions for *de novo* methylation by DNA methyltransferases (DMTase) and associated chromatin factors. RNA-directed DNA methylation is characterized by methylation (m) of cytosines in all sequence contexts (CNG, CG, CNN, where N is A, T, or C).

RNAs that contain promoter sequences can target homologous promoter elements for methylation and induce TGS *(7–13)*.

To study the mechanism of RNA-mediated TGS and promoter methylation, we have established several two-component transgene systems in the model plant *Arabidopsis thaliana (7,13)*. The systems consist of two unlinked transgene loci: a silencer locus and a target locus. The silencer locus contains an inverted DNA repeat of target promoter sequences that is transcribed to produce a double-stranded (hairpin) RNA. The hairpin RNA is processed by Dicer-like activity into short RNAs *(14)*, leading ultimately to *de novo* methylation and TGS of a homologous promoter at the target locus (*see* **Fig. 1**). After briefly outlining the features of transgene constructs that can initiate RNA-mediated TGS, we describe techniques used in our lab for the isolation and detection of dsRNAs and short RNAs active in this silencing pathway *(7–9,13,14)*.

2. Materials

2.1. Chemicals and Solutions

1. 1 *M* Tris-HCl, pH 8.0 (Sigma 7-9) (Sigma, cat. no. T1378).
2. 1 *M* Tris-HCl, pH 8.5.
3. 1 *M* PIPES/NaOH buffer, pH 6.4 (Sigma, cat. no. P9291).
4. 0.5 *M* Phosphate buffer, pH 7.2.
5. Na_2CO_3/$NaHCO_3$, 120 m*M*/80 m*M*: Dissolve 636 mg sodium carbonate (Sigma, cat. no. S7795) and 336 mg sodium bicarbonate (Sigma, cat. no. S6297) in double distilled water to a final volume of 50 mL. Keep in a tightly closed 50-mL plastic tube at room temperature. Always make fresh.
6. 0.5 *M* Na_2EDTA/NaOH, pH 8.0 (EDTA = Titriplex, Merck, cat. no. K31678818).

7. 20% SDS (sodium dodecyl sulfate).
8. 10 mM Tris-HCl, pH 8, 0.1% SDS.
9. 40% Polyethylene glycol MW 8000 (dissolved in water) (Sigma, cat. no. P2139).
10. 3 M Sodium acetate (NaOAc), pH 5.
11. 5 M NaCl.
12. 0.2 M Dithiothreitol (DTT) (Bio-Rad, electrophoresis grade, cat. no. 161-0611).
13. 40% Ammonium peroxodisulfate (Merck, cat. no. K19701601).
14. Diethylpyrocarbonate (DEPC) (Sigma, cat. no. D5758).
15. Concentrated bleach (sodium hypochlorite solution) (Roth, cat. no. 9062.3).
16. Chloroform.
17. Distilled phenol (phenol p.a. from Merck, cat. no. 1.00206.0250; distill in standard organic chemistry apparatus, with air cooling only, in chemical fume hood; boiling temperature of phenol-182°C. Avoid skin contact.)
18. 20X SSC: 3 M NaCl, 0.3 M sodium citrate.
19. High-salt mix: 0.8 M sodium citrate, 1.2 M NaCl.
20. 10X TBE: 121.1 g Tris-HCl (Sigma 7-9), 51.3 g boric acid, 3.72 g Na$_2$EDTA (Titriplex); dissolve in distilled water; final volume-1 L.
21. Tris-buffer-saturated phenol–chloroform: 100 g distilled phenol, 96 mL chloroform, 4 mL isoamyl alcohol, 100 mg 8-hydroxyquinoline (Sigma, cat. no. H-6378); equilibrated four times with an equal volume of 100 mM Tris-HCl, pH 8.
22. Gel loading buffer: 96% formamide, 20 mM EDTA, 1 mg/mL bromphenol blue.
23. 70% Ethanol.
24. Absolute ethanol.
25. Isoamyl alcohol.
26. Trizol reagent (Invitrogen Life Technologies, cat. no. 15596-018).
27. Urea (Invitrogen, cat. no. 15505-027).
28. Acrylamide:bis-acrylamide 37.5 : 1 (AppliChem, cat. no. A1577.0500).
29. TEMED (*N,N,N′,N′*-tetramethylethylenediamine) (Bio-Rad, cat. no. 161-0801).
30. Bromophenol blue (Merck, cat. no. 8122).
31. Deionized formamide: Pour 1 L formamide (Merck, cat. no. K31880384) into a baked 1-L glass beaker containing a sterile magnetic stirring bar, add 10 g resin [AG501-X8 (D), Bio-Rad, cat. no. 142-6425], cover with aluminum foil to minimize air contact, and stir until the resistance reaches a stable level of 1 kΩ (measure with a conductivity meter). Add another 5 g of resin and continue stirring until the resistance reaches a stable level of 5 kΩ. Filter through Whatman No. 1 filter paper, aliquot into 50-mL plastic tubes with screw caps, and store at −80°C.
32. 10 mg/mL Ethidium bromide stock (dissolved in double-distilled water) (Merck, cat. no. K26175015).
33. Double-Distilled water.
34. Liquid nitrogen.
35. Chromic acid (Merck, cat. no. 1.02499.1000).

36. Hybridization solution: 125 mM sodium phosphate buffer, pH 7.2 (10 mL of a 0.5 M stock), 250 mM sodium chloride (2 mL of 5 M NaCl), 7% SDS (2.8 g of SDS in 40 mL final volume), and 50% deionized formamide (20 mL).
37. Hamilton's homogenization solution (50 mL): 2.5 mL of 1 M Tris-HCl, pH 8.5, 1 mL of 0.5 M Na$_2$EDTA/NaOH, pH 8, 1 mL of 5 M NaCl, 5 mL of 20% SDS, and 40.5 mL of DEPC-treated double-distilled H$_2$O.
38. RNAseA buffer (200 mL): 20 mM Tris-HCl, pH 7.5 (4 mL of 1 M stock), 60 mM NaCl (2.4 mL of 5 M stock), 5 mM EDTA (2 mL of 0.5 M stock), 10 µg/mL of RNase A (200 µL of 10 mg/mL stock), and distilled water to 200 mL.

2.2. Supplies

1. Vinyl gloves (Kimberly Clark, cat. no. 9982).
2. Grinding spheres (Plant DNA isolation kit [Roche, cat. no. 1667319]).
3. Miracloth (Calbiochem, cat. no. 475855).
4. 3MM paper (Whatman, cat. no. 3030917).
5. Zeta-Probe GT nylon membrane (Bio-Rad, cat. no. 162-0197).
6. Sephadex G50 (Pharmacia, cat. no. 17-0573-02).
7. Kodak X-ray film (Sigma, cat. no. F5513-50EA).
8. Hybridization bags (Roche, cat. no. 1666649).
9. Parafilm.
10. Saran Wrap.

2.3. Centrifuges and Tubes

1. 1.5-mL Eppendorf tubes.
2. Eppendorf centrifuge 5415 D.
3. Cooled tabletop centrifuge for 1.5-mL tubes capable of approx 10,000g centrifugation force (e.g., Sigma1-15K at 13,000 rpm).
4. 50-mL plastic conical tubes, sterile (Greiner Bio-one, cat. no. 227261).
5. 15-mL Corex tubes plus rubber adaptors.
6. 30-mL Corex tubes plus rubber adaptors.
7. High-speed refrigerated centrifuge (e.g., Beckman J2-21).
8. High-speed swing-out rotor suitable for 15-mL and 30-mL Corex tubes plus rubber adaptors (e.g., Beckman JS13.1).

2.4. Incubators, Ovens

1. 42°C incubator.
2. 64°C incubator.
3. Heating block (e.g., Eppendorf Thermostat 5320).
4. 80°C vacuum oven.
5. Baking oven (180°C).

2.5. Equipment

1. GeneQuant ultraviolet (UV) spectrophotometer (Pharmacia, cat. no. 80-2190-98).
2. Polyacrylamide gel electrophoresis apparatus (vertical gel electrophoresis system; Biometra, cat. no. V15-17).

3. UV transilluminator.
4. Trans-blot SD semidry transfer cell (Bio-Rad, cat. no. 170-3848).
5. Gilson Pipetman (20 µL, 200 µL, 1 mL, 10 mL), with appropriate autoclaved tips.
6. Plastic dessicator attached to vacuum pump for drying nucleic acid pellets.
7. Hybridization bag sealer (e.g., Audion Elektro, Sealboy 235).

2.6. Enzymes, Radioactive dNTPs/NTPs, Kits, Nucleic Acids

1. RNase One (Promega, cat. no. M 426 C).
2. Megaprimed DNA probe synthesis (Amersham kit RPN 1605).
3. NTP set, 100 mM solutions (Amersham, cat. no. 27-2025-01).
4. [α-^{32}P]dATP (New England Nuclear, cat. no. NEG012H or equivalent).
5. [α-^{32}P]UTP (New England Nuclear, cat. no. NEG507H or equivalent).
6. 0.155 to 0.177-kb RNA size standards (Invitrogen Life Technologies, cat. no. 15623-010).
7. RNasin (Roche, cat. no. 3 335399).
8. RNA polymerase (T7, T3, SP6 as appropriate) (Roche: cat. nos. T7, 881767; T3, 1031163: SP6, 810 274).
9. DNaseI, RNase-free (Amersham, cat. no. 27-0514-01).
10. RNase A (10 mg/mL stock solution in water) (Sigma, cat. no. R-5500).
11. tRNA (1 µg/mL stock solution in double-distilled water) (Roche, cat. no. 109517).
12. pGEM-T easy vector (Promega, cat. no. A1360).
13. DNA oligonucleotides for transcribing short RNA standards (Biolegio or a comparable supplier).

3. Methods

The methods described below outline (1) a transgene approach for inducing RNA-mediated TGS via promoter dsRNAs, (2) isolation of dsRNA and small RNAs, (3) polyacrylamide gel electrophoresis (PAGE) of dsRNA and small RNAs, (4) transfer of RNAs from gel to hybridization membrane by electroblotting, and (5) hybridization conditions to detect dsRNA and small RNAs.

3.1. Transgene Strategy for Inducing RNA-Mediated Transcriptional Gene Silencing

The basic requirement for RNA-mediated TGS is a dsRNA that contains promoter sequences. This is synthesized most efficiently by transcribing an inverted DNA repeat of target promoter sequences. Factors to consider when assembling a silencing construct for inducing RNA-mediated TGS are (1) the susceptibility of the target promoter to this type of silencing, (2) the length of the inverted repeat and spacer, (3) the region of target promoter to be covered by the RNA trigger, (4) the characteristics of the promoter transcribing the inverted repeat, and (5) the need for transcription termination sequences.

3.1.1. Susceptibility of the Target Promoter

Although a large number of plant promoters have not yet been examined for susceptibility to RNA-mediated TGS, those tested so far appear to be silenced effectively by this approach. This includes transgene promoters (constitutive and tissue-specific) as well as endogenous gene promoters. Various target promoters have been silenced successfully in different plant species, including *Arabidopsis thaliana*, *Nicotiana tabacum* (tobacco), and *Petunia hybrida* *(7–13)*. Efficient downregulation (up to 50-fold) of even strong, tissue-specific promoters has been achieved using this technique *(13)*. It is conceivable that sequence composition (i.e., number and density of CpG, CpNpG, and CpNpN nucleotide groups) could affect sensitivity to RNA-mediated TGS. For example, a promoter that is rich in CpG dinucleotides, which maintain methylation much more reliably than non-CpGs in the absence of the RNA trigger *(7,10)*, might be more persistently silenced by RNA-mediated TGS than a promoter that is deficient in CpGs. More studies are required to assess this possibility and to determine specific features of target promoters that enhance or reduce susceptibility.

3.1.2. Length of Inverted Repeat and Spacer

The silencer locus contains a transcribed inverted repeat of target promoter sequences. From studies carried out so far, the optimal length of each copy of the inverted repeat appears to be 200–400 bp (basepairs), with an intervening spacer region comprising several hundred basepairs of unrelated sequence *(7,13)*. Although a systematic analysis has not yet been conducted, it is likely that longer sequences in the inverted repeat will not always be transferred intact into host chromosomes. A generous spacer is necessary to ensure stability of the inverted repeat construct during cloning steps in bacteria. In our silencing systems, short RNAs originate from the entire dsRNA, indicating efficient processing and subsequently full coverage of the complete DNA sequence that is targeted *(14)*.

3.1.3. Region of Target Promoter to Be Covered by the RNA Trigger

Constructs used successfully in our lab have targeted the region around the transcription start site and sequences upstream. Addition of the 5′-untranslated region into the silencing construct does not appear to interfere with RNA-mediated TGS but is not necessary to achieve efficient silencing. Any regions containing cytosine residues that are known to be critical for silencing when methylated should be included in the inverted repeat sequence.

3.1.4. Characteristics of Promoter Transcribing the Inverted Repeat

Using a strong constitutive promoter to transcribe the inverted repeat of promoter sequences results in abundant dsRNA and short RNA processing products *(7,10,13)*. For this purpose, we normally use the nominally constitutive 35S promoter of cauliflower mosaic virus, which produces around 1–10 pg of dsRNA per 100 μg total RNA. A tissue-specific promoter can also be employed to transcribe the inverted repeat in selected cell types, but this might not always achieve the desired result. For example, we have used a strong embryo-specific promoter to drive transcription of an inverted repeat containing sequences of a distinct but related embryo-specific promoter. Instead of uniform, complete silencing of the target promoter in embryos, which had been expected based on the individual expression patterns of the two promoters, a random, mottled pattern of silencing was frequently observed (*see* **Fig. 2, left**). A more predictable pattern of silencing of the embryo-specific target promoter was achieved using the nominally constitutive 35S promoter (*see* **Fig. 2, right**). Although the basis of the uneven silencing observed with the embryo-specific transcribing promoter is not yet known, it is possible that competition for common embryo transcription factors stochastically disturbs the activity of the transcribing promoter. To ensure the most consistent silencing, regardless of the tissue specificity of the target promoter, we recommend using a strong constitutive promoter to transcribe the inverted repeat.

3.1.5. Need for Transcription Termination Sequences

Whether to include transcription termination sequences or polyadenylation signals might depend on the system under investigation. Plants have at least one nuclear-localized Dicer-like activity *(14)*, which is thought to process dsRNAs to short RNAs in the nucleus. Accordingly, we routinely avoid placing a transcriptional terminator downstream of the transcribed promoter inverted repeat. This presumably retains the dsRNAs in the nucleus, where they ultimately exert their effect. Further analysis is required to assess the use of RNA-mediated TGS for mammals, where the single Dicer protein is cytoplasmic *(15)*, because promoter dsRNAs might provoke an interferon response in these organisms. For animals that lack an interferon response, information about the subcellular localization of Dicer activities should help to decide whether or not to include transcription termination sequences downstream of the transcribed inverted repeat.

3.2. Detection of dsRNA

This protocol was optimized to detect hairpin RNAs that have a perfect stem of about 200–300 nucleotides and are present at a concentration of about 1–10 pg per 100 μg of total RNA. Such RNAs have been used successfully to induce RNA-mediated TGS and *de novo* methylation of target promoters (*see* **Subheading 3.1.**). We have not systematically analyzed whether the procedure will also work for dsRNAs that are larger or are present at lower concentrations.

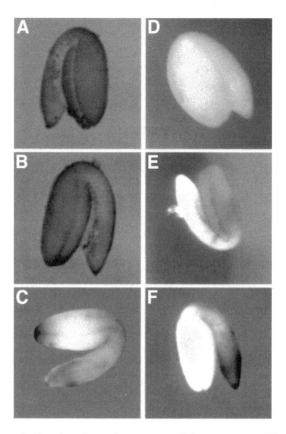

Fig. 2. Pattern of silencing depends on transcribing promoter. The target embryo-specific promoter (**A**) and a related embryo-specific promoter (**B**) both drive strong and uniform expression of a *GUS* reporter gene (dark color) in plant embryos. Surprisingly, when the related embryo-specific promoter is used to transcribe an inverted repeat of the target promoter, instead of uniform and complete silencing [which would result in a completely white embryo (**D**)], incomplete and stochastic silencing of the target promoter is observed (**C**). By contrast, when a nominally constitute viral promoter, which drives *GUS* expression primarily in cotyledons (embryonic leaves), (**E**) is used to transcribe the inverted repeat of the target promoter, silencing occurs in a predictable pattern, which is a mirror image of the pattern of activity of the viral promoter (high in cotyledons, weak in root tip) (**F**).

To selectively enrich for dsRNAs, total RNA (*see* **Subheading 3.2.1.**) is treated with a single-strand-specific ribonuclease (RNase) (*see* **Subheading 3.2.3.**) and then separated by gel electrophoresis (*see* **Subheading 3.2.4.**). After transfer to a membrane (*see* **Subheading 3.2.5.**), the dsRNA of interest is detected by hybridization (*see* **Subheading 3.2.6.**).

3.2.1. Isolation of Total RNA from Plant Tissues

In our laboratory, total RNA is extracted from young tobacco and *Arabidopsis* leaves using the Trizol reagent according to the manufacturer's instructions. However, total RNA prepared with any other method should work as well provided that it does not contain RNase One inhibitors (e.g., SDS). All solutions and plasticware (Eppendorf tubes, pipet tips, etc.) should be RNase-free and sterilized by autoclaving. Glass Corex tubes should be cleaned overnight with chromic acid, rinsed thoroughly with tap water and, finally, with distilled water, capped with aluminum foil, and baked overnight in a 180°C oven. To make double-distilled water that is RNase-free for use in enzyme reactions, add DEPC (250 µL/L water) before autoclaving. Vinyl gloves should be worn when carrying out RNA isolation procedures, enzyme reactions, and blot hybridization steps. Total RNA is treated with RNase One to enrich for dsRNA (*see* **Subheading 3.2.3.**). Before describing that procedure, we outline a method for making synthetic dsRNA, which can be used to assess the efficiency of the RNase One reaction and to generate size markers.

3.2.2. Synthetic dsRNA Size Markers

When investigating a dsRNA encoded by a transgene inverted repeat of known length, a synthetic size standard can be made by annealing sense and antisense RNAs that have been transcribed in vitro. The sequence of interest must be cloned into a plasmid (e.g., pGEM-T easy), which allows in vitro transcription of both insert DNA strands from a promoter of a DNA-dependent RNA polymerase (T3, T7, or SP6).

3.2.2.1. In Vitro Transcription of Sense and Antisense RNAs

1. Cut plasmid containing DNA sequences homologous to dsRNA with appropriate enzymes to allow runoff transcripts from T3, T7, or SP6 RNA polymerase.
2. Set up separate 50-µL reactions for sense and antisense RNA synthesis in two 1.5-mL Eppendorf tubes by adding the following: 2 µL plasmid digest (approx 1 µg plasmid DNA), 34 µL DEPC-treated double-distilled water, 5 µL 10X in vitro transcription buffer (supplied with RNA polymerase); 2.5 µL of 0.2 M DTT; 2 µL RNasin (40 U/µL), 4 µL of 25 mM ATP, CTP, GTP, UTP mix (made by mixing equal volumes of the four 100 mM stock solutions in the NTP set), 1 µL appropriate RNA polymerase (20 U/µL).
3. Incubate at 37°C for 2 h.
4. Add 2.5 µL DNase I (20 U/µL), RNase-free, and incubate at 37°C for 5 min; then add 1 mL Trizol reagent, vortex, add 200 µL chloroform, vortex, and incubate at room temperature for 2–3 min.
5. Centrifuge for 15 min at 4°C, 13,000 rpm (Sigma 1-15K centrifuge); then transfer the upper phase to a new Eppendorf tube.

6. Add 0.5 vol of isopropanol and 0.5 vol of 0.8 *M* sodium citrate, 1.2 *M* of NaCl (high-salt mix), vortex, and incubate 15 min at room temperature.
7. Centrifuge as in **step 5**. Discard supernatant and wash pellet with 1 mL of 70% ethanol; centrifuge as in **step 5**.
8. Discard supernatant and vacuum-dry for 10 min.
9. Dissolve each pellet in 10 µL DEPC-treated double-distilled water; quantify with GeneQuant (*see* **Subheading 3.3.2.**). The expected yield for each reaction is several microgram sense or antisense RNA.
10. Aliquots can be used immediately for annealing (*see* **Subheading 3.2.2.2.**).
11. As controls for sufficient digestion of single-stranded RNA in the RNase One reaction (*see* **Subheading 3.2.3.**), make separate 1 pg/µL stock solutions for sense and antisense RNA in double-distilled water containing 1 µg/µL tRNA as the carrier and store at −80°C.
12. Store remaining sense and antisense RNA at −80°C (solutions most stable if left undiluted).

3.2.2.2. ANNEALING OF SENSE AND ANTISENSE TRANSCRIPTS

1. Put the following in a 1.5-mL Eppendorf tube: 0.1 µg sense RNA, 0.1 µg antisense RNA, 20 µL of 5X RNase protection buffer (0.2 *M* PIPES, pH 6.4, 2 *M* NaOAc, 5 m*M* EDTA), and 80 µL deionized formamide.
2. Incubate 5 min at 80°C, switch off heating block, allow to slowly cool to room temperature in the heating block for 5 h, and store at −80°C until use.
3. Make 1 pg/µL stock in double-distilled water containing 1 µg/µL tRNA as a carrier and store at −80°C.

3.2.3. RNase One Treatment

For each RNase One reaction (equals one gel lane), 50 or 100 µg of total plant RNA are digested using 1–2 U RNase One per 10 µg RNA (*see* **Note 1**).

Controls employing synthetic RNAs (*see* **Subheading 3.2.2.1.**) can be processed in parallel: 0.3 pg and 1 pg of separate sense or antisense RNA are digested with RNase One in a background of 50 or 100 µg tRNA. The absence of a signal after hybridization indicates sufficient digestion of single-stranded RNA. Then, 0.3 pg and 1 pg of synthetic dsRNA are digested with RNase One in a background of 50 or 100 µg of tRNA. A single well-defined band of the expected size indicates the successful enrichment for dsRNA and provides a measure to estimate the amount of dsRNA in the accompanying plant RNA samples (*see* **Fig. 3, right**).

1. To treat total plant RNA with RNase One, put into a sterile 1.5-mL Eppendorf tube RNA dissolved in DEPC-treated double-distilled water and adjusted to 90 µL.
2. Add 10 µL of 10X RNase One reaction buffer (provided with the enzyme by the supplier). Mix well by flicking at the tube, briefly (10 s), then spin in an Eppendorf centrifuge at room temperature to collect the solution at the bottom of the tube.
3. Add RNase One (supplied as 7–10 U/µL), mix well, briefly spin as in **step 2**, and incubate 10 min at 37°C.

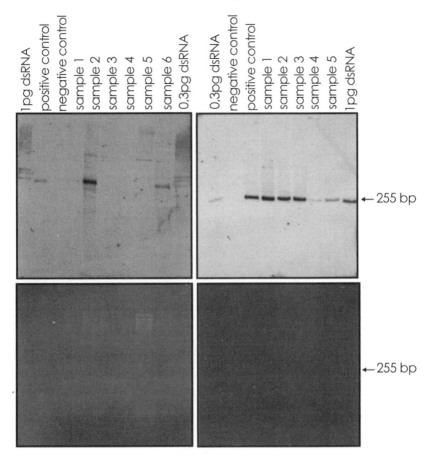

Fig. 3. Detection of dsRNA by RNase One digestion. **Left**: Incomplete RNase One digestion, leading to weak and/or multiple bands in addition to the one of the expected size (255 bp) on the Northern blot **(top)**. **Right**: Optimal digestion produces a single band of the expected size and the quantitative controls (0.3 pg and 1 pg) display the desired ratio of signal strength. Note that this blot is derived from an ethidium bromide-stained gel in which no RNA was visible **(bottom, right)** as compared to the stained gel for the incomplete RNase One treatment, on which stained bands were still apparent **(bottom, left)**. For further information, *see* **Note 1**.

4. Add 5 μL of 20% SDS (stops the reaction).
5. Extract with Tris-buffer-saturated phenol–chloroform (add 150 μL Tris-buffer-saturated phenol–chloroform, vortex for 30 s, centrifuge 5 min in an Eppendorf centrifuge at room temperature to separate phases, and transfer the upper aqueous phase to a sterile 1.5-mL Eppendorf tube).
6. Extract with chloroform (add 150 μL chloroform and treat as in **step 5**; transfer the upper aqueous phase to a sterile 1.5-mL Eppendorf tube).

7. Precipitate RNA from aqueous phase: Add 12 µL of 3 *M* NaOAc, pH 5, then add 300 µL absolute ethanol, mix well by vortexing, and incubate >30 min at −20°C.
8. Centrifuge 15 min at 4°C, 13,000 rpm (Sigma 1–15K centrifuge), discard the supernatant, and wash the sediment with 1 mL of 70% ethanol.
9. Centrifuge 15 min at 4°C, 13,000 rpm (Sigma 1–15K centrifuge), discard the supernatant, and vacuum-dry the sediment for 10 min.
10. Dissolve RNA in 10 µL of 0.2% SDS and store at −20°C or use directly for PAGE (*see* **Subheading 3.2.4.**).

3.2.4. PAGE of dsRNA

To separate RNase One digestion products (*see* **Subheading 3.2.3.**), we typically use a 5% polyacrylamide gel containing 7 *M* urea (useful range about 100–1000 nucleotides). A suitable size standard for this experiment is the 0.155- to 1.77-kb RNA standard (5 µg per lane) from Invitrogen. If the expected size and sequence of the RNase One digestion product are known (e.g., from a hairpin RNA encoded by a transgene inverted repeat of known length), a synthetic dsRNA can be made and used as a hybridization control and size standard (*see* **Subheading 3.2.2.**).

Electrophoresis on 5% polyacrylamide–7 *M* urea gel (0.225X TBE):

1. Size: 18 cm × 13 cm × 1.5 mm, comb for 12 wells, about 10-mm well width; glass plates are cleaned with dishwashing detergent and 70% ethanol.
2. Gel: Add to a 250-mL Erlenmeyer flask the following: 28.8 g urea (final concentration, 7 *M*), 28.5 mL double-distilled water, 7.5 mL of 40% acrylamide : *bis*-acrylamide 37.5 : 1 (final percent = 5%) and 1.35 mL of 10X TBE (final, 0.225X).
3. Heat (64°C incubator) until the urea is completely dissolved (swirl flask to mix contents, then filter through Miracloth into a 100-mL Erlenmeyer flask, and let stand at room temperature for 30–45 min to cool.
4. Add 62.5 µL of 40% ammonium peroxodisulfate and 40 µL TEMED; mix well and pour gel.
5. Allow to polymerize overnight (cover with Saran Wrap to avoid dessication), then, set up in gel box with 0.225X TBE running buffer.
6. Prepare the RNA samples: To the RNase One–treated RNA sample dissolved in 10 µL of 0.2% SDS (*see* **Subheading 3.2.3.**), add 40 µL gel loading buffer, denature 5 min at 95°C on heating block, place on ice for 5 min, then load samples to gel wells (*see* **Note 2**).
7. Carry out the electrophoresis: Set limits of the power supply to 800 V, 25 mA, 10 W and run until the bromophenol blue reaches the lower edge of of the gel (time: 60 min [bromophenol blue has about same mobility as tRNA of 70 nucleotides] or longer [e.g., 90 min, optimal result for RNA of 300 nucleotides]).
8. Stain the gel for 10 min in 0.5X TBE containing 1 µg/mL ethidium bromide and then wash 20 min in 0.5X TBE. At this point, the gel can be optionally photographed under UV transillumination (*see* **Fig. 3, bottom**) and can now be used to transfer the RNA to hybridization membrane (*see* **Subheading 3.2.5.**).

3.2.5. Transfer of RNA to Hybridization Membrane by Electroblotting

Whereas the transfer of RNA from agarose gels to hybridization membrane can be carried out by standard capillary blotting, the transfer from polyacrylamide gels requires electroblotting.

1. Setup: Bio-Rad Trans-Blot SD semidry transfer cell in cold room; all steps at 4°C. Place onto transfer cell from bottom (anode) to top (cathode) the following: three sheets of Whatman 3MM, 20 cm × 15 cm, soaked with 0.5X TBE (*see* **Note 3**), a piece of Bio-Rad Zeta-Probe GT nylon membrane, 20 cm × 15 cm, soaked with 0.5X TBE, the gel, equilibrated with 0.5X TBE, and, finally, three sheets of Whatman 3MM paper, 20 cm × 15 cm, soaked with 0.5X TBE.
2. Transfer step: Transfer RNA to the membrane at 10 V for 1 h, and then rinse the membrane briefly in 0.5X TBE. Rub gently but firmly with a gloved finger to remove particles that can contribute to background hybridization. Briefly air-dry the membrane. The positions of the bands can be optionally visualized under UV transillumination, and the positions of the bands of the 0.155- to 1.77-kb RNA size standards can be marked with a pencil.
3. Place the membrane between two sheets of Whatman 3MM paper and vacuum-dry at 80°C for 2 h to crosslink the RNA to the membrane.

3.2.6. Hybridization with ^{32}P-Labeled DNA Probes

To detect dsRNAs of interest, ^{32}P-labeled "megaprimed" DNA probes are used. Details of probe synthesis and hybridization conditions suitable for the membranes used for electroblotting are given in **Subheading 3.2.6.1.** and **Subheading 3.2.6.2.**, respectively.

3.2.6.1. "MEGAPRIMED" DNA PROBE SYNTHESIS

For this, the Amersham kit RPN 1605 is used. Detailed information is available on the Amersham-Pharmacia website.

For each 15-cm × 20-cm membrane, two 50-μL reactions are performed:

1. To a sterile 1.5 mL Eppendorf tube, add the following: 1 μL template DNA (25 ng/μL) (*see* **Note 4**), 5 μL primer solution (supplied in kit), 20 μL DEPC-treated double distilled water.
2. Denature 5 min at 95°C on a heating block; incubate 10 min on ice.
3. Add 12 μL dNTPs (minus dATP) (supplied in kit), 5 μL (50 μCi) [α-^{32}P]dATP, 5 μL of 10X reaction buffer (supplied in kit), 2 μL of 1 U/μL Klenow DNA polymerase (supplied in kit) and incubate for 2 h at 37°C.
4. Pool reactions (final volume to 100 μL) and remove unincorporated ^{32}P by gel filtration on a 1-mL column of Sephadex G50 resin equilibrated in 10 mM Tris-HCl, pH 8, 0.1% SDS.
5. Denature the probe for 5 min at 95°C in a heating block, followed by incubating for 10 min on ice.

3.2.6.2. Hybridization Reaction (dsRNA)

1. Prewet the membrane in double-distilled water, place it in a hybridization bag, and prehybridize membrane in 40 mL hybridization solution at 42°C for 4–5 h (*see* **Note 5** for details of the preparation of the hybridization solution).
2. Replace the hybridization solution and add freshly denatured probe (*see* **Note 6**).
3. Hybridize at 42°C overnight.
4. Remove the filter from the bag and place in appropriate tray and wash:
   a. One time, 15 min in 2X SSC, 0.1% SDS at room temperature.
   b. One time, 15 min in 0.5X SSC, 0.1% SDS at room temperature.
   c. Two times, 15 min in 0.1X SSC, 0.1% SDS at 64°C.
5. Expose to X-ray film (*see* **Fig. 3, top**).

3.3. Detection of Small RNAs

Preparations enriched for low-molecular-weight RNA (about 10–100 nucleotides) are obtained according to the following protocol (*see* **Subheading 3.3.1.**), which is a modified version of one provided by Andrew Hamilton (*16; A. Hamilton, personal communication*). Also described are procedures for quantifying small RNAs (*see* **Subheading 3.3.2.**), synthesis of synthetic small RNA markers (*see* **Subheading 3.3.3.**), PAGE of small RNAs (*see* **Subheading 3.3.4.**), and hybridization conditions and probe synthesis (*see* **Subheading 3.3.5.**).

3.3.1. Small RNA Isolation

1. Collect 1 g fresh leaf tissue in a sterile 50-mL conical plastic tube containing five grinding spheres (cleaned in concentrated bleach and then rinsed thoroughly with DEPC-treated double-distilled water). Add liquid nitrogen to freeze tissue. Wait until liquid nitrogen evaporates and grind frozen tissue to powder by vortexing for 1 min.
2. Suspend frozen powder in 5 mL Hamilton's homogenization solution, vortex, and put on ice.
3. Add 5 mL Tris-buffer-saturated phenol–chloroform; cover tube with parafilm, and vortex. (*Caution*: Avoid skin contact with phenol. Wash in stream of running tap water if exposed.)
4. Incubate 5 min on ice.
5. Transfer to a clean 15-mL Corex centrifugation tube (baked overnight at 180°C); centrifuge in Beckman JS13.1 rotor, 9000 rpm (about 12,500g) for 15 min at 4°C.
6. Using a 10-mL Gilson pipetman and sterile 10-mL tips, transfer the upper aqueous phase (approx 5 mL) to a clean 15-mL Corex tube, add 5 mL Tris-buffer-saturated phenol–chloroform, cover with parafilm, and vortex to mix.
7. Incubate 5 min on ice.
8. Centrifuge in JS 13.1 rotor, 9000 rpm (about 12,500g) at 4°C for 15 min.
9. Collect aqueous supernatant (approx 5 mL) in a clean 30-mL Corex tube.

10. Add 0.5 mL of 3 M NaOAc, pH 5, and 15 mL absolute ethanol. Cover with parafilm and mix well. Incubate at least 2 h at −20°C.
11. Centrifuge in JS 13.1 rotor as in **step 8**.
12. Discard supernatant and wash sediment with 10 mL of 70% ethanol.
13. Centrifuge as in **step 8**.
14. Vacuum-dry sediment for 10 min.
15. Dissolve in 3.87 mL DEPC-treated double-distilled H_2O (cap tube with aluminum foil and incubate for 15 min at 64°C in incubator; vortex briefly after 7 min) (*see* **Note 7**).
16. Add 0.5 mL of 5 M NaCl and 0.63 mL of 40% polyethylene glycol, molecular weight = 8000; incubate on ice for 30 min; centrifuge as in **step 8**.
17. *Arabidopsis* only: Transfer supernatant to a clean 15-mL Corex tube, add 5 mL Tris-buffer-saturated phenol–chloroform, vortex, and centrifuge as in **step 8** (*see* **Note 8**).
18. Save supernatant and transfer to a clean 30-mL Corex tube. Add 0.5 mL of 3 M NaOAc and 15 mL absolute ethanol. Cover with parafilm and mix well. Incubate for 2 h at −20°C. Centrifuge as in **step 8**. Discard supernatant and wash with 10 mL of 70% ethanol. Centrifuge as in **step 8**. Discard supernatant and vacuum-dry for 10 min.
19. Dissolve in 50–100 μL DEPC-treated double-distilled H_2O (cap tube with aluminum foil and place at 64°C for 10 min); mix by pipeting up and down with a 200-μL Gilson pipetman; briefly centrifuge to sediment any remaining particulate material; transfer supernatant to autoclaved 1.5-mL Eppendorf tube.
20. Store at −80°C.

3.3.2. Quantification of Small RNAs

1. Mix 1 μL RNA solution (**step 19** of **Subheading 3.3.1.**) + 3 μL DEPC-treated double-distilled H_2O; measure absorbance at 260 nm using the Pharmacia GeneQuant UV spectrophotometer (*see* **Note 9**).
2. The yield is typically about 70–140 μg for *Arabidopsis* and 250–300 μg for tobacco. After quantification, samples are ready for PAGE (*see* **Subheading 3.3.4.1.**; per lane, use 20 μL of RNA dissolved in DEPC-treated double-distilled water [approx 30–40 μg for *Arabidopsis* and 50–80 μg for tobacco]).

3.3.3. Synthetic Small RNA Markers

For size markers of known sequence to run on the small RNA gel, sense and antisense short RNAs of the desired size (e.g., 17, 21, and 24 bases) can be synthesized from the appropriate DNA oligonucleotides using T7 RNA polymerase directed in vitro transcription *(17)*.

3.3.3.1. ANNEALING OF OLIGONUCLEOTIDES

Run a separate reaction for each length and polarity.

1. Obtain lyophilized DNA oligonucleotides from standard custom synthesis and dissolve to a final concentration of 1 nmol/μL in DEPC-treated double-distilled water.

2. Add 1 nmol (1 µL of stock) of each, the T7 promoter oligonucleotide and the oligonucleotide for the short RNA to be synthesized (*see* **Note 10**), to 50 µL of 10 m*M* Tris-HCl, 1 m*M* Na₂EDTA, pH 8.0, in a 1.5-mL Eppendorf tube.
3. Place sample in 95°C heating block for 2 min.
4. Switch off heating block and allow samples to cool down slowly to room temperature overnight.

3.3.3.2. IN VITRO TRANSCRIPTION OF SMALL RNAS (100 µL REACTION)

Run a separate reaction for each length and polarity.

1. 54 µL of DEPC-treated double-distilled water.
2. 10 µL of 10X in vitro transcription buffer (supplied with T7 RNA polymerase).
3. 5 µL of 0.2 *M* DTT.
4. 20 µL of DNA oligonucleotide templates.
5. 4 µL of 25 m*M* NTP mix (made from NTP set, 100 m*M* solutions).
6. 2.5 µL of RNasin (40 U/µL).
7. 5 µL of T7 RNA polymerase (20 U/µL).
8. Incubate at 37°C for 2 h.
9. Add 1 µL of Dnase I (10 U/µL), RNase-free.
10. Incubate for 15 min at 37°C.
11. Extract with Tris-buffer-saturated phenol–chloroform (add 150 µL of Tris-buffer-saturated phenol–chloroform, vortex for 30 s, centrifuge for 5 min in an Eppendorf centrifuge at room temperature to separate the phases, and transfer the upper aqueous phase to a sterile 1.5-µL Eppendorf tube).
12. Extract with chloroform (add 150 µL of chloroform, treat as in **step 11**, and transfer the upper aqueous phase to a sterile 1.5-mL Eppendorf tube).
13. Add 12 µL of 3 *M* NaOAc, pH 5.0, and 300 µL of ethanol, mix well, and incubate for 2 h at –20°C.
14. Centrifuge for 15 min at 4°C, 13,000 rpm (Sigma 1-15K centrifuge).
15. Vacuum-dry the pellet for 10 min; dissolve in 50 µL of DEPC-treated double-distilled water.
16. Quantify each short RNA separately with Pharmacia GeneQuant UV spectrophotometer (*see* **Subheading 3.3.2.**); dilute to obtain separate "sense" and "antisense" stock solutions containing 1 pmol/µL each of the 17, 21, and 24 nucleotide short RNAs in DEPC-treated double distilled water (*see* **Note 11**).
17. Store at –80°C until use.
18. Add 20 µL of 17 + 21 + 24 sense or antisense stock solution (sense or antisense according to the polarity of probe to be used) to 80 µL of 96% deionized formamide, 20 m*M* EDTA, 1 mg/mL bromophenol blue and load on gel (*see* **Fig. 4**).

3.3.4. PAGE of Small RNAs

Small RNA preparations are separated on a 15% polyacrylamide gel containing 7 *M* urea (useful range about 10–100 nucleotides) using essentially the same gel system as described for dsRNA (*see* **Subheading 3.2.4.**). Small

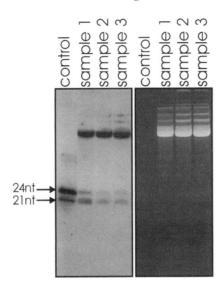

Fig. 4. Detection of small RNAs. On the stained gel (**right**), several bands are visible in sample lanes containing preparations from plant sources. They presumably represent 4S tRNA (approx 80 nucleotides) and 5S rRNA (approx 120 nucleotides). These RNA species can serve as loading controls. After electroblotting and hybridization to detect NOSpro short RNAs (**left**) (*see* **Subheading 3.3.**), the stained bands show background hybridization but also visible are the NOSpro short RNAs (21, 22, and 24 nucleotides) in the three sample lanes. The control lane contains a mixture of 17-, 21-, and 24-nt NOSpro RNAs synthesized in vitro (*see* **Note 11**). Only the two larger sizes are visible on this exposure.

RNA size standards of the sequence of interest are synthesized in vitro (*see* **Subheading 3.3.3.**).

Electrophoresis on 15% polyacrylamide–7 *M* urea gel (0.225X TBE):

1. Size: 18 cm × 13 cm × 1.5 mm, comb for 12 wells, about 10-mm well width; glass plates cleaned with dishwashing detergent and 70% ethanol.
2. Place the following in a 250-mL Erlenmeyer flask: 28.8 g of urea (final concentration, 7 *M*), 13.5 mL double-distilled water, 22.5 mL of 40% acrylamide : *bis*-acrylamide 37.5 : 1 (final, 15%), and 1.35 mL of 10X TBE (final, 0.225X).
3. Heat (64°C incubator) until the urea is completely dissolved (swirl flask to mix contents).
4. Filter through Miracloth into a 100-mL Erlenmeyer flask and let stand at room temperature for 30–45 min to cool.
5. Combine 62.5 µL of 40% ammonium peroxodisulfate and 40 µL of TEMED. Mix well, pour gel and allow to polymerize overnight. Cover with Saran Wrap to avoid dessication.
6. Set up in gel box with 0.225X TBE running buffer.

7. Dissolve 50–100 µg RNA in 20 µL of 0.2% SDS, add 80 µL of gel loading buffer, denature 5 min at 95°C in heating block, and place on ice for 5 min.
8. Load samples to gel wells (*see* **Note 2**).
9. For electrophoresis, set limits to 800 V, 25 mA, 10 W and run until bromophenol blue reaches the lower edge of the gel (75 min, bromophenol blue has about same mobility as RNA of 10 nucleotides).
10. Stain the gel for 10 min in 0.5X TBE containing 1 µg/mL ethidium bromide. It is also required to equilibrate the gel to 0.5X TBE for electroblotting.
11. Wash 20 min in 0.5X TBE. At this point, an optional photo can be taken under UV transillumination (*see* **Fig. 4, right**).
12. Transfer RNA to hybridization membrane (*see* **Subheading 3.2.5.**).

3.3.5. Hybridization with Fragmented ^{32}P-Labeled RNA Probes

For the detection of small RNAs, fragmented ^{32}P-labeled in vitro transcripts are used. Depending on which polarity of short RNA is to be detected on the blot, either a sense or antisense probe is made. Some purposes might require two blots probed with either a sense or antisense probe, respectively. Details of probe synthesis and hybridization conditions are given in **Subheading 3.3.5.1.** and **Subheading 3.3.5.2.**, respectively.

3.3.5.1. FRAGMENTED RNA PROBE SYNTHESIS

1. For each 15-cm X 20-cm membrane, one 20-µL reaction is performed, containing 1 µL of a 1 µg/µL linearized template DNA for in vitro transcription (*see* **Note 12**), 2 µL of 10X in vitro transcription buffer (supplied with RNA polymerase), 1.5 µL of ATP, CTP, GTP mix, 6.7 mM for each nucleotide (made from NTP set, 100 mM solutions), 1 µL of 0.2 M DTT, 12.5 µL (125 µCi) of [α-^{32}P]UTP, 1 µL of RNasin 40 U/µL and, 1 µL of RNA polymerase 20 U/µL (T7, T3, SP6 as appropriate).
2. Incubate for 2 h at 37°C.
3. Add 1 µL of DNase I, RNase-free 10 U/µL and incubate 15 min at 37°C.
4. Add 300 µL of Na_2CO_3/$NaHCO_3$ 120 mM/80 mM and incubate at 60°C for 2.5–3 h (time depends on probe length; for formula, *see* **Note 13**).
5. Add 20 µL of 3 M NaOAc, pH 5.0.
6. Add probe to hybridization buffer without denaturation.

3.3.5.2. HYBRIDIZATION REACTION (SMALL RNAS)

1. Prewet the membrane in double-distilled water, transfer into a hybridization bag, and prehybridize in 40 mL of hybridization solution at 42°C for 4–5 h.
2. Replace hybridization solution, add probe without denaturation (*see* **Note 6**), and hybridize at 42°C overnight.
3. Wash two times for 30 min with 2X SSC and 0.2% SDS at room temperature.
4. Expose to X-ray film overnight.
5. After the first overnight exposure, incubate the membrane for 1 h at 37°C in 200 mL of RNAseA buffer to reduce unspecific hybridization signals.

6. Briefly (2 min) rinse the membrane in 200 mL of 50 m*M* sodium phosphate buffer, pH 7.2.
7. Expose to X-ray film (*see* **Fig. 4**).

4. Notes

1. Some pilot experiments using synthetic dsRNA as well as the corresponding sense and antisense single-stranded RNA controls should be performed for each individual dsRNA to determine the optimal RNase One concentration and/or in order to avoid underdigestion or overdigestion and loss of signal strength after hybridization. **Figure 3** shows two experiments: one with incomplete digestion, in which bands are still visible on the stained gel (bottom, left), and one with complete (but not excessive) digestion, in which the stained gel appears empty (bottom, right). Both resulted in successful detection of the expected 255-bp dsRNA (sample lanes and positive control lane), but the "complete digestion" blot is much cleaner (top, right), lacking the additional larger bands seen on the "incomplete digestion" blot (top, left). In the experiment shown here to detect nopaline synthase promoter (NOSpro) dsRNA, the sense and antisense transcripts used to make the synthetic dsRNA were not perfectly matching but had some unique sequences at the 5′ ends that are derived from the multiple cloning site of the pGEM or pBC vectors during in vitro transcription. For typical control lanes, 1-pg or 0.3-pg synthetic annealed sense and antisense transcripts were digested with RNase One in a background of 50 or 100 μg of tRNA. This simulates the situation when detecting a small amount of natural dsNOSpro RNA derived from the silencer locus in the background of 50 or 100 μg of total RNA extracted from plants. In a suboptimal digestion reaction (**Fig. 3**; top, left), undigested sense and antisense transcripts somewhat "larger" than the one for the dsRNA are still visible on the blot (three bands total; 1 pg dsRNA and 0.3 pg dsRNA); only with optimal digestion (**Fig. 3**; top, right), a single band representing the perfectly double-stranded part annealed sense and antisense transcript remained (0.3 pg dsRNA and 1 pg dsRNA). This could serve as a size marker and quantitative measure for NOSpro dsRNA in various plant RNA preparations (**Fig. 3**; right, sample lanes 1–5).
2. Before loading gel, rinse slots using a buffer-filled syringe. To avoid band "smiling effects", load empty lanes adjacent to actual RNA lanes with an equivalent amount of gel loading buffer (96% deionized formamide, 20 m*M* EDTA, 1 mg/mL bromophenol blue).
3. The easiest way to attract the gel to the hybridization membrane is to let the gel float in 0.5X TBE and to bring the membrane under the gel. The membrane plus attached gel is then carefully lifted from the buffer tray and placed on the three presoaked sheets of Whatman 3 MM paper. On top of the gel, three additional presoaked pieces of Whatman 3MM paper are placed. A glass pipet is rolled carefully over the stack in order to remove trapped air bubbles that could interfere with RNA transfer.
4. An appropriate template is a plasmid fragment excised and extracted from an agarose gel (using QIAquick Gel Extraction kit [Qiagen, 28706] according to the

manufacturer's instructions) and then dissolved in double-distilled water to a concentration of 25 ng/mL.

5. To make the 40 mL hybridization solution, place 2.8 g SDS in an autoclaved 100-mL capped Erlenmeyer flask containing a magnetic stir bar, add 6 mL double-distilled water, and stir on a magnetic stirrer at room temperature until the SDS is dissolved. Then add, in the following order, 20 mL deionized formamide, 10 mL of 0.5 *M* sodium phosphate buffer, pH 7.2, and 2 mL of 5 *M* NaCl. Continue stirring for approx 5 min; then prewarm the hybridization solution by placing the flask in a 42°C incubator before adding it to the prewetted filter in the hybridization bag.

6. Make sure to remove all larger air bubbles from the hybridization bag by gently moving and pressing the bag to the open side before sealing. This can also be done by rolling a glass pipet over the filled hybridization bag to push air bubbles out at an open corner of the bag.

7. Insoluble material can remain after this step, but usually the small RNAs are completely dissolved.

8. This additional extraction step has been found useful for *Arabidopsis* small RNA preparations to remove poorly soluble material. Omitting this step can lead to band distortions during polyacrylamide gel electrophoresis.

9. Centrifuge briefly to sediment insoluble material before quantifying RNA.

10. For the synthesis of T7-generated short RNAs from the sequence of interest, an 18-mer oligonucleotide comprising the T7 promoter (T7pro, below) is annealed to the 3' part of a 34- to 41-mer oligonucleotide (depending on desired final size of short RNA) with the complementary sequence of the T7 promoter downstream of the target sequence (reading the sequence 5' to 3'). This forms a double-stranded DNA region comprising the T7 promoter sequence that is recognized by T7 RNA polymerase. In vitro transcription is initiated with the last G of the T7pro 18mer as the first nucleotide of the transcript. Transcription proceeds along the single stranded unique part of the 34- to 41-mer oligonucleotide (making an RNA complementary to the single-stranded DNA) until the end is reached and the T7 RNA polymerase "runs off" its template *(17)*. As examples, we list here the sense (S) and antisense (AS) oligonucleotides used for the nopaline synthase promoter (Np) short RNA size markers (17, 21, and 24 bases). These contain a sequence from the Np *(18)* (not underlined) and the complementary T7pro 18mer at the 3' ends (underlined). The Np sequences in the oligonucleotides shown here are one nucleotide shorter than the resulting small RNA since the first G of the T7 transcript is complementary to the adjacent base (not included in the oligonucleotide) in the Np sequence. This provides a short RNA of the desired length with full homology to the target Np. In addition to choosing a region where there is a C adjacent to the last nucleotide, a second rule of thumb when selecting sequences for oligonucleotides is to have an intermediate GC content (40–60%) for efficient hybridization with the probe. Note that the 17-base RNA marker does not produce strong signals under the hybridization conditions used (*see* **Fig. 4**), probably indicating that a sequence of this length is at the lower limit of hybrid stability.

T7pro:
5′ TAA TAC GAC TCA CTA TAG 3′
NpS17+T7:
5′ CAA TAA TGG TTT CTG A<u>CT ATA GTG AGT CGT ATT A</u> 3′
NpS21+T7:
5′ CGC GCA ATA ATG GTT TCT GA<u>C TAT AGT GAG TCG TAT TA</u> 3′
NpS24+T7:
5′ GAA CGC GCA ATA ATG GTT TCT GA<u>C TAT AGT GAG TCG TAT TA</u> 3′
NpAS17+T7:
5′ TCA GAA ACC ATT ATT G<u>CT ATA GTG AGT CGT ATT A</u> 3′
NpAS21+T7:
5′ TAC GTC AGA AAC CAT TAT TG<u>C TAT AGT GAG TCG TAT TA</u> 3′
NpAS24+T7:
5′ ACA TAC GTC AGA AAC CAT TAT TG<u>C TAT AGT GAG TCG TAT TA</u> 3′

11. Calculate: approx 300 g/mol per nucleotide: "17," 5100 g/mol; "21," 6300 g/mol; "24," 7200 g/mol. Therefore, 1 pmol: "17," 5.1 ng; "21," 6.3 ng; "24," 7.2 ng. Dilute 10 ng/µL ≅ 1 pmol/µL. Load on gel 20 µL (≅ 20 pmol) plus 80 µL 96% formamide, 20 mM EDTA, 1 mg/mL bromophenol blue.

12. An appropriate template is cloned in pGEM or other appropriate vector providing promoter sequences for in vitro transcription. The vector is cut with an appropriate restriction enzyme such that either a sense or antisense transcript will be made.

13. The incubation time can be calculated according to the formula $t = (L_i{-}L_f)/(k \times L_i \times L_f)$, where t is time (in min), L_i is the initial length of probe (in kb), L_f is the final length of the probe (in kb), and k is a rate constant of 0.11/kb/min.

Acknowledgments

The authors thank Dr. Andrew Hamilton for providing a protocol for isolating small RNAs. Work on RNA-mediated silencing in our lab is supported by grants from the Austrian Fonds zur Fîrderung der wissenschaftlichen Forschung (grant no. P15611-B07) and the European Union (contract no. HPRN-CT-2002-00257).

References

1. Finnegan, E. J. and Matzke, M. (2003) The small RNA world. *J. Cell Sci.* **116,** 4689–4693.

2. Volpe, T., Kidner, C., Hall, I., Teng, G., Grewal, S., and Martienssen, R. (2002) Regulation of heterochromatin silencing and histone H3 lysine-9 methylation by RNAi. *Science* **297,** 1833–1837.

3. Hall, I., Shankaranarayana, G., Noma, K., Ayoub, A., Cohen, S., and Grewal, S. (2002) Establishment and maintenance of a heterochromatic domain. *Science* **297,** 2232–2237.

4. Zilberman, D., Cao, X., and Jacobsen, S. (2003) ARGONAUTE4 control of locus-specific siRNA accumulation and DNA and histone methylation. *Science* **299,** 716–719.

5. Pal-Bhadra, M., Leibovitch, B., Gandhi, S., et al. (2004) Heterochromatic silencing and HP1 localization in *Drosophila* are dependent on the RNAi machinery. *Science* **303,** 669–672.

6. Wassenegger, M. (2000) RNA-directed DNA methylation. *Plant Mol. Biol.* **43,** 203–220.

7. Aufsatz, W., Mette, M. F., van der Winden, J., Matzke, A., and Matzke, M. (2002) RNA-directed DNA methylation in *Arabidopsis. Proc. Natl. Acad. Sci. USA* **99,** 16,499–16,506.

8. Mette, M. F., Aufsatz, W., van der Winden, J., Matzke, M., and Matzke, A. (2000) Transcriptional silencing and promoter methylation triggered by dsRNAs. *EMBO J.* **19,** 5194–5201.

9. Aufsatz, W., Mette, M. F., van der Winden, J., Matzke, M., and Matzke, A. (2002) HDA6, a putative histone deacetylase needed to enhance CG methylation induced by dsRNAs. *EMBO J.* **21,** 6832–6841.

10. Jones, L., Ratcliff, F., and Baulcombe, D. (2001) RNA-directed transcriptional silencing in plants can be inherited independently of the RNA trigger and requires Met1 for maintenance. *Curr. Biol.* **11,** 747–757.

11. Sijen, T., Vijn, A., Rebocho, A., et al. (2001) Transcriptional and posttranscriptional gene silencing are mechanistically related. *Curr. Biol.* **11,** 436–440.

12. Melquist, S. and Bender, J. (2003) Transcription from an upstream promoter controls methylation signaling from an inverted repeat of endogenous genes in *Arabidopsis. Genes Dev.* **17,** 2036–2047.

13. Kanno, T., Mette, M. F., Kreil, D.P., Aufsatz, W., Matzke, M., and Matzke, A. J. M. (2004) Involvement of putative SNF2 chromatin remodeling protein DRD1 in RNA-directed DNA methylation. *Curr. Biol.* **14,** 801–805.

14. Papp, I., Mette, M. F., Aufsatz, W., et al. (2003) Evidence for nuclear processing of plant microRNA and short interfering RNA precursors. *Plant Physiol.* **132,** 1382–1390.

15. Billy, E., Brondani, V., Zhang, H., Muller, U., and Filipowicz, W. (2001) Specific interference with gene expression induced by long, dsRNAs in mouse embryonal carcinoma cells. *Proc. Natl. Acad. Sci. USA* **98,** 14,428–14,433.

16. Hamilton, A. and Baulcombe, D. (1999) A species of small antisense RNA in post-transcriptional gene silencing in plants. *Science* **286,** 950–952.

17. Donzé, O. and Picard, D. (2002) RNA interference in mammalian cells using siRNAs synthesized with T7 RNA polymerase. *Nucleic Acids Res.* **30,** e46.

18. Depicker A., Stachel, S., Dhaese, P., Zambryski, P., and Goodman, H. (1982) Nopaline synthase: transcript mapping and DNA sequence. *J. Mol. Appl. Genet.* **1,** 5561–5573.

7

Strategies for the Design of Random siRNA Libraries and the Selection of anti-GFP siRNAs

Mouldy Sioud and Sébastien Wälchli

1. Introduction

Various methods have been devised to elucidate gene function in a high-throughput format. With the potential to silence any gene once the sequence is known, small, interfering RNAs (siRNAs) have been considered ideal for functional analysis and gene target validation *(1–3)*. Furthermore, the technology has unprecedented target flexibility. By performing RNAi on a large scale using siRNA libraries, this reverse-genetic method can essentially be used as a forward-genetic screening tool. In this respect, genomewide siRNA screens in *Caenorhabditis elegans* using libraries of in vitro-transcribed long double-stranded RNAs (dsRNAs) have been useful in gene discovery and functional annotation in various processes, including embryonic development *(4,5)*.

To increase the speed of RNA interference (RNAi) screening in mammalian cells, several groups have designed methods for the high-throughput development of hairpin-producing plasmid libraries *(6–8)*. These libraries were either constructed from double-stranded cDNAs, which were cleaved into multiple DNA fragments and then cloned into appropriate vectors *(6,7)* or by the creation of a large collection of retroviral vectors encoding a large number of distinct hairpin siRNAs *(8)*. Among the application of siRNA libraries would be the search for new target genes critical for various cellular processes such a metastasis and drug resistance. Moreover, the same approach can be used for the selection of effective siRNAs for therapeutic applications. As with any genetic approach, however, the power of each RNAi screen will depend on the nature of the libraries and the appropriate choice of the functional readout. In view of the high efficacy of siRNA in vitro and in vivo, we

From: *Methods in Molecular Biology, vol. 309: RNA Silencing: Methods and Protocols*
Edited by: G. G. Carmichael © Humana Press Inc., Totowa, NJ

have reasoned that siRNA random libraries may also be useful for functional gene discovery. Previously, random ribozymes and antisense libraries have been used to identify novel functional genes *(9,10)*. In this chapter, we outline two strategies for the construction of random siRNA libraries and their potential applications.

2. Materials

2.1. Buffer Solution

1. 10X TBE buffer: 890 mM Tris-borate, pH 8.3, 20 mM EDTA.
2. 10X deoxynucleotides stock mixture: 5 mM each dNTP.
3. 10X polymerase chain reaction (PCR) buffer: 500 mM NaCl, 10 mM Tris-HCl, pH 7.4, 1.5 mM MgCl$_2$.
4. 10X restriction enzyme buffers (supplied by the manufacturers).

2.2. Enzymes and Reagents

1. T4 DNA ligase (1 U/µL).
2. DNA polymerase.
3. Tap DNA polymerase.
4. dNTPs (10 mM each).
5. Lipofectamin, USA 2000 (Invitrogene).
6. Plasmid miniprep kit (Qiagen).
7. Plasmid maxiprep kit.
8. Luria–Bertani (LB)/ampicillin plates.
9. Fetal calf serum (FCS).
10. Growth medium (appropriate for cell type).
11. Chloroform/isoamyl alcohol (24 : 1, v/v).
12. 10% sodium dodecyl sulphate (SDS).
13. 30% acrylamide/0.8% *N,N'* methylene *bis*-acrylamide.
14. Phenol.

2.3. Vectors

1. pcDNA3 (Invitrogen).
2. pEGFP-N3 (Invitrogen).
3. GeneBuster™ vector (Genordia, Sweden).
4. pcMV-DsRed (Invitrogen).

2.4. Oligonucleatides

1. DNA oligonucleotides (Invitrogen).
2. T7 and T3 DNA primers carrying appropriate restriction sites.

2.5. Cells

1. HEK 293 cells.
2. *Escherichia coli* competent cells, DH5α (Invitrogen), LB medium.

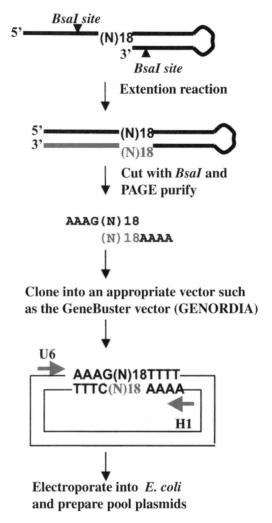

Fig. 1. Schematic diagram for the construction of the dual random siRNA library.

3. Methods

3.1. Construction of a Dual Random siRNA Library

1. The procedure of constructing the dual siRNA library is summarized in **Fig. 1**. The strategy employs a self-priming oligonucleotide that contains two recognition sequences GGTCTC for *Bsa*I restriction enzyme. Also, it contains internal 18 degenerated bases, denoted as N, which represents A, T, C, or G. The designed oligonucleotide is self-annealing and priming (5′GAATAT**GGTCTC**GAAAA-(N)18CTTTAGAGACCATATGCCAAAAGGCATAT**GGTCTC** 3′). In addition, we have used a non-self-priming oligonucleotide and a reverse primer (*see* **Note 1**).

2. The designed oligonucleotides are converted into dsDNA by T4 DNA polymerase and in the presence of 0.5 mM of each dNTP, and then the resulting dsDNA is digested, with *Bsa*I leaving the random siRNA sequences flanked by four nucleotides AAAA/GAAA 5′ overhangs (*see* **Fig.1**).

3. The cleaved DNA is purified by 15% agarose gels and then ligated into GeneBuster vector™ (Genordia) that contains four TTTT/CTTT 3′overhangs as recommended by the manufacturer's instructions. The principle of the vector is identical to that published by Zheng et al. *(11)*.

4. Once the ligation reaction is completed, electroporate the mixture into *E. coli* DH5α, add 1 mL SOC medium, and incubate for 1 h at 37°C with agitation.

5. Plate 200 µL of cells on prewarmed LB agar plates containing 50 µg/mL ampicillin.

6. Incubate plates overnight at 37°C.

7. Collect cells from the LB agar plates by scraping and purify plasmids using Qiagen maxiprep kits, which provide high-quality plasmid DNA for transfection of eukaryotic cells. This plasmid preparation constitutes the random siRNA pool that should be stored at –20°C until use.

3.2. Subcloning of the siRNA Library into the pcDNA3 Vector

Although the constructed library can be used for functional analysis in eukaryotic cells, it is preferable to subclone it into a eukaryotic expression vector. Several choices of expression vectors are available. We chose the pcDNA3 vector (Invitrogen).

1. To generate the siRNA cassettes, set up the following 100 µL PCR reaction: 81 µL water, 10 µL 10X Buffer, 2 µL dNTPs (5 mM each), 2 µl *Bgl*II-T7 primer (20 µM), 2 µL *Not*I-T3 primer (20 µM), 2 µL pooled plasmids (0.2 µg).

2. Add 2 U of Tap polymerase and carry out the PCR reaction using the following parameters: 1 min at 94°C, followed by 20 cycles consisting of 30 s at 52°C and 30 s at 72°C. A final elongation step at 72°C for 5 min is carried out, and the reaction mixture is cooled to 4°C. Perform three PCR reactions.

3. Check 5 µL of the PCR reaction on 1.2% agarose gel for the presence of the expected PCR product. The siRNA cassette should be around 600 bp in length.

4. Extract the reaction mixture with phenol, followed by extraction with chloroform/isoamyl alcohol (24 : 1), and collect the DNA by ethanol precipitation and centrifugation.

5. Digest about 4 µg of the amplified DNA with the two restriction enzymes whose recognition sequences were engineered into the T7 and T3 primers (e.g., *Bgl*II and *Not*I) and then purify the cleaved DNA by a 1.2% agarose gel.

6. Subsequent to purification, ligate the DNA into, for example, *Bgl*II and *Not*I-cleaved pcDNA3 vector.

7. Electroporate the ligation mixture into *E. coli* DH5α and plate out cells on LB agar plates containing 50 µg/mL ampicillin.

8. Collect bacteria and prepare plasmids using Qiagen maxiprep kits. DNA plasmid at this stage is pure enough for transfecting eukaryotic cells (*see* **Note 2**).

3.3. Selection of Anti-GFP siRNAs from Random siRNA Libraries

The embryonic kidney HEK 293 T cells are maintained in RPMI 1640, supplemented with 10% FCS and antibiotics.

1. On the day before transfection, cells are trypsinized and 10^6 cells are plated in tissue culture dishes in 5 mL complete medium (RPMI supplemented with 10% FCS).
2. To select siRNAs against green fluorescent protein (GFP), cells are transfected with the siRNA library (10 µg) and pEGFP-N3 plasmid encoding GFP protein (1 µg). To visualize the transfected cell population by flow cytometry, the pcMV-DsRed (0.2 µg), encoding the *Discosoma* sp. red fluorescent protein, is included as a selection marker for transfection.
3. After 24 h transfection time, negative cells that express the red fluorescence but not the GFP are sorted by flow cytometry. This step is expected to enrich for truly transfected cells that have received GFP and active siRNA sequences. For an illustration, *see* **Fig. 2**.
4. Plasmid-expressing siRNAs are rescued from the sorted cells (GFP negative, DsRed positive) by alkaline lysis/SDS method, precipitated with ethanol, and then transformed into *E. coli* DHα5.
5. The following day, plasmids are prepared using Qiagen maxiprep kits and the selection step was repeated by transfecting HEK 293 T-cells as described above.
6. After the desired rounds of selection, single colonies appearing on LB ampicilin plates are picked up; plasmid prepared using Qiagen miniprep and siRNA sequences are determined.

3.4. Data Analysis

In order to detect changes in the desired phenotype, several rounds of selection may be needed. To enrich for anti-GFP siRNAs and/or miRNAs (*see* **Note 3**), the selection protocol was repeated five times. After the last round of selection, the percentage of GFP negative cells and DsRed positive cells increased significantly. Indeed, flow cytometry analysis shows that GFP was more attenuated in cells transfected with the selected libraries (**Fig. 2D**, gate P2) than in cells transformed with the parental library (**Fig. 2B**, gate P2), indicating that our selection protocol does enrich for anti-GFP siRNAs *(12)*. The present approach, which requires no prior sequence information, should provide an additional tool for functional gene analysis and the selection of effective siRNAs or miRNAs (*see* **Note 4**). An additional strategy for the expression of hairpin type siRNA library is outlined in **Fig. 3**.

1. In the first step, the self-priming oligonucleotide is extended with DNA polymerase to generate dsDNA.
2. After primer extension, purify the dsDNA with 15% polyacrylamide gel.
3. Dissolve the sample in PCR reaction mixture, add the reverse primer, and incubate the sample at 94°C for 5 min.

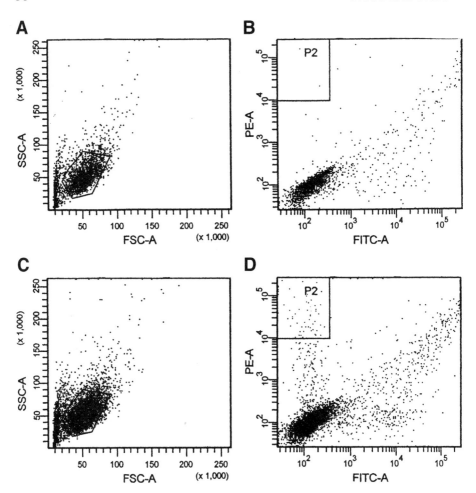

Fig. 2. Enrichment of anti-GFP siRNAs as detected by flow cytometry. To evaluate the effectiveness of the designed library, the green fluorescent protein is used as a working model. In this experimental protocol, HEK 293 T-cells are transfected with the random dual siRNA library in combination with pEGFP-N3 plasmid encoding GFP and pcMV- DsRed encoding the red protein. Cells were harvested 16 h after transfection and analyzed by flow cytometry. Cells that are identified as negative for GFP and positive for Ds-Red are sorted and plasmids rescued. Data are from the first round of selection (**A,B**) and the fourth round of selection (**C,D**).

Fig. 3. (*opposite page*) Strategy for the design of hairpin-type random siRNA libraries. The expression of the siRNAs from the U6 promoter would generate siRNAs with extra sequences that correspond to the linker used used for priming. Experimental data indicate that this extra sequence has no effects on siRNA activity. On the contrary, it enhanced the activity of siRNAs (*see* **Note 5**).

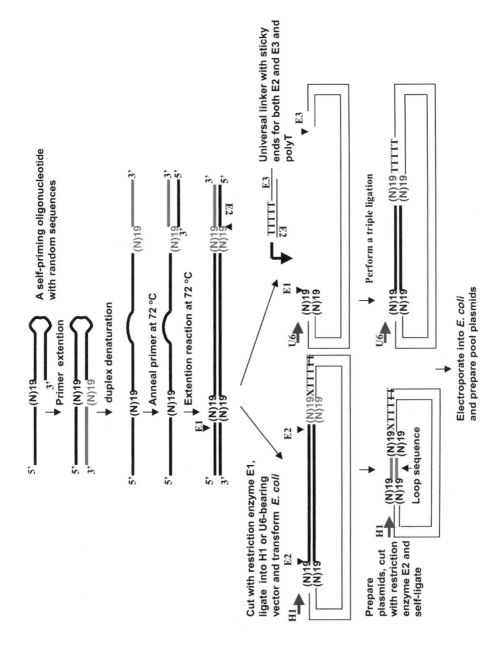

89

4. Place the sample at 4°C, add 2 U Tap polymerase, and then incubate the sample at 72°C for 10 min. The key here is that complementary sense and and antisense DNA sequences can be generated by this strategy. Digest the filled DNA with the appropriate restriction enzymes and then ligate into the appropriate expression vector.

4. Notes

1. For some unknown reasons, self-priming oligonucleotides are not efficiently cleaved by *Bsa*I restriction enzyme. Therefore, we have also used a non-self-priming oligonucleotide and a reverse primer (5′ GAATAAGGTCTCGAAAG(N)18TTT-AGAGACCATATGCC 3′; 5′ GGCATATGGTCTCTAAAA 3′). After annealing and primer extention, dsDNA is processed as in **Subheading 3.**

2. A quick evaluation of whether the pool of random siRNAs possesses silencing activity should be made by transfecting cells expressing a reporter gene such as GFP. Silencing activity is indicated by the knockdown of the reporter gene. However, the silencing effect will vary according to the frequency of each siRNA within the library.

3. By using random sequences, it might be possible to select siRNAs that function as microRNAs; that is, only the translation is arrested. Indeed, some of the selected anti-GFP molecules are partially homologous to GFP mRNA. However, an siRNA targeting the following site was selected (5′ GGCACAAGCUGGAGUACAA 3′).

4. Another application of the random siRNA libraries would be the selection of effective siRNA or microRNA against known target genes. Indeed, by fusing any target gene with a reporter gene such as GFP, siRNA targeting the fusion mRNA can be easily quantitated. After selection, sequence analysis of individual clones will indicate which part of the fusion transcript is targeted.

5. As illustrated in Fig. 3, the self-priming oligonuleotide with random sequences contains a self-priming linker, which will be part of the expressed siRNA sequences. To test whether the addition of such extra sequence would affect siRNA activity in human cells, a siRNA targeting the GFP with an additional sequence (5′-GCAAGCUGACCCUGAAGUUC**UCCCGGG**-3′) was expressed in HEK-293 cells under the U6 promoter. The expressed siRNA inhibited GFP expression. The inhibition was even more pronounced than the parental siRNA (5′-GCAAGCUGACCCUGAAGUUC-3′). Thus, the addition of extra sequences does not affect siRNA activity and therefore the construction of siRNA libraries with self-priming is feasible.

Acknowledgments

This work was supported by the Norwegian Cancer Society. We thank Dr. Anne Dybwad for critical reading of the manuscript.

References

1. Elbashir, S. M., Harborth, M. J., Ledecknel, W., Yalcin, A., Weber, K, and Tuchl, T. (2001) Duplexes of 21-nucleotide RNAs mediate RNA interference in cultured mammalian cells. *Nature* **411,** 494–498.

2. Hannon, G. J. (2002) RNA interference. *Nature* **418,** 244–251.
3. Sioud, M. (2004) Therapeutic siRNAs. *Trends Pharmacol. Sci.* **25,** 22–28.
4. Kamath, R. S., Fraser, A. G., Dong, Y., et al. (2003) Systematic functional analysis of the *Caenorhabditis elegans* genome using RNAi. *Nature* **421,** 231–237.
5. Kamath, R. S. and Ahringer, J. (2003) Genome-wide RNAi screening in *Caenorhabditis elegans. Methods* **30,** 313–321.
6. T. S., Myers, J. W., and Blau, H. M. (2004) Restriction enzyme-generated siRNA (REGS) vectors and libraries. *Nat. Genet.* **36,** 183–189.
7. Luo, B., Heard, A. D., and Lodish, H. F. (2004) Small interfering RNA production by enzymatic engineering of DNA (SPEED). *Proc. Natl. Acad. Sci. USA* **101,** 5313–5314.
8. Berns, K., Hijmans, E. M., Mullenders, J., et al. (2004) A large-scale RNAi screen in human cells identifies new components of the p53 pathway. *Nature* **428,** 431–437.
9. De Backer, M. D., Nelissen, B., Logghe, M., et al. (2001) An antisense-based functional genomics approach for identification of genes critical for growth of *Candida albicans. Nat. Biotechnol.* **19,** 235–241.
10. Kawasaki, H., Onuki, R., Suyama, E., and Taira, K. (2002) Identification of genes that function in the TNF-α-mediated apoptotic pathway using randomized hybrid ribozyme libraries. *Nat. Biotechnol.* **20,** 376–380.
11. Zheng, L., Batalov, S., Zhou, D., Orth, A., Ding, S., and Schultz, P. G. (2004) An approach to genomewide screens of expressed small interfering RNAs in mammalian cells. *Proc. Natl. Acad. Sci. USA* **101,** 135–140.
12. Sioud, M. (2004) Therapeutic potential of small interfering RNAs. *Drugs of the Future,* **29,** 741–750.

8

Silencing Gene Expression with Dicer-Generated siRNA Pools

Jason W. Myers and James E. Ferrell, Jr.

1. Introduction

With the explosion of genomic information, there is an increasing need to analyze gene function in a high-throughput fashion. This makes reverse genetic approaches extremely attractive; however, in most mammalian and vertebrate systems it has been difficult and time-consuming to develop a cellular model deficient in one or more proteins. The discovery of RNA interference (RNAi) has made loss-of-function studies relatively quick and easy and amenable to high-throughput formats *(1–3)*. Double-stranded RNAs (dsRNAs) 21 nucleotides in length, known as small interfering RNAs (siRNAs), are introduced into the cytosol, where they are unwound *(4)*, allowing the antisense strand to interact in a sequence-specific manner with the complementary mRNA *(5)*. Binding of the antisense strand to the target mRNA triggers cleavage of the mRNA *(6)*. siRNA-mediated gene silencing is very specific, most likely because of the high specificity of nucleotide base-pairing.

The design of effective siRNAs has been a major obstacle in siRNA-mediated gene silencing, as there is great variability among siRNAs in terms of their effectiveness. We reasoned that the tedious process of siRNA design could be eliminated if a simple method to prepare a large number of distinct siRNAs could be developed. We then realized that nature had already designed a method and it simply had to be recapitulated. Dicer, an RNase III family enzyme, cleaves dsRNA into a pool of siRNAs suitable for gene silencing. Therefore, we *(7)* and others *(8–10)* have produced a recombinant version of Dicer (r-Dicer) and used it to digest in vitro–transcribed dsRNAs into a complex pool of siRNAs (d-siRNAs) *(7)*. Nearly every pool of d-siRNAs is capable of eliciting

From: *Methods in Molecular Biology, vol. 309: RNA Silencing: Methods and Protocols*
Edited by: G. G. Carmichael © Humana Press Inc., Totowa, NJ

specific gene silencing *(11)*. This approach, in vitro dicing, eliminates the need to identify an individual effective siRNA and has proven to be very useful for transiently silencing many endogenous genes in several types of cultured cell *(7,11–13)*.

Unlike *Caenorhabditis elegans*, *Drosophilia melanogaster*, or other invertebrates, RNAi-mediated gene silencing in mammalian cells must be accomplished with an siRNA because dsRNAs greater than 30 basepairs are cytotoxic *(14–18)*. There are many types of siRNA to choose from: chemically synthesized siRNAs *(5)*, in vitro–transcribed siRNAs *(19)*, short-hairpin RNAs (shRNAs) *(20–23)*, or siRNAs *(24,25)* that are encoded from DNA in vivo. The shRNAs are equivalent to siRNAs once the loop is removed by the cellular Dicer. All of these siRNAs are, in general, used as a single siRNA targeting a single gene, but in some cases, three to five different siRNAs, all targeting the same gene, are pooled to increase the odds of attaining efficient gene silencing. In vitro dicing pushes this pooling concept further; a diced pool of siRNAs typically contains hundreds of different oligonucleotides, all targeting the same gene (Gong et al., unpublished data).

Each technique has advantages and disadvantages, but in vitro dicing is simple, cost-effective, and scalable to high-throughput formats, more so than many of the individual siRNA approaches even when considering that d-siRNAs are not capable of providing stable gene suppression. The major benefit of in vitro dicing is that it eliminates the need to design and test siRNAs for efficacy and potency. This both simplifies and decreases the cost of gene silencing experiments. Nearly every pool of d-siRNAs elicits specific gene silencing *(11)*, whereas typically only one out of three individual siRNAs is a potent gene silencer *(26,27)*. Thus, to silence one gene, at least three individual siRNAs must be designed and tested, thereby increasing the cost and amount of work. This is especially a problem if a large collection of genes or even a genome is to be silenced, because the number of assays will be three times greater if siRNAs are used instead of d-siRNAs. Several groups have designed a clever scheme to generate libraries of shRNA *(28–30)* or siRNA *(31)* expression vectors from a cDNA library. This process reduces the cost but does not eliminate the need to find potent gene silencers and, therefore, may be best suited to forward genetic screens.

In vitro dicing is suitable for high-throughput reverse genetic screens for several reasons. Although siRNA selection criteria have been improved *(4,26,27)*, the potency of individual siRNAs can vary over three orders of magnitude, 0.1 nM to 100 nM, making high-throughput silencing very complicated. If the concentration is too low, many individual siRNAs will not be effective, and if it is too high, silencing will be accompanied by off-target and nonspecific affects. For example, in many screens, siRNAs are being used at 100 n*M* (nmoles of siRNA per liter of culture volume), a concentration that will likely provoke off-target and nonspecific silencing, but if the concentration is lowered, many of the targets will not be silenced. The lack of a standard concentration for attaining specific and efficient gene silencing will result in many false-positive and false-negative results. However, nearly every pool of d-siRNAs is effective and specific between 1 and 30 nM and the potency of different pools of d-siRNAs targeting different genes varies by only twofold to threefold. Thus, for screening applications, d-siRNAs can be used at a standard concentration, thereby improving the quality of the resulting data.

The specificity of siRNA-mediated gene silencing has been scrutinized *(32–34)* and it has been argued that a pool of siRNAs, like d-siRNAs or siRNAs generated by bacterial RNase III digestion, will further increase the possibility of off-target gene silencing. However, the opposite may be argued as well. A complex mixture of siRNAs will certainly contain more off-target matches than a single siRNA would, particularly if the pool consists of short siRNAs (12–15 nucleotides) like those produced by RNase III digestion. However, the concentration of any individual siRNA in a complex d-siRNA pool is quite low—perhaps approx 1/300th of the total siRNA concentration (Gong et al., unpublished data)—and so, unless many different members of the pool share the same off-target effects, the effects should be insignificant. In support of this view, the gene silencing achieved in *C. elegans* by the complex pools of d-siRNAs produced from large dsRNAs in vivo has proven to be quite specific in worms; time and time again the RNAi-induced phenotype reiterates the null phenotype *(35)*. Indeed, analysis of the silencing of closely related gene pairs indicates that d-siRNAs are likely to be highly specific. The hypothesis is that if the gene with the highest degree of nucleotide identity to the target is not silenced when the target is silenced, then it is unlikely that more distantly related genes would be silenced. For example, only the target gene is silenced when we analyze closely related gene pairs like cyclin B1 and cyclin B2, cyclin-dependent kinase 1 (Cdk1) and cyclin-dependent kinase 2 (Cdk2) (Myers et al., unpublished data), and cdc25C and cdc25A *(7)*, suggesting that off-target gene silencing is not a likely consequence of d-siRNA-mediated gene suppression. It will be of interest to test the specificity of d-siRNAs globally through expression profiling using DNA microarrays.

This chapter describes the materials and methods necessary to produce r-Dicer and d-siRNAs for gene silencing experiments in any given system capable of siRNA-mediated gene silencing. The basic approach is as follows. The r-Dicer is expressed in insect cells using a common insect cell expression system. After partial purification, the r-Dicer is used to cleave long dsRNAs into a complex pool of d-siRNAs. The long dsRNAs are in vitro transcribed from PCR (polymerase chain reaction) products. Finally, the d-siRNAs are purified away from all reaction components and introduced into the cells or organisms via common delivery tactics. The power of this approach is that it is very efficient: it requires only moderate hands-on time, it takes only 5–7 d to produce d-siRNAs and assay for gene silencing, and the odds of attaining high levels of specific gene silencing are very good. Finally, in vitro dicing was developed for high-throughput loss-of-function screening. Thus, the methods to produce d-siRNAs and attain gene silencing in a 96-well format are also presented.

2. Materials

2.1. Bacmid Preparation

1. RNeasy® Kit, including Qiashredder™ (Qiagen, Valencia, CA; cat. nos. 74104 and 79654, respectively); β-mercaptoethanol not included.
2. TE, pH 8.0: 10 mM Tris-HCl, pH 8.0, 1 mM EDTA (ethylenediamine tetraacetic acid), pH 8.0; made with RNase-free reagents.
3. Agarose (Invitrogen, Carlsbad, CA, cat. no. 11510-019) or equivalent.
4. 50X TAE: 2 M Tris-acetate, 0.1 M EDTA.
5. 10X TBE: 0.89 M Tris base, 0.89 M boric acid, 0.02 M EDTA.
6. Ethidium bromide (Invitrogen, cat. no. 15585-001) or equivalent.
7. 2X RNA loading buffer: 95% formamide, 0.025% xylene cyanol, 0.025% bromophenol blue, 18 mM EDTA, and 0.025% sodium dodecyl sulfate (SDS); made with RNase-free reagents and stored at −20°C.
8. RNase-free H$_2$O (Invitrogen, cat. no. 10977-015); or treat nanopure H$_2$O with 0.1% DEPC (**item 9**) for 6 h to overnight in a fume hood with continuous mixing and then autoclave for 30 min to inactivate DEPC. Incomplete activation can inhibit enzymes used in subsequent manipulations. Store at room temperature.
9. DEPC (diethyl pyrocarbonate) (Sigma-Aldrich, St. Louis, MO; cat. no. D5758); very toxic and should be handled in a fume hood; moisture causes hydrolysis to ethanol and carbon dioxide, thus stock solution should be stored under airtight conditions (e.g., wrap lid in parafilm and place in a desiccator) and at 4°C.
10. Superscript™ RTIII First-Strand Synthesis System for RT-PCR (Invitrogen, cat. no. 18080-051); includes 5X reaction buffer, 0.1 M DTT (dithiothreitol), 10 mM dNTPs, 50 μM oligo(dT)$_{20}$, RNaseOUT™ (40 units/μL), Superscript™ RTIII (200 units/μL), and RNase-free H$_2$O and *E. coli* RNase H (2 units/μL); stored at −20°C.
11. Nuclease-free 0.2-mL thin-wall PCR tube (E and K Scientific, Campbell CA; cat. no. 690002) or equivalent.
12. Platinum® *Pfx* polymerase (Invitrogen, cat. no. 11708-013); 10X *Pfx* Amplification Buffer and 50 mM MgSO$_4$ are included; stored at −20°C.
13. Oligonucleotide primers, 10-nmol scale, desalted (Invitrogen).
14. 10X DNA loading buffer: 50% glycerol, 100 mM EDTA, 0.25% bromophenol blue, 0.25% xylene cyanol; stored at room temperature.
15. 1-kb DNA ladder (Invitrogen, cat. no. 15615-016) or equivalent diluted with 10X DNA loading buffer (**item 14**); stored at −20°C (long term) or 4°C for several months.
16. QIAquick® PCR Purification Kit (Qiagen, cat. no. 28104).

17. *Not*I, *Xho*I, and *Bam*HI restriction enzymes (New England Biolabs, Beverly, MA; R0189S, R0146S, and R0136S, respectively); 10X reaction buffers and BSA (bovine serum albumin) are included; stored at −20°C.
18. Bac-to-Bac® HT Vector Kit (Invitrogen, cat. no. 10584-027); many components and various storage conditions.
19. Nuclease-free 1.5-mL microcentrifuge tubes (E and K Scientific, cat. no. 280150) or equivalent.
20. QIAquick® Gel Extraction Kit (Qiagen, cat. no. 28704).
21. High DNA Mass Ladder (Invitrogen, cat. no. 10496-013) or equivalent diluted with 10X DNA loading buffer (**item 14**); stored at −20°C (long term) or 4°C for several months.
22. T4 DNA ligase (New England Biolabs, cat. no. M0202S); 10X buffer is included; stored at −20°C.
23. LB (Luria–Bertani) ampicillin plates: 10 g bacto-tryptone, 5 g bacto-yeast extract, 10g NaCl, and 15 g bacto-agar per liter containing 100 µg/mL ampicillin (**item 24**) *(36)*; stored at 4°C and protected from light.
24. Ampicillin; dissolved in H_2O at 50 mg/mL *(36)*; should be filter-sterilized (0.22 µm) and aliquoted into 500 µL; stored at −20°C.
25. One Shot® TOP10 Chemically Competent Cells (Invitrogen, cat. no. C4040-50); stored at −80°C.
26. SOC (Invitrogen, cat. no. 15544-034) or 20 g bacto-tryptone, 5 g bacto-yeast extract, 0.5 g NaCl per liter plus 20 m*M* glucose *(36)*; stored at room temperature.
27. 3- or 5-mm Glass beads (Fisher, Suwanee, GA; cat. no. 11-312A or 11-312C, respectively).
28. LB medium: 10 g bacto-tryptone, 5 g bacto-yeast extract, and 10g NaCl per liter *(36)*; stored at room temperature.
29. 14-mL Polypropylene culture tubes (Becton Dickinson, Franklin Lakes, NJ; cat. no. 352059; distributed by Fisher Scientific, cat. no. 431144).
30. QIAprep® Miniprep Kit (Qiagen, cat. no. 27106).
31. Max Efficiency® chemically competent DH10Bac™ cells (Invitrogen, cat. no. 10361-012); stored at −80°C.
32. LB plates for bacmid growth: 10 g bacto-tryptone, 5 g bacto-yeast extract, 10g NaCl and 15 g bacto-agar per liter *(36)* containing 50 µg/mL kanamycin (**item 33**), 10 µg/mL tetracycline (**item 34**), 7 µg/mL gentamycin (**item 35**), 100 µg/mL bluo-gal (**item 36**), and 40 µg/mL isopropylthio-β-galactoside (IPTG) (**item 37**); stored at 4°C and protected from light.
33. Kanamycin; dissolved in H_2O at 10 mg/mL *(36)*; should be filter-sterilized (0.22 µm) and aliquoted into 500 µL; stored at −20°C.
34. Tetracycline; dissolved in ethanol at 5 mg/mL *(36)*; store at −20°C protected from light; must be used in medium without magnesium ions; LB is fine.
35. Gentamycin (Sigma-Aldrich, cat. no. G1272); provided at 10 mg/mL in H_2O; stored at 4°C.
36. Bluo-gal; dissolved in DMSO (dimethyl sulfoxide) at 20 mg/mL; should be dissolved in and stored in a glass or polypropylene tube; stored at −20°C and protected from light.

37. IPTG; dissolved in H_2O at 200 mg/mL; should be filter-sterilized (0.22 μm) and aliquoted into 1 mL; stored at –20°C.
38. Solution I: 15 mM Tris-HCl, pH 8.0, 10 mM EDTA, 100 μg/mL RNase A; filter-sterilized and stored at 4°C.
39. RNase A 100 mg/mL (Qiagen, cat. no. 19101); stored at room temperature.
40. Solution II: 0.2 N NaOH, 1% SDS; filter-sterilized and stored at room temperature.
41. 3 M Potassium acetate, pH 5.5; autoclaved and stored at 4°C.
42. Nuclease-free 2.0-mL microcentrifuge tube (E and K Scientific, cat. no. 280200) or equivalent.
43. Isopropanol.
44. 70% Ethanol; dilute 100% ethanol with H_2O.
45. TE, pH 8.0 (10 mM Tris-HCl, pH 8.0, 1 mM EDTA, pH 8.0); should be sterile but does not need to be RNase free as in **item 2**. If same TE is used for RNA and DNA, be sure that TE remains RNase-free.
46. Platinum® Taq high-fidelity (Invitrogen, cat. no. 11304-011); 10X amplification buffer and 50 mM $MgSO_4$ are included; stored at –20°C.
47. 30% (v/v) glycerol solution prepared by dilution in H_2O.

2.2. Baculovirus Production

1. Sf-900 II SFM (Invitrogen, cat. no. 10902-096); stored at 4°C and protected from light.
2. Sf9 cells, serum-free adapted (Invitrogen, cat. no. 11496-015); stored in vapor phase of, or submerged in, liquid nitrogen.
3. 125-mL disposable Erlenmeyer flasks with vented caps (Corning; distributed by Fisher Scientific, cat. no. 431143) or equivalent.
4. Trypan blue (Invitrogen, cat. no. 15250-061) or equivalent.
5. Bright-Line® hemacytometer (Reichert, Buffalo, NY; distributed by Fisher Scientific, cat. no. 026715) or equivalent.
6. 250-mL disposable Erlenmeyer flasks with vented caps (Corning; distributed by Fisher Scientific, cat. no. 431144) or equivalent.
7. Tissue-culture-grade DMSO (Sigma-Aldrich, cat. no. D2650); stored at room temperature.
8. CryoTube® vials (Nunc; distributed by Fisher Scientific, cat. no. 377267) or equivalent.
9. Six-well tissue culture plates (Becton Dickinson; distributed by Fisher Scientific, cat. no. 353046) or equivalent.
10. Grace's unsupplemented medium (Invitrogen, cat. no. 11595-030); stored at 4°C and protected from light.
11. 5-mL Polypropylene round-bottom tube (Becton Dickinson; distributed by Fisher Scientific, cat. no. 352063).
12. Cellfectin® (Invitrogen, Carlsbad, CA; 10362-010); stored at 4°C.
13. Fetal bovine serum (FBS) (Invitrogen, cat. no. 26140-079); stored in aliquots (5–30 mL) at –20°C
14. 10-cm Tissue culture dish (Becton Dickinson; distributed by Fisher Scientific, cat. no. 353003) or equivalent.

2.3. Expression and Purification of r-Dicer

1. T7 Tag® Antibody HRP conjugate (Novagen, Madison, WI; cat. no. 69048-3); stored at –20°C.
2. 24-Well tissue culture plates (Becton Dickinson; distributed by Fisher Scientific, cat. no. 353047) or equivalent.
3. D-PBS (Dulbecco's phosphate-buffered saline) (Invitrogen, cat. no. 14040-182); stored at room temperature.
4. 6X SDS Laemmli sample buffer: 0.35 M Tris-HCl, pH 6.8, 36% glycerol, 10% SDS, 0.65 M DTT, 0.012% bromophenol blue; stored at –20°C; make 1X by dilution with H_2O.
5. 40% acrylamide (Fisher Scientific, cat. no. BP14021); stored at 4°C; equivalent acceptable.
6. 2% *bis*-acrylamide (Fisher Scientific, cat. no. BP1404250); stored at 4°C; equivalent acceptable.
7. 40% 29 : 1 Acrylamide : *bis*-acrylamide (Fisher Scientific, cat. no. BP 14081); stored at 4°C; equivalent acceptable.
8. 4X Resolving gel buffer: 1.5 M Tris-HCl, 0.4% SDS (w/v), pH 8.8 *(36)*.
9. 4X Stacking gel buffer: 0.5 M Tris-HCl, 0.4% SDS (w/v), pH 6.8 *(36)*.
10. 10% APS (ammonium persulfate) (w/v); aliquoted into 10 mL; short-term storage at 4°C and long-term storage at –20°C.
11. TEMED (Invitrogen, cat. no. 14424-010); stored at 4°C; equivalent acceptable.
12. 10X SDS-PAGE running buffer: 0.25 M Tris-HCl, 1.92 M glycine, 1% SDS (w/v), pH 8.6.
13. Immobilon-P PVDF (polyvinylidene fluoride) membrane (Millipore; distributed by Fisher Scientific, cat. no. IPVH 00010) or equivalent.
14. 10X Transfer buffer: 0.2 M Tris-Hcl, 1.5 M glcyine, pH 8.2; used at 1X with 10% methanol.
15. Antibody diluent: D-PBS (**item 3**) containing 1% nonfat milk powder (**item 16**) and 0.01% TWEEN® 20 (**item 17**).
16. Nonfat milk powder (generic store brand).
17. TWEEN® 20 (Sigma-Aldrich, St. Louis, MO; P1379).
18. Immun-Star™ HRP Chemiluminescent Kit (Bio-Rad, Hercules, CA; cat. no. 170-5040); contains one bottle each of Luminol/enhancer solution and peroxide buffer; stored at 4°C.
19. α-Extraction buffer pH 8.0: 50 mM phosphate buffer, 300 mM NaCl, 1% NP-40 (NP-40 alternative; Calbiochem, La Jolla, CA; cat. no. 492106); store at 4°C; 50 mM $NaPO_4$ at pH 8.0 is made by mixing the appropriate ratios of 1 M monobasic and 1 M dibasic sodium phosphate with other components as described in **ref**. *36* (*see* **Note 1**). Solution containing protease inhibitors must be prepared fresh for lysis: Add protease inhibitors to 10 mL of α-extraction buffer, 50 µL of 200 mM PMSF (**item 20**), 1 µL each of 5.5 mg/mL leupeptin (**item 21**), 5.5 mg/mL pepstatin A (**item 22**), and 6.3 mg/mL aprotinin (**item 23**).
20. 200 mM PMSF (phenylmethyl sulfonylfluoride) dissolved in 95% ethanol (Sigma-Aldrich. cat. no. P-7626); store at 4°C; equivalent acceptable. PMSF is stable in

aqueous solutions for only approx 30 min and should be added last; inhibition of proteases is irreversible. PMSF is very toxic and can penetrate skin once dissolved in ethanol.

21. 5.5 mg/mL leupeptin dissolved in nanopure H_2O (Sigma-Aldrich. cat. no. L-2023); store at $-20°C$; equivalent acceptable.

22. 5.5 mg/mL pepstatin A dissolved in DMSO (Sigma-Aldrich. cat. no. P-4265); store at $-20°C$; equivalent acceptable.

23. 6.3 mg/mL aprotinin (Sigma-Aldrich, cat. no. A-6279); store at $4°C$; equivalent acceptable.

24. Talon™ resin (Clontech, Palo Alto, CA; cat. no. 8901-2); store at $4°C$.

25. β-Wash buffer pH 8.0: 50 mM phosphate buffer, 500 mM NaCl, 1% NP-40 (NP-40 alternative; Calbiochem, cat. no. 492106); store at $4°C$. 50 mM NaPO$_4$ at pH 8.0 is made by mixing the appropriate ratios of 1 M monobasic and 1 M dibasic sodium phosphate with other components as described in **ref. 36** (*see* **Note 1**).

26. δ-Wash buffer pH 8.0: 50 mM phosphate buffer, 500 mM NaCl; store at $4°C$. 50 mM NaPO$_4$ at pH 8.0 is made by mixing the appropriate ratios of 1 M monobasic and 1 M dibasic sodium phosphate with other components as described in **ref. 36** (*see* **Note 1**).

27. γ-Elution buffer pH 8.0: 50 mM phosphate buffer, 500 mM NaCl, 150 mM imidazole; store at $4°C$. 50 mM NaPO$_4$ at pH 8.0 is made by mixing the appropriate ratios of 1 M monobasic and 1 M dibasic sodium phosphate with other components as described in **ref. 36** (*see* **Note 1**).

28. Slide-A-Lyzer® Dialysis Cassette: 3- to 15-mL capacity; 10,000 molecular weight cutoff (MWCO) (Pierce Biotechnology, Rockford, IL; cat. no. 66810).

29. Dialysis buffer pH 8.0: 60 mM HEPES, pH 8.0, 500 mM NaCl, 0.1 M EDTA; make fresh and chill at $4°C$ prior to use.

30. 50 mM MgCl$_2$; prepare with RNase-free components or treat with DEPC and autoclave. Make sure DEPC is completely hydrolyzed so that r-Dicer is not inhibited; store at room temperature.

31. Reaction dilution buffer: 60 mM HEPES, pH 8.0, 500 mM NaCl, 0.1 mM EDTA; prepare with RNase-free components or treat with DEPC and autoclave. Make sure DEPC is completely hydrolyzed so that r-Dicer is not inhibited; store at room temperature.

32. 10X RNA loading buffer: 50% glycerol, 100 mM EDTA, 0.25% bromophenol blue, 0.25% xylene cyanol. Make sure this buffer is prepared with RNase-free reagents; it cannot be autoclaved and, therefore, cannot be treated with DEPC; store at room temperature.

2.4. Production of In Vitro Transcription Templates

Low DNA Mass Ladder (Invitrogen, cat. no. 10068-013) or equivalent diluted with 10X DNA loading buffer; stored at $-20°C$ (long term) or $4°C$ for several months.

2.5. In Vitro Transcription and Dicing

MEGAscript™ (Ambion, Autsin, TX; cat. no. 1334); includes all necessary components; stored at –20°C.

2.6. Purification of d-siRNAs

1. Micro-to-Midi™ Total RNA Purification System (Invitrogen, cat. no. 12183-018).
2. Lysis buffer: component of Micro-to-Midi Total RNA Purification System (**item 1**); β-mercaptoethanol must be added before use to 1% (v/v); prepare fresh.
3. β-Mercaptoethanol (Sigma-Aldrich, cat. no. M6250); store at 4°C.
4. RNase-free isopropanol; ACS grade.
5. RNase-free spin cartridge: component of Micro-to-Midi Total RNA Purification System (**item 1**).
6. 1X Wash buffer II: component of Micro-to-Midi Total RNA Purification System (**item 1**); make fresh by diluting the wash buffer II (5X) with 100% ethanol **Do not dilute 5X Buffer with H$_2$O.**
7. 100% Ethanol; ACS grade.

2.7. Gene Suppression in Tissue Culture Cells

1. HeLa cells (American Tissue Type Collection, Manassas, VA; CCL-2); stored in vapor phase of, or submerged in, liquid nitrogen.
2. Complete growth medium: DMEM (**item 3**) containing 10% FBS and 1X penicillin–streptomycin–glutamine (**item 4**); stored at 4°C and protected from light (bottle can be wrapped in aluminum foil or, preferably, light should be blocked from entering the incubator); usable for 1 mo.
3. DMEM (Dulbecco's modified Eagle's medium) (Invitrogen, cat. no. 11995-65); stored at 4°C and protected from light.
4. Penicillin–streptomycin–glutamine (100X) (Invitrogen, cat. no. 10378-016); stored in 6-mL aliquots at –20°C and protected from light.
5. Trypsin–EDTA (1X) (Invitrogen, cat. no. 25300-054); stored at 4°C (short term) or at –20°C (long term).
6. Genesilencer™ (Gene Therapy Systems, San Diego, CA; cat. no. T500750); kit contains "siRNA diluent" and "Genesilencer™ siRNA transfection reagent"; store at 4°C.
7. ART (aerosol-resistant tips) (Molecular BioProducts; distributed by Fisher Scientific, cat. nos. Art 10 reach, 2140; Art 20P, 2149P; Art 200, 2069; Art 1000, 2279) or equivalent.
8. 2X Medium: DMEM (**item 3**) containing 20% FBS and 2X penicillin–streptomycin–glutamine (**item 4**); stored at 4°C and protected from light (bottle can be wrapped in aluminum foil or, preferably, light should be blocked from entering the incubator); stable for 1 mo.
9. OPTI-MEM® I (Invitrogen, cat. no. 31985-070); stored at 4°C.
10. Hoechst 33342 (Invitrogen, cat. no. H3570); stored at 4°C.

2.8. Assaying and Troubleshooting Gene Silencing

1. PARIS™ (Ambion, cat. no. 1921); kit contains many components stored at both 4°C and room temperature; note that different lysis buffers are used for fractionating nucleoplasm and cytoplasm.
2. M-PER® Mammalian protein extraction reagent (Pierce Biotechnology, cat. no. 78501); stored at room temperature.
3. Protease inhibitors used at a final concentration of 1 mM PMSF (**Subheading 2.3., item 20**), 30 µg/mL aprotinin (**Subheading 2.3., item 23**), 10 µg/mL each of leupeptin (**Subheading 2.3., item 21**), pepstatin A (**Subheading 2.3., item 22**), and chymostatin (**item 4**).
4. 5.5 mg/mL Chymostatin dissolved in DMSO (**Subheading 2.2., item 7**) (Sigma-Aldrich, cat. no. C7268); store at –20°C; equivalent acceptable.
5. Phosphatase inhibitors used at a final concentration of 50 mM β-glycerophosphate (**item 6**), 10 mM sodium fluoride (**item 7**), and 1 mM sodium orthovanadate (**item 8**).
6. 0.5 M β-glycerophosphate, dissolved in H_2O (Sigma-Aldrich, cat. no. G6251); store at 4°C; equivalent acceptable.
7. 1 M Sodium fluoride, dissolved in H_2O (Sigma-Aldrich, cat. no. S7920); store at 4°C; equivalent acceptable.
8. 100 mM Sodium orthovanadate dissolved in H_2O (Sigma-Aldrich, cat. no. A6508); store at –20°C; equivalent acceptable; requires activation for maximal inhibition of phosphatases *(37)*.
9. BCA Protein Assay Kit (Pierce Biotechnology, cat. no. 23225); kit contains two components; store at room temperature.
10. 96-Well plates (Costar; distributed by Fisher Scientific, cat. no. 3595).
11. Pre-Diluted Protein Assay Standards: Bovine Serum Albumin (BSA) Set (Pierce Biotechnology, cat. no. 23208); kit contains various concentrations of BSA; store at 4°C.
12. ZyMax™ HRP-conjugated secondary antibodies (1.5 mg/mL) (Zymed, San Francisco, CA; rabbit–anti-goat, cat. no. 81-1620; rabbit–anti-mouse, cat. no. 81-6720; goat–anti-rabbit, cat. no. 81-6120); store at 4°C.
13. Anti-Actin antibody (Santa Cruz Biotechnology, Santa Cruz, CA; goat–anti-Actin, cat. no. SC-1616, recognizes human Actin; mouse–anti-Actin, cat. no. SC-8432, recognizes human Actin); store at 4°C.
14. Phospho-specific eIF2α antibody (Biosource International, Carmillo, CA; cat. no. 44-728G); store at –20°C.
15. *Silencer*™ siRNA Labeling Kit–Cy3 (Ambion, cat. no. 1632); store at –20°C.
16. *Silencer*™ siRNA Labeling Kit–FAM (Ambion, cat. no. 1634); store at –20°C.

2.9. High-Throughput Dicing

1. Multichannel pipets: Pipet-Lite™ with LTS™ (Rainin, Oakland, CA, cat. nos. L12-10, L12-20, L12-200, L12-300, and L12-1200); equivalents are acceptable, but these pipets are ideal for high-throughput applications.

2. Multichannel pipet tips (Rainin, cat. no. SS-L10, SS-L250, SS-L300, and RT-L1200); required for use with LTS pipets; sterile tips and barrier tips are available and denoted with an "F" after the product number.

3. Multichannel reservoir basins (Labcor Products Inc., Concord, ON, Canada; distributed by Fisher Scientific, cat. no. 13681101.PK) or equivalent.

4. Platinum® PCR Supermix (Invitrogen, cat. nos. 11306-081 for 5000 reactions; also available in 96-well format, cat. nos. 11306-065 or 11306-073); only the primers and the template must be added to the supermix, thus saving time and preventing error.

5. Nuclease-free 96-well plates (Denville Scientific Inc., Metuchen, NJ; cat. no. C-18082).

6. Polarseal; foil adhesive tape for sealing 96-well plates (E and K Scientific, cat. no. T592100); works very well for sealing plates stored at −80°C; or equivalent.

7. MEGAscript™ can be purchased in bulk quantity (Ambion, inquire for catalog number); includes all necessary components.

8. Concert™ 96 RNA Purification Kit (Invitrogen, cat. no. 12173-011); each kit contains reagents sufficient to purify 192 d-siRNAs (*see* **items 10–14**).

9. 1-mL Deep well, 96-well plate (Axygen; distributed by VWR Scientific, West Chester, PA; cat. no. 10011-940); if 2-mL deep well required, order cat. no. 10011-942.

10. Lysis buffer: component of Concert™ 96 RNA Purification Kit (**item 8**); β-mercaptoethanol must be added before use to 1% (v/v); prepare fresh.

11. Filter binding plate: component of Concert 96 RNA Purification Kit (**item 8**).

12. 1X Wash buffer II: component of Concert 96 RNA Purification Kit (**item 8**); make fresh by diluting the wash buffer II (5X) with 100% ethanol **Do not dilute 5X Buffer with H_2O.**

13. RNase-free H_2O: component of Concert 96 RNA Purification Kit (**item 8**).

14. Nuclease-free 96-well collection plate: component of Concert 96 RNA Purification Kit (**item 8**).

15. Ultraviolet (UV)-plate, 96-well (Corning; distributed by Fisher Scientific, cat. no. 3635); low absorbance for UV and, therefore, appropriate for determining concentration of nucleic acid.

16. RediPlate™ 96 RiboGreen® RNA Quantitation Kit (Invitrogen, cat. no. R32700); other packaging available: cat. nos. R-11490 and R-11491.

17. 96-well tissue culture plates (Costar; distributed by Fisher Scientific, cat. nos. 3603 or 3904 depending on packaging); ideal for microscopy.

18. 96-well tissue culture plates (Costar; distributed by Fisher Scientific, cat. no. 3614); ultrathin well bottoms are ideal for high-resolution microscopy; sterile lids are ordered separately (Costar; distributed by Fisher Scientific, cat. no. 3931).

19. GeNunc™ 12-well modules (Nunc; distributed by PGC Scientifics, Frederick, MA; cat. no. 232034); nuclease-free; for support, frames are available, cat. no. (232042).

3. Methods

The methods presented here describe how to (1) construct the pFastbac-HTC-Dicer construct and Dicer bacmid, (2) produce high-titer baculoviral stocks for expression of the r-Dicer, (3) express and purify the r-Dicer, (4) design and produce templates for in vitro transcription, (5) set up in vitro transcription and dicing reactions, (6) purify d-siRNAs away from contaminating proteins and reaction components, (7) transfect d-siRNAs into cultured cells, (8) assay and troubleshoot gene silencing, and (9) perform high-throughput in vitro dicing. Many of the methods used here require general molecular biology skills. If the instructions provided here or by the manufacturer of the suggested kits are insufficient please refer to **ref. 36.** On the other hand, if you are highly skilled in molecular biology, you may want to disregard the detail in **Subheading 3.1.** The time to production of r-Dicer will be shortened if you begin culturing Sf9 cells (*see* **Subheading 3.2., steps 1** and **2**) while preparing the Dicer bacmid (*see* **Subheading 3.1.**). This will provide ample time to practice appropriate culturing techniques and ensure that Sf9 cells are healthy and growing properly. Finally, for more information regarding baculovirus biology, see published reference sources *(38,39)*.

3.1. Bacmid Preparation

This section describes (1) the pFastbac™-HTC vector and general methodology of the Bac-to-Bac® expression system, (2) the cloning of the *Homo sapiens* Dicer cDNA into pFastbac-HTC, (3) transposition of the pFastbac-HTC-Dicer into DH10Bac cells to produce the Dicer bacmid, and (4) selection of positive recombinants and purification of the Dicer bacmid. Many of the methods described in this section are derived from our interpretation of the protocol for the Bac-to-Bac expression system. If a more detailed explanation is desired or for additional troubleshooting information, please refer to the instructions provided by the manufacturer (Bac-to-Bac® Baculovirus Expression System Version D April 6, 2004, 10359) (http://www.invitrogen.com/content/sfs/manuals/bactobac_man.pdf).

3.1.1. pFastbac-HTC and the Bac-to-Bac Expression System

Insect cells are a good expression system for large-scale production of enzymes like r-Dicer. Unlike bacteria (*Escherichia coli*) and yeast (*Pichia pastoris*), insect cells are capable of expressing very large proteins in soluble form as well as providing some degree of posttranslational modification. It is very simple to express recombinant enzymes in insect cells. A stock of high-titer baculovirus is produced and used to infect insect cells, forcing them to expresses the gene of interest. We have had good experience using the Bac-to-Bac® Expression System (Invitrogen). This system allows a recombinant baculovirus *(40)* to be selected without plaque purification, a technique that can be difficult and time-consuming. The general approach is as follows. A cDNA encoding the gene of interest is cloned into a pFastbac vector, and a recombinant bacmid containing the gene is made in the DH10Bac cells and the bacmid is subsequently used to produce recombinant baculovirus as described in **Subheading 3.2.**

In the pFastbac vector *(41)*, the gene of interest is flanked by a polyhedrin promoter *(38)*, a gentamycin-resistance gene, and the right and left recombination arms of the Tn7 transposon *(42)* (*see* **Fig. 1**). The DH10Bac cells contain a bacmid (bMON14272) *(42)* that is selected by kanamycin resistance. Once the pFastbac vector containing the gene of interest is transformed into the DH10Bac cells, the gene is transposed onto the bacmid, making a recombinant bacmid *(40,42)*. The transposition occurs because the bacmid contains mini-*att*Tn7 attachment sites appropriate for a transposase encoded by a tetracycline-resistant helper plasmid (pMON7124) *(46)*. The insertion of the gene of interest disrupts a LacZα peptide. Thus, positive recombinants are selected as white colonies that are gentamycin, kanamycin, and tetracycline resistant. The recombinant bacmid is propagated in the DH10Bac cells, isolated, and used to make recombinant baculovirus.

The relevant vector implemented here is pFastbac-HTC (*see* **Fig. 1**). The pFastbac HT vectors enable the user to add a 6X-His tag to the N-terminus of the gene of interest. The vectors are available in all three reading frames, HTA, HTB, and HTC. The reading frame is relative to the start codon located upstream of the 6X-His tag and is shifted at the *Bam*HI site.

3.1.2. Cloning of Homo sapiens Dicer into pFastbac-HTC

The *Homo sapiens* Dicer cDNA (NM_177438.1) *(8–10,47–49)* is amplified by PCR (polymerase chain reaction) from HeLa cDNA, digested with *Not*I and *Xho*I restriction enzymes, ligated into pFastbac-HTC, and transformed into TOP10 bacterial cells.

3.1.2.1. PREPARATION OF TOTAL RNA FROM HELA CELLS

1. Isolate total RNA from a 10-cm dish of HeLa cells growing in mid-log phase, approx 3×10^6 cells or approx 60% confluency, with the RNeasy® Mini Kit as directed by the manufacturer (*see* **Note 2**).
2. Determine concentration and purity by measuring absorbance at 260 and 280 nm after dilution of the RNA at 1 : 250 in TE, pH 8.0 *(36)*. 1 OD_{260} = 40 µg/mL, so calculate the concentration of the RNA with the equation

$$[RNA] = (A_{260})(40 \text{ µg/mL})(250)$$

Calculate the purity or relative absence of protein contamination with the equation

$$Purity = A_{260}/A_{280} \quad (\text{should be} \geq 2)$$

The typical yield should be approx 100 µg and the elution volume is 100 µL, so the concentration should be approx 1 µg/µL.

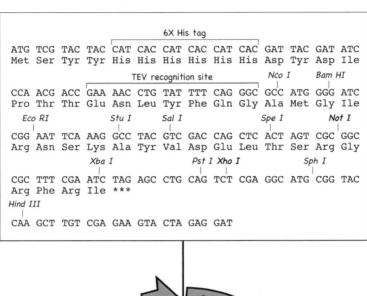

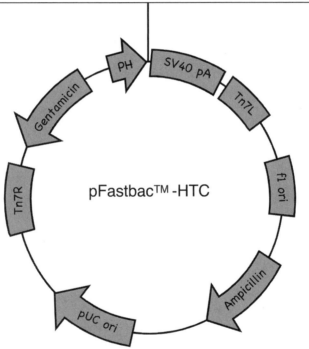

Fig. 1. Schematic of the multiple-cloning site and functional elements present in the pFastbac-HTC vector (adapted from Invitrogen, Carlsbad, CA). The Dicer coding region is cloned into the pFastbac-HTC vector at the *Not*I and *Xho*I sites (in bold) in the correct reading frame with the 6X-His (histidine) amino-terminal tag and downstream of the TEV (tobacco etch virus) protease recognition site *(54,55)*. The ampicillin-resistance gene and pUC ori (pUC origin) select for and allow amplification in bacteria. The right (Tn7R) and left (Tn7R) mini-Tn7 elements flank the Dicer gene to facilitate site-directed transposition into the bacmid (bMON14272) *(42)*.

3. The integrity of the mRNA can be estimated from the integrity of the 28S and 18S rRNA (ribosomal RNA) species. The simplest method to determine the integrity of rRNA is by agarose gel electrophoresis *(36)*. Traditionally, denaturing agarose gels, either glyoxal or formaldehyde, are used for electrophoresis of RNA. However, in this case, a good estimate of quality can be obtained by using a 1% native agarose gel so long as the rRNA is properly denatured prior to electrophoresis. The agarose gel should be buffered with either 1X TAE or 1X TBE. Ethidium bromide at a final concentration of 0.5 μg/mL should be added to the gel prior to casting. Be sure that the all solutions and electrophoresis equipment are RNase-free.

4. Mix 1 μg of RNA with an equal volume of 2X RNA loading buffer. If RNA is concentrated, 1 μg can be diluted with RNase-free H_2O to 5 μL and then 5 μL of 2X RNA loading buffer should be added. Heat mixture at 95°C for 5 min and load directly onto the gel; there is no need to place on ice. Visualize the 28S and 18S rRNA bands with UV light and photograph. It can be assumed that the mRNA is intact if the 28S and 18S bands are sharp and free of smearing and the intensity of the 28S rRNA is 1.5- to 2-fold greater than the intensity of the 18S rRNA. Smearing below either band indicates degradation of the rRNA and presumably the mRNA (for more information on RNA integrity, *see* **ref. 50**). If only one intense band is seen, the rRNA was not properly denatured. Make sure that the RNA sample is heated thoroughly in the 2X RNA loading buffer.

3.1.2.2. PREPARATION OF HELA CDNA

1. Produce single-stranded cDNA from the total RNA (**Subheading 3.1.2.1., Item 2**) with the Superscript RTIII First-Strand Synthesis System for RT-PCR as described below from the manufacturer's protocol.

2. Add the following components to a RNase-free PCR tube: 50 μ*M* of 1 μL Oligo(dT)$_{20}$, 5 μg Total RNA (up to 11 μL can be added), 1 μL of 10 m*M* dNTP mix (10 m*M* of each dNTP), and RNase-free H_2O to 13 μL.

3. Heat mixture to 65°C for 5 min and chill on ice for 2 min.

4. Centrifuge briefly and add the following: 4 μL of 5X first-strand buffer, 1 μL of 0.1 *M* DTT, 40 units/μL of 1 μL RNaseOut, and 200 units/μL of 1 μL Superscript RT III.

5. Mix by gently pipetting up and down, centrifuge briefly, and incubate at 50°C for 60 min.

6. Inactivate the enzyme by heating to 70°C for 15 min.

7. Add 1 μL of *E. coli* RNase H (2 units/μL) and incubate at 37°C for 20 min. The cDNA does not need to be purified; it can be added directly to the PCR.

Fig. 1. (*Continued*) The gentamycin-resistance gene provides a means for selection of the recombinant bacmid. The PH (polyhedrin) promoter *(38)* is upstream of the 6X-His element and Dicer coding region to provide high-level expression in insect cells. The SV40 (simian virus 40) poly adenylation signal *(56)* is downstream of the Dicer coding region and permits efficient transcription termination and poly adenylation of the Dicer mRNA.

3.1.2.3. PCR CONDITIONS FOR PRODUCING THE DICER INSERT

The T7 epitope tag (not to be confused with the T7 polymerase promoter) must be added to the N-terminus of the Dicer coding region. This is done with two rounds of PCR. First, the full-length Dicer insert is amplified so that the T7 epitope will be added to the N-terminus. Then, a second round of PCR, with a different sense strand primer, is used to add the *Not*I restriction site 5′ of the T7 epitope. To decrease the odds of introducing a deleterious mutation, the number of amplification cycles is kept to a minimum.

1. Platinum *Pfx* DNA polymerase is used to produce the Dicer insert, with the following primers :

Sense 1
5′-**ACTGGTGGACAGCCAATGGGT**ATGAAAAGCCCTGCTTTGCAAC-3′
Sense 2
5′-GAGAGT**GCGGCCGC**<u>ATGGCTAGCATGACTGGTGGACAGCCAA-TGG</u>-3′
Antisense
5′-CCGCCG**CTCGAG**TCAGCTATTGGGAACCTGAGG-3′

The gene-specific sequence used to attain the Dicer cDNA is from the accession number NM_177438.1. The T7 epitope (ATGGCTAGCATGACTGGTG-GACAGCAAATGGGT) is broken into pieces and shown in bold with the underline. The *Not*I (GCGGCCGC) and *Xho*I (CTCGAG) restriction sites, in bold, are added to flank the Dicer gene for subsequent subcloning into pFastbac-HTC (*see* **Fig. 1**). The bases at the 5′ end of the sense 2 and the antisense primers are just filler to ensure efficient digestion close to the end of the fragment. The T_m for the primers is calculated at 25 mM salt and 0.3 μM primer concentration. In the gene-specific region of the sense 1 primer, which corresponds to the Dicer sequence, the T_m is 51.2°C, and for the entire primer, after the first round of amplification, the T_m is 63.3°C. For the sense 2 primer, the T_ms are 52.3°C and 67.1°C, respectively. For the antisense primer, the T_ms are 50.7°C and 63.6°C, respectively.

2. In a PCR tube, combine the following components and overlay with mineral oil if necessary: 5.0 µL of 10X *Pfx* amplification buffer, 1.5 µL of 10 mM dNTP mixture, 1.0 µL of 50 mM MgSO$_4$, 1.5 µL of 10 µM sense 1 primer, 1.5 µL of 10 µM antisense primer, 2.0 µL cDNA, 1.0 µL Platinum *Pfx* DNA polymerase (2.5 units/µL), 36.5 µL DNase-free H$_2$O.

3. Heat the reaction at 94°C for 2.5 min to denature the template and activate the enzyme. The Platinum technology provides an inactivating antibody against the polymerase and it must be denatured in order for the polymerase to work. Two annealing temperatures are used to increase the specificity: 55°C is used for the first three rounds of amplification, then the annealing temperature is changed to 60°C. The lower temperature allows the gene-specific region to prime the cDNA. However, after the first two rounds, the entire primer will match the template; thus, the annealing temperature is raised to improve specificity.

4. Follow the 2-min incubation with a three-step cycle; use three rounds of amplification: denature: 94°C for 15 s; anneal: 55°C for 30 s; extend: 68°C for 6 min (1 min per kilobase). Then switch the annealing temperature to 60°C and proceed with 12 more rounds of amplification: denature: 94°C for 15 s; anneal: 60°C for 30 s; extend: 68°C for 6 min (1 min per kilobase); maintain at 4°C after cycling.

5. You can run a portion of the PCR on an agarose gel; however, because only 15 rounds of amplification have been performed, it may be difficult to see a band. Thus, running a gel is really not necessary at this point.

6. Set up a second PCR to add th e *Not*I restriction site: 5.0 μL of 10X *Pfx* amplification buffer, 1.5 μL of 10 m*M* dNTP mixture, 1.0 μL of 50 m*M* MgSO$_4$, 1.5 μL of 10 μ*M* sense 2 primer, 1.5 μL of 10 μ*M* antisense primer, 2.0 μL first PCR reaction, 1.0 μL Platinum *Pfx* DNA polymerase (2.5 units/μL), and 36.5 μL DNase-free H$_2$O.

7. Heat the reaction at 94°C for 2.5 min to denature the template and activate the enzyme. Follow with a three-step cycle; use 20 rounds of amplification. As in **step 4**, two annealing temperatures will be used.

8. Follow the 2.5-min incubation with a three-step cycle; use three rounds of amplification: denature: 94°C for 15 s; anneal: 55°C for 30 s; extend: 68°C for 6 min (1 min per kilobase). Then switch the annealing temperature to 60°C and proceed with 17 more rounds of amplification: denature: 94°C for 15 s, anneal: 60°C for 30 s, extend: 68°C for 6 min (1 min per kilobase), maintain at 4°C after cycling.

9. Analyze PCR by running 5 μL on a 0.8% agarose containing ethidium bromide (*see* **Subheading 3.1.2.1., step 3**). Use the 10X DNA loading buffer. A good size marker is the 1-kb DNA ladder. The correct band is 5795 bp. For troubleshooting information, *see* **Note 3**.

10. If the correct band is present, or abundant enough to clone relative to contaminating bands, purify DNA away from the enzyme and reaction components using the QIAquick PCR purification kit as directed by the manufacturer. Elute in 30 μL EB (10 m*M* Tris-Hcl, pH 8.5; provided in the kit) so that the subsequent enzymatic steps can be performed in a smaller reaction volume. To verify recovery of the insert, run another 0.8% agarose gel (*see* **step 9**). The amount of DNA to load can be extrapolated from the previous gel.

3.1.2.4. Digestion and Gel Purification of the pFastbac-HTC Vector and the Dicer Insert

1. To digest the Dicer insert with *Not*I and *Xho*I, set up the following reaction and incubate at 37°C for 2 h (*see* **Note 4**): 0.5–1.0 μg of 25 μL DNA insert, 4 μL of 10X NEB buffer 3, 4 μL of 10X BSA, 2 μL *Not*I (10 units/μL), 1 μL *Xho*I (20 units/μL), 4 μL H$_2$O.

2. To digest the pFastbac-HTC vector set up the following reaction and incubate at 37°C for 2 h (*see* **Note 4**): 2 μL of pFastbac-HTC 0.5 μg/μL, 4 μL of 10X NEB buffer 3, 4 μL of 10X BSA, 10 units/μL of 2 μL *Not*I, 20 units/μL of 1 μL *Xho*I, of 27 μL H$_2$O.

3. After the 2-h incubation, add 5 μL of 10X DNA loading buffer to each reaction and load the 45 μL in a large well of a 0.8% agarose gel and allow sufficient time

for good resolution (e.g., run the bromophenol dye marker to the bottom of an 8-cm gel). Run gel as described previously.

4. Visualize bands with UV light, photograph, and excise bands with a sharp scalpel or razor blade. Avoid a large excess of agarose by carefully trimming it away from the DNA. Place excised bands in a microcentrifuge tube. The linearized vector should migrate at 4858 bp and the digested insert at 5776 bp. It will be impossible to determine if the insert is digested or if the vector is digested with both enzymes.

5. Extract DNA with the QIAquick Gel Extraction Kit per manufacturer's suggestions (*see* **Note 5**). Elute in 30 μL EB (10 mM Tris-HCl, pH 8.5; included in kit).

6. Run a 0.8% agarose gel to assess quality and quantity of the linearized vector and digested insert. The High DNA Mass Ladder is good for estimating the concentration of the digested vector and insert.

3.1.2.5. LIGATION OF THE DICER INSERT INTO THE pFASTBAC-HTC VECTOR

The digested Dicer fragment is ligated into the digested pFastbac-HTC vector with T4 DNA ligase as described by the manufacturer (*see* **Note 6**). Mix the following components in a microcentrifuge tube, pipet up and down, centrifuge briefly, and incubate at room temperature for 20 min. After setting up the ligation reaction, thaw One Shot TOP10 Chemically Competent Cells on ice for the subsequent transformation (*see* **Subheading 3.1.2.6.**). Thaw one tube per ligation reaction—4 tubes in total if ligations are set up as described as follows:

Vector-only control
50 ng of pFastbac-HTC digested with *Not*I and *Xho*I, 2 μL of 10X T4 DNA ligase buffer, 1 μL T4 DNA ligase (400 units/μL), H₂O to final volume of 20 μL.

Vector plus insert
50 ng of pFastbac-HTC digested with *Not*I and *Xho*I, 0.5-, 1-, and 3-fold molar excess of digested Dicer insert, 2 μL of 10X T4 DNA ligase buffer, 1 μL T4 DNA ligase (400 units/μL), H₂O to final volume of 20 μL.

3.1.2.6. TRANSFORMATION OF THE DICER-pFASTBAC-HTC CONSTRUCT INTO TOP 10 CELLS

1. Other strains of bacteria or cells you made can be used instead of the commercially available competent cells. Other forms of transformation can also be used where appropriate.

2. Place LB plates containing 100 μg/mL ampicillin in an incubator at 37°C. This allows plates to warm to 37°C while performing the transformation and recovery. Prewarm two LB plates per transformation.

3. Mix one-half of the ligation into 50 μL of One Shot TOP10 Chemically Competent Cells. Do not pipet up and down. Save the remaining ligation at −20°C.

4. Incubate on ice for 30 min.
5. Heat shock the cells for 30 s at 42°C.
6. Transfer tubes immediately to ice.
7. Add 250 µL of SOC media and shake for 1 h at 37°C and 225 rpm.
8. Evenly spread 60 µL (one-fifth) of the transformation on one prewarmed LB plate and the remaining 240 µL (four-fifths) on the second prewarmed LB plate. Glass beads or conventional plate spreaders could be used. Using 2 vol of the transformation will ensure that one of the plates will have well-isolated single colonies for selection (*see* **Subheading 3.1.2.7.**).
9. Return plates to the 37°C incubator and allow bacteria to grow for 16 h.
10. Count colonies and determine likelihood of subcloning success (*see* **Note 7**).

3.1.2.7. ISOLATE DNA FROM COLONIES AND SCREEN FOR PFASTBAC-HTC-DICER CLONES

1. Select colonies and grow overnight in 3 mL LB containing 100 µg/mL ampicillin. Four to ten colonies should be a sufficient number to obtain a positive subclone (*see* **Note 8**). Place cultures in a shaking incubator at 37°C and 250 rpm and grow for approx 16 h. Disposable 14-mL culture tubes work well. Refer to QIAprep Miniprep Kit protocol for more information.
2. Isolate DNA with the QIAprep Miniprep Kit as directed by the manufacturer.
3. Digest each sample with either *Not*I or *Bam*HI for 30 min to 2 h at 37°C. Mix as described below, gently pipet up and down, and centrifuge briefly. A master mix can be prepared if a large number of colonies were selected: For *Not*I: 100–200 ng of 1 µL miniprep DNA, 2 µL of 10X NEB buffer 3, 2 µL of 10X BSA, 1 µL *Not*I (10 units/µL), 14 µL H_2O. For *Ban*HI: 100–200 ng 1 µL miniprep DNA, 2 µL of 10X *Bam*HI buffer, 2 µL of 10X BSA, 0.5 µL *Bam*HI (20 units/µL), 14.5 µL H_2O.
4. Add 2 µL of 10X DNA loading buffer to each digest and electrophorese on a 0.8% agarose gel as described previously. If the Dicer insert was ligated into the pFastbac-HTC vector, then the *Not*I digest will linearize the vector and the size should be 10,605 bp. However, if the Dicer insert is not present, the linearized vector will migrate at 4858 bp. Likewise for the *Bam*HI digest; if the insert is present, two bands will be generated: one at 6020 bp and the other at 4585 bp. If it is just vector, the band will migrate at 4858 bp. The linearized pFastbac-HTC vector should be run as a control for size. Any other pattern of bands suggests a problem with vector and insert preparation, such as introduction of mutations into the insert. Cloning steps (**Subheadings 3.1.2.3.–3.1.2.7.**) should be repeated. The presence of the insert can also be confirmed by running undigested miniprep DNA and undigested pFastbac-HTC on a 0.8% agarose gel because the vector plus insert will migrate as a very large supercoiled band, much larger than the supercoiled vector. If undigested miniprep DNA is used, be sure to run undigested pFastbac-HTC vector as a control.
5. If the correct bands are seen after the digestion sequence, then subclone to ensure that it does not contain any deleterious mutations. Before proceeding, the Dicer

cDNA should be sequenced rigorously because it is such a large gene and was amplified by PCR (*see* **Note 9**).

3.1.3. Transposition of the Dicer cDNA in DH10Bac Cells

The transposition of the Dicer gene onto the bacmid (*see* **Subheading 3.1.1.**) takes place after a simple transformation of the pFastbac-HTC-Dicer construct into chemically competent DH10Bac cells. The transformation is performed as suggested by the manufacturer; see the Bac-to-Bac instruction manual for additional details and troubleshooting information.

1. Thaw one tube, 100 μL,Cof MAX Efficiency DH10Bac cells on ice, approx 20–30 min. Heat a water bath to 42°C.
2. Place DH10Bac cells into a prechilled 14-mL tube. The tube is important because of variation in heat transfer between different tubes (e.g., if the tubes are different in thickness or in composition).
3. Add 5 μL (200 pg/μL) of the sequence-verified pFastbac-HTC-Dicer construct to the cells and mix gently, but do not pipet up and down.
4. Incubate on ice for 30 min.
5. Heat shock the cells at 42°C for 45 s.
6. Transfer the tubes quickly back to ice for 2 min.
7. Add 900 μL of S.O.C. medium.
8. Place tube in a shaking incubator set at 37°C and 225 rpm. Allow cultures to incubate for 4 h. This incubation might seem long relative to normal recovery incubations used in routine transformations. However, this extended incubation is critical because the bacteria must deal with replication of a very large bacmid.
9. During the last hour of the incubation place four LB plates containing 50 μg/mL kanamycin, 10 μg/mL tetracycline, 7 μg/mL gentamycin, 100 μg/mL bluo-gal, and 40 μg/mL IPTG in a 37°C incubator. Be sure that the plates were prepared with Luria broth and not Lennox L broth because use of Lennox L results in fewer colonies and less intense color.
10. At the end of the 4-h recovery, plate 1, 10, 100, or approx 890 μL of the cells on the prewarmed LB plates. In order to efficiently spread the small volumes of cells, add 99 μL of SOC to the 1 μL of cells and 90 μL to the 10 μL. It is not necessary to add any medium to the 100 μL of cells, but in order to plate the 890 μL of cells, the volume must be reduced. Transfer the remainder of the cells (all 890 μL) to a microcentrifuge tube and spin for 2 min at 1000*g*. Aspirate the majority of the medium, leaving about 100 μL. Gently resuspend the cells by pipteting up and down and then transfer to one of the LB plates. Also, transfer the other 3 vol of cells, each in 100 μL, each to a different LB plate. Spread cells with glass beads or a conventional plate spreader.
11. Place plates in a 37°C incubator and check for colony growth at 24 h. As a general rule, it will take 48 h until colonies are the appropriate size for selection, but very small colonies might be visible at 24 h.

3.1.4. Selection of a Recombinant Bacmid

Recombinant bacmids are first selected by blue/white screening of colonies. The bacmid DNA is then isolated with traditional alkaline lysis. General miniprep kits do not work well because the bacmid is so large. There are kits available for isolation of bacmid DNA but are not necessary. Finally, PCR screening is used to verify that the Dicer gene was transposed onto the bacmid.

3.1.4.1. SELECTION OF WHITE COLONIES AND PREPARATION FOR BACMID ISOLATION

1. To avoid false positives, select well-isolated, large, white colonies. Avoid colonies that appear gray or darker in the center because they might contain a mixture of bacteria that harbor both bacmids with and without the insert. Viewing the colonies under a dissecting microscope against a white background facilitates the selection process.
2. Four colonies should be restreaked on LB plates containing 50 µg/mL kanamycin, 10 µg/mL tetracycline, 7 µg/mL gentamycin, 100 µg/mL bluo-gal, and 40 µg/mL IPTG in a 37°C incubator. Colonies should be visible after overnight incubation at 37°C, in contrast to the previous 48-h incubation. This will confirm selection of bacmids containing the insert.
3. From the restreaked colonies, again select four colonies and grow in 3 mL of LB containing 50 µg/mL kanamycin, 10 µg/mL tetracycline, 7 µg/mL gentamycin. Grow liquid cultures overnight to stationary phase at 37°C and 250 rpm; because the bacmid is so large, bacterial growth is slower than normal.
4. A glycerol stock of the recombinant bacmid should be made. Take one LB plate-containing 50 µg/mL kanamycin, 10 µg/mL tetracycline, 7 µg/mL gentamycin, 100 µg/mL bluo-gal, and 40 µg/mL IPTG and divide into four quadrants using a Sharpie to label the bottom of the plate. Before isolating bacmid DNA, use a pipet tip or inoculation loop to streak a portion of each of the four cultures in one of the quadrants of the LB plate. Place the plates in the 37°C incubator overnight and store at 4°C once the colonies are visible. These colonies are essential for making a glycerol stock (*see* **step 6** in **Subheading 3.1.4.3.**), which ensures that once a verified bacmid is found, transposition will not have to be repeated. Note that the bacmid cannot easily be transformed into bacteria, so a glycerol stock is very important.

3.1.4.2. ISOLATION OF BACMID DNA

Bacmid DNA can be isolated with a modified alkaline lysis protocol (developed by Invitrogen through communication with Peter deJong). The protocol that we use is described here. Additional information can be found in the Bac-to-Bac instruction manual.

1. Make sure solution I contains 100 µg/mL RNase A. RNase A can be added to the stock solution if all of the solution will be used within 6 mo. Alternatively, the

RNase A can be added fresh to a small amount of solution I prior to isolation of DNA.

2. Transfer 1.5 mL of bacterial culture to a microcentrifuge tube and centrifuge at 14,000g for 1 min to pellet the cells.

3. Remove the supernatant with an aspirator or pipet.

4. Resuspend the pellet in 300 µL of solution I. Gently vortex or pipet up and down to resuspend.

5. Add 300 µL of solution II and gently mix by inverting the tube five times. Incubate at room temperature for 5 min. The solution should become translucent and viscous.

6. To precipitate cellular protein and genomic DNA add 300 µL of 3 M potassium acetate, pH 5.5, and gently mix by inverting the tube several times. Precipitate is white and thick. Place on ice for 5 min.

7. Centrifuge for 10 min at 14,000g and room temperature.

8. Avoid the precipitate and carefully transfer the supernatant to a 2-mL microcentrifuge tube containing 800 µL of isopropanol, mix gently, and incubate on ice for 10 min. This is a good stopping point if necessary because the precipitate can be stored overnight at –20°C.

9. To pellet the bacmid DNA-precipitate, centrifuge for 15 min at 14,000g at room temperature.

10. Carefully remove the supernatant without disturbing the pellet. Note that the pellet is often translucent and difficult to see. Holding the tube at the correct angle toward a light provides sufficient contrast to identify the location of the pellet. Keep track of the tube orientation during centrifugation so that the pellet can be identified.

11. Overlay the pellet with 500 µL of 70% ethanol to wash the pellet.

12. Centrifuge for 5 min at 14,000g and room temperature.

13. Repeat **steps 10** and **11**. This second wash is not optional; it is essential to remove contaminating proteins and salts, thereby improving the purity of the bacmid for the subsequent transfection.

14. Carefully remove the 70% ethanol. Centrifuge briefly, and using a very fine pipet tip (e.g., gel loading tip), carefully remove all traces of 70% ethanol. Air-dry the pellet for 5–10 min, taking care not to overdry. It is always difficult to set a time on drying a nucleic acid pellet, because of differences in the accuracy with which the trace amounts of ethanol are removed. A good way to tell if the pellet is dry is to note when it becomes very translucent and difficult to see. Do not use a Speedvac to dry the pellet.

15. Resuspend the pellet in 40 µL TE. Just overlay the pellet and allow it to dissolve for 30–45 min at room temperature. The tube can be gently tapped and centrifuged briefly. Do not pipet up and down or vortex because this will shear the bacmid DNA.

16. Dilute the bacmid DNA 1 : 200 and 1 : 400 in TE and measure absorbance at 260 and 280 nm (*see* **Subheading 3.1.2.1.**, **step 2**). Use the following conversion (1 OD_{260} = 50 µg/mL):

$$[DNA] = (A_{260})(50 \text{ μg/mL})(\text{dilution factor}), \text{Dilution factor} = 200 \text{ or } 400$$

A typical yield should be 10–20 ng and the elution volume is 40 μL, so the concentration should be about 0.25–0.5 μg/μL.

Calculate the purity or relative absence of protein contamination with the following equation:

$$\text{Purity} = A_{260}/A_{280} \text{ should be } \geq 1.8$$

Lower ratios will result in decreased transfection efficiencies, poor cell health, and lower-quality virions (*see* **Subheading 3.2.2.**). Store the DNA at 4°C until verification of the transposition is completed.

3.1.4.3. CONFIRMATION OF TRANSPOSITION

The easiest and most reliable method to confirm that Dicer was transposed onto the bacmid is PCR. The bacmid is around 135 kbp, much too large to use restriction digest analysis for verification. A low-percentage agarose gel can be used, but interpretation of the results is often difficult, unreliable, and not reassuring. The Tn7 transposition sites are flanked by the M13 forward and reverse primer sites, making PCR analysis simple. Also, a combination of one M13 primer and a gene-specific primer could be used, however, two gene-specific primers should not be used because residual pFastbac-HTC-Dicer DNA could result in a false positive.

1. M13 forward primer: 5′-GTTTTCCCAGTCACGAC; M13 reverse primer: 5′-CAGGAAACAGCTATGAC (T_m is approx 50°C).
2. Use Platinum *Taq* high-fidelity polymerase for the PCR. Mix the following in a PCR tube and heat for 2.5 min at 94°C. This initial step is important to denature the template and activate the enzyme; the Platinum enzymes contain an inactivating antibody that must be denatured to achieve full polymerase activity: 5 μL of 10X high-fidelity buffer, 1 μL of 10 m*M* dNTPs, 2 μL of 50 m*M* MgSO$_4$, 2 μL of 10 μ*M* M13 forward, 2 μL of 10 μ*M* M13 reverse, 1 μL of 100 ng/μL bacmid DNA (generally a 1 : 4 dilution), 0.25 μL Platinum *Taq* high-fidelity, H$_2$O to 50 μL.
3. Use the following cycling parameters: denature = 94°C for 30 s; anneal = 55°C for 30 s; extend = 68°C for 8.5 min. The extension time is 1 min per kb; 8.5 min in this case because Dicer is approx 6 kb; and there is approx 2.5 kb of additional bases flanking the Dicer insert between the two M13 primer sites (*see* **Note 10**).
4. Run 5 μL of the PCR reaction on a 0.8% agarose gel as described previously. Use the 10x DNA loading buffer. The 1-kbp ladder is a good size marker. If Dicer was transposed onto the bacmid, then the band will migrate at 8206 bp (5776 bp + 2430 bp); a lack of transposition will result in a band at approx 300 bp (*see* **Note 11**). Because the expected band is so large, the gel will have to run for a long time in order to accurately compare the large band with the bands in the 1-kbp ladder.

The bromophenol blue dye can be run nearly to the bottom of an 8-cm gel, but keep in mind that if transposition did not occur, a 300-bp band will be present and this band will run below the bromophenol dye.

5. Determine which bacmids are positive—likely to be all that were chosen because of blue/white selection. It is best to aliquot the DNA for storage. An aliquot size of 5 µL is good. Store half of the aliquots at –20°C for long-term storage and the other half at 4°C for transfection (*see* **Subheading 3.2.3.**). Note that repeated freeze–thaw cycles can result in shearing of the bacmid DNA. Mechanical stress will cause shearing of the bacmid DNA; thus, never vortex or pipet up and down.

6. Make a glycerol stock of the DH10Bac cells that harbor the Dicer bacmid. Before isolating bacmid DNA, a portion of the cultures were streaked on an LB plate (*see* **step 4** in **Subheading 3.1.4.1.**). From a colony that was confirmed to have been transposed with dicer inoculate 3 mL of LB containing 50 µg/mL kanamycin, 10 µg/mL tetracycline, and 7 µg/mL gentamycin. Grow liquid cultures overnight to stationary phase at 37°C and 250 rpm. Remove 1 mL of culture and mix with 1 mL of 30% glycerol and mix very well by inverting the tube several times and then snap-freeze in liquid nitrogen and store at –80°C.

3.2. Production of High-Titer Viral Stocks

In order to express r-Dicer, a high-titer baculovirus stock must be produced in Sf9 cells. This section describes how to (1) initiate a Sf9 culture, (2) maintain a Sf9 culture, (3) produce frozen Sf9 cell stocks, (4) produce a P1 viral stock by transfecting the Dicer bacmid into Sf9 cells, (5) generate a P2 viral stock from amplification of the P1 viral stock, and (6) amplify the P2 viral stock to produce a large volume of high-titer P3 viral stock to use for r-Dicer production. As in **Subheading 3.1.**, the methods described in this subsection are our interpretation of the Bac-to-Bac instruction manual. Please refer to the manual for additional details and troubleshooting information.

3.2.1. Starting an Sf9 Culture

The easiest way to culture Sf9 cells is in suspension, preferably in disposable flasks with vented filter caps. The preferred growth medium is Sf900 II SFM; refer to the Bac-to-Bac Expression System manual for alternate growth medias. Cells double more frequently without antibiotics. Thus, it is better to grow without the antibiotics (*see* **Note 12**). The cells grow best at 28 + 0.5°C, shaking at 135 rpm. Note that CO_2 and humidity are not required. Self-contained shaking incubators (Innova models) work best because temperature fluctuation is minimal; however, cells can be grown on a benchtop shaker if the room temperature can be set to 28 ± 0.5°C with minimal fluctuation (*see* **Note 13**). The incubator should be capable of heating and cooling because the optimal growth temperature, 28 ± 0.5°C, is very close to room temperature. The incubator should be kept clean and sterilized once a month to avoid fungal or bacterial contamination (*see* **Note 14**). Cells can be grown in adherent cultures if a shaker or shaking incubator is not available; large tissue culture plates or flasks are optimal (*see* **Note 15**). However, adherent cultures require more work both to maintain the culture and to produce protein; thus, they are used only when necessary, as described here.

1. Sf9 cells, already adapted to grow in suspension, can be purchased from Invitrogen. Cells are supplied as 1.5×10^7 viable cells in 1.5 mL.
2. In a sterilized tissue culture hood add 27 mL of Sf900 II SFM medium (**Subheading 2.2., item 1**) to a 125-mL culture flask. Avoid bubbles by holding the flask at about a 45° angle and dispensing the medium down the side of the flask. Place the flask in a shaking incubator at 28 ± 0.5°C for 30 min to equilibrate the medium to the appropriate temperature (*see* **Note 16**).
3. Thaw cells as quickly as possible in a 37°C water bath; remove when about one-third of solid material remains. Thaw this last bit of solid by inverting the tube several times. Heating the cells excessively at 37°C will result in cell death, making it impossible to continue with the culture.
4. Clean the outside of the tube with 70% ethanol and move to a sterilized tissue culture hood.
5. Resuspend the cells by gently inverting the tube. Do not mix by pipeting up and down because it will shear the cells. Add the cells to the 125-mL flask containing 27 mL of Sf900 II SFM prewarmed to 28°C (**step 2**). Density is approx 8×10^5 viable cells per milliliter.
6. Let the flask shake at 135 rpm and 28 ± 0.5°C. Be sure to loosen the lid one-quarter turn to allow high levels of aeration. Loosen the cap even if the filtered caps are being used so that aeration is not the limiting factor.
7. Let the cells grow for 3 d and then determine density and viability. Move the flask to a sterile hood, remove cap, tilt flask about 45° to one side (side from which you will pipet), and transfer 1 mL from the upper portion of the culture to a microcentrifuge tube. Recap the flask, making sure to loosen one-quarter turn, and return to incubator. The cells will settle to the bottom of the flask and form clumps if left in the hood while cells are being counted.
8. Move the cells in the microcentrifuge tube to the benchtop and transfer 100 µL of the 1-mL cell suspension to another microcentrifuge tube containing 100 µL of trypan blue. Mix by gently flicking the tube. Work quickly to ensure that the cells do not settle to the bottom of the tubes. If cells settle, resuspend by gently flicking the tube. Do not pipet up and down because this can damage the cell wall and interfere with the accuracy of measuring cell viability.
9. Transfer 10 µL of the trypan blue-stained culture to a hemacytometerNote that 10 µL is common to most hemacytometers, but refer to the specific directions for your hemacytometer. Count the number of viable and dead cells based on the fact that the viable cells exclude the trypan blue and are transparent, and the dead cells are stained blue. Cells should be counted in four quadrants.

10. To determine the viable cell density, calculate the average cell number of viable cells per quadrant and account for the twofold dilution. This number is then multiplied by 1×10^4 and the resulting number is the viable cell density in cells per milliliter. Again, the conversion is common to most hemacytometers, but refer to the specific directions for your hemacytometer. Also, determine the viability of the culture by calculating the percentage of cells that are viable (*see* **Note 17**).

11. It is very likely that 3 d after initiation, the density of the freshly started culture will be only $(6–7) \times 10^5$ viable cells per milliliter with a viability of 60–70%. The growth rate and viability are much lower than seen in a healthy culture (*see* **Note 18**). Even though the density is low, the cells must be diluted in order to improve cell health. Do not worry about the cell health and growth rate; they will improve over the next two passages, as long as the cells are split every 3 d. If the density and viability are much higher, $(2–3) \times 10^6$ and >90%, that is excellent; proceed to the next step.

12. Seed a fresh 125-mL flask at 3×10^5 viable cells per milliliter (*see* **Note 19**). It is best to make the volume of the culture 50 mL because the cells seem to grow better in a larger volume. It is likely that you will be diluting the culture only twofold because the original culture volume was about 28 mL and the density is likely to be $(6–7) \times 10^5$ viable cells per milliliter. Note that the maximum culture volume in a 125-mL flask is 50 mL. (*see* **Table 1**).

13. Again, after 3 d of growth, determine cell density and viability. It is very likely that the density will again be $(6–7) \times 10^5$ viable cells per milliliter, but the viability should have improved to 70–80%. You should split the cells to make a 100-mL culture in a 250-mL disposable flask. It is a good to have a large volume of cells for making frozen cell stocks (*see* **Subheading 3.2.3.**). The seeding density should again be 3×10^5 viable cells per milliliter. If the density and viability are higher, that is excellent; seed the cells at 3×10^5 viable cells per milliliter.

14. By the third passage the growth rate and viability should be increasing; for example, after 3 d, the density should be $(1–2) \times 10^6$ viable cells per milliliter and the viability should be >85%. Again, split the cells to a density of 3×10^5 viable cells per milliliter. Be sure to maintain a 100-mL culture for making frozen cell stocks; ideally, three to five 100-mL cultures should be grown to a density of $(2–3) \times 10^6$ viable cells per milliliter for preparing frozen cell stocks (*see* **Subheading 3.2.3**); however, frozen stocks should be made only if the cells are growing well. If the growth rate and viability are low, *see* **Note 20**.

15. After the third passage and up until the sixth or seventh passage, the strategy changes. The viability should be improving, so now the focus is on the growth rate. At this point, the growth rate may be slow; a density of $(2–3) \times 10^6$ viable cells per milliliter may not be attained until 5–7 d after seeding at 3×10^5 viable cells per milliliter. This is fine as long as the viability is not decreasing. Let the cells grow until mid-log phase growth is attained. During this extended growth period, be sure to count cells and check the viability every day. To improve the growth rate, cells should be seeded at a density of 5×10^5 viable cells per milliliter. If the growth rate and viability remain low, see **Note 21**.

Table 1
Recommended Culture Conditions for Sf9 Suspension Culture

Flask size	Culture volume	Seeding density	Mid-log phase density
125 mL	35–50 mL	$(3–5) \times 10^5$ viable cells per milliliter	$(2–3) \times 10^6$ viable cells per milliliter
250 mL	75–100 mL	$(3–5) \times 10^5$ viable cells per milliliter	$(2–3) \times 10^6$ viable cells per milliliter
500 mL	150–250 mL	$(3–5) \times 10^5$ viable cells per milliliter	$(2–3) \times 10^6$ viable cells per milliliter
1 L	300–400 mL	$(3–5) \times 10^5$ viable cells per milliliter	$(2–3) \times 10^6$ viable cells per milliliter
2 L	600–800 mL	$(3–5) \times 10^5$ viable cells per milliliter	$(2–3) \times 10^6$ viable cells per milliliter
3 L	900–1200 mL	$(3–5) \times 10^5$ viable cells per milliliter	$(2–3) \times 10^6$ viable cells per milliliter

Note: Listed here are the suggested cell densities and culture volumes for subculturing Sf9 cells in the relevant flask. To ensure optimal cell growth, the maximum culture volume per flask volume should not be exceeded. Furthermore, irregardless of flask size cells should be grown to mid-log phase density before subculturing and should not be seeded at a density less than 3×10^5 viable cells per milliliter.

16. After approximately five to seven passages, the growth rate should be optimal and the cells can be used to produce a virus or protein.
17. A log of cell growth should be maintained. At each passage, keep track of the date, the viable cell density, the viability, and notes regarding general cell health and morphology as well as the volumes to which the cells are split. Note the date of initiation and the lot number of the cells that were purchased. It is also important to keep track of the lot number of the Sf900 II SFM medium that is used, in case a bad lot is encountered.

3.2.2. Maintaining an Sf9 Culture

Once a freshly started culture stabilizes, the viability will be >95% and will be able to reach mid-log phase growth in 2–4 d after splitting. To maintain the culture, use the following procedure.

1. Grow the cells until they reach mid-log phase at a density of $(2–3) \times 10^6$ viable cells per milliliter. We favor densities closer to 3×10^6 viable cells per milliliter because the viability is usually higher. The viability should always be >95%. Cell density and viability are determined as described previously (*see* **Note 17**). If the cells are seeded at 3×10^5 viable cells per milliliter it will take 3–4 d for cells to reach a density of $(2–3) \times 10^6$ viable cells per milliliter. If cells are seeded at 5×10^5 viable cells per milliliter, it will take only 2 d to reach a density of $(2–3) \times 10^6$ viable cells per milliliter. It is not a problem if cells reach a density greater than $(2–3) \times 10^6$ viable cells

per milliliter but the growth rate might decrease over the next 3 d. Maximum and minimum volumes of medium for each flask size are listed in **Table 1**. If the growth rate and viability decrease, try to rescue the culture as described in **Note 21**.

2. In order to eliminate unwanted metabolic products, the cells must be resuspended in fresh medium every 3 wk. Count the cells and determine the viability as described previously. Transfer the culture to a sterile 50-mL conical tube, note the volume, and spin at 500g and room temperature for 5 min.

3. Return to the sterile hood and remove the medium with a pipet; do not aspirate. Recap the tube and gently tap the bottom so that the cells are pushed up the walls of the tube. Remove the cap and transfer fresh Sf900 II SFM medium to the conical tube; add medium so that is runs along the wall of the tube. Use the same volume of medium that the cells were in before centrifugation, so that the previous estimation of cell density is still accurate. Gently resuspend the cells by inverting the tube several times. In some instances, it might be difficult to resuspend the last bit of cells, but simply inverting the tube several more times or gentle tapping of the tube will finish the job.

4. Use a portion of the fresh cell suspension to seed a flask at $(3–5) \times 10^5$ viable cells per milliliter. You might see a few small clumps when counting cells at the next passage. The clumps will not affect cell growth or infection and should disseminate by the next passage. If clumps persist, make sure shaking is at 135 rpm and that cells are not centrifuged for longer than 5 min or at speeds greater than 500g.

5. A log of cell growth should be maintained as described in **Subheading 3.2.1., step 15**. Cells should be maintained for only 30 passages. Infectivity, viability, and reliability will decrease after 30 passages. It is important to start a new culture before the 30 passages has been reached.

3.2.3. Making Frozen Sf9 Cell Stocks

If you purchased cells, you will need to make frozen cell stocks. If you already have frozen cell stocks, you can skip this subsection until you have used up most of your own frozen cell stocks. It is very important to make a large number of frozen cell stocks because cultures are maintained for only 30 passages. It is also very important to make the frozen cell stocks at the lowest possible passage number. Cells are frozen in 46.25% fresh Sf900 II SFM medium, 46.25% conditioned Sf900 II SFM medium (medium the cells have been growing in for 2–4 d), and 7.5% DMSO.

1. As soon as the cells you purchased stabilize, cell stocks should be made. The frozen cell stocks can be made once the fresh culture has a viability >95% and has attained mid-log phase growth, a density of $(2–3) \times 10^6$ viable cells per milliliter. This should be after three or four passages. Even if the freshly started culture has viability >95% and the growth rate is ideal, cell stocks should not be made until cells have been passed three or four times.

2. If storage space is not limited, it is worthwhile to make 50 frozen cell stocks. For safety, store an equal number of stocks in two independent cell freezers. Cells can be stored submerged in liquid nitrogen or the vapor phase.

3. Frozen cell stocks are 1 mL of cells at a density of 1.5×10^7 viable cells per milliliter. Four 100-mL cultures at a density of 2×10^6 viable cells per milliliter will be enough cells to make 50 frozen cell stocks.

4. Once the cells have stabilized, seed cells at a density of $(3–5) \times 10^5$ viable cells per milliliter in 100 mL in a 250-mL flask. You will need four flasks.

5. After 2–4 d, determine the cell density and viability. The viability must be >95% and the cells must be in mid-log phase growth, a density of $(2–3) \times 10^6$ viable cells per milliliter. It is alright to allow the cells to grow up to 7 d to reach mid-log phase and this might be the case because it is a fresh culture. Determine the cell density and viability of one of the cultures. You will need 7.5×10^8 viable cells (50 mL of 1.5×10^7 viable cells per milliliter). Portion the appropriate amount of cells (e.g., 300 mL of cells at 2.5×10^6 viable cells per milliliter) into sterile 50-mL conical tubes. Spin the cells into a pellet at $500g$ for 5 min at room temperature.

6. Return to the tissue culture hood and use a sterile pipet to remove the medium. Remember to save approx 30 mL of the conditioned growth medium in a sterile 50-mL conical tube to make the freezing medium. Leave about 5 mL of medium on top of each cell pellet so the cells do not dry out while you are making the cell-freezing medium.

7. Make the freezing medium by mixing 23.2 mL of conditioned medium with 23.2 mL fresh Sf900 II SFM and 3.25 mL DMSO. The volume can be scaled up or down accordingly to the number of cell stocks you are making.

8. Working with one 50-mL Falcon tube at a time, remove the last 5 mL of medium with a sterile pipet. Recap the tube and gently tap the bottom of the tube to disperse the cells on the walls of the tube. Dispense approx 10 mL of cell-freezing medium along the side of the tube. Recap and invert several times to resuspend the cells, taking care to eliminate all cell clumps; this may require gently tapping the bottom of the tube. Repeat until all cell pellets are resuspended.

9. Combine all of the cells pellets into one 50-mL conical tube and dispense 1-mL aliquots into cryopreservation vials. It is important to work very quickly.

10. Place the cells at 4°C for 30 min, then at –20°C for 30 min, then at –80°C overnight, and, finally, store submerged or in the vapor phase of a cell freezer containing liquid nitrogen.

11. After 1 wk, start a culture from a frozen stock to evaluate the quality of the frozen cell stocks (*see* **Subheading 3.2.1.**). Cells should reach normal growth rate and viability of >95% after three to four passages.

3.2.4. Producing a P1 Viral Stock

The initial P1 viral stock is prepared by transfecting the bacmid into Sf9 cells and isolating the supernatant after 3–5 d. It is a good idea to use a six-well plate (**Subheading 2.2., item 9**) and to transfect four or five of the six wells. Leave one well untransfected to assess the efficiency of transfection. Perform each step of the transfection in a sterilized tissue culture hood.

1. Grow 50 mL of Sf9 cells to $(2–3) \times 10^6$ viable cells per milliliter and, viability at ≥97%.
2. Plate 9×10^5 cells per well, in each well of a 6-well dish by diluting the culture to 4.5×10^5 cells per milliliter with Sf900II SFM and adding 2 mL to each well.
3. Let the cells attach for 30 min to 1 h in an incubator at $28 ± 0.5°C$. The incubator does not need either CO_2 or humidity. If the cells are very healthy, they will attach in about 15 min, however, wait at least 30 min.
4. Bacmid DNA and transfection reagents can be combined and can incubate while the cells are attaching. For each transfection, add 1µg of bacmid DNA to 100 µL of unsupplemented Grace's medium in a sterile 5-mL polypropylene round-bottom tube. Be sure to use unsupplemented Grace's medium without FBS (fetal bovine serum) and antibiotics for optimal transfection and viral production. For the untransfected well, use 100 µL of unsupplemented Grace's medium.
5. For each transfection, add 6 µL of Cellfectin to a second sterile round-bottom tube containing 100 µL of unsupplemented Grace's medium. Be sure to mix Cellfectin by inverting five times before taking the 6-µL aliquot. Alternatively, a master mix of Cellfectin and unsupplemented Grace's medium can be prepared. If six transfections are required, then add 7 vol of Cellfectin and 7 vol of unsupplemented Grace's medium to the round-bottom tube. Use 7 vol so that there will be enough of the mixture.
6. Add the Cellfectin-unsupplemented Grace's medium mixture to the bacmid-unsupplemented Grace's medium mixture; 106 µL of the master mix. Mix gently by flicking the tube and incubate for 30 min. The sample for the untransfected well should also contain Cellfectin and 200 µL of unsupplemented Grace's medium, but no DNA.
7. About 5 min before the end of the 30-min incubation in **step 6**, aspirate the medium from the cells and add 2 mL of unsupplemented Grace's medium.
8. Working with one tube at a time, add 0.8 mL of unsupplemented Grace's medium to the Cellfectin–bacmid mixture and mix gently by flicking the tube. Aspirate the 2 mL of unsupplemented Grace's medium from the appropriate well and overlay the cells with the 1 mL of transfection mix. Repeat this process for the remaining wells.
9. Place the cells in the incubator at $28 + 0.5°C$ for 5 h.
10. Aspirate the transfection mix and add 2 mL of Sf900 II SFM.
11. Incubate the cells for at least 3 d, at which time you should see signs of viral infection (see **Table 2**). If the cells do not show signs of infection, wait for an additional 1–2 d, but do not exceed 5 d. Longer incubation times compromise the quality of the virus.
12. Once late stages of infection have been reached, the P1 viral stock is isolated. The virus is released from the cells and is, therefore, in the medium. In a sterilized tissue culture hood, remove the medium (approx 1.5 mL) and transfer to a sterile 2-mL tube. Centrifuge for 5 min at $500g$ at room temperature to remove cells and large debris.
13. Return to the hood and transfer the virus-containing supernatant to a new sterile 2-mL tube. Add sterile FBS to 2% (v/v). This prevents proteases from destroying viral coat proteins. Store the virus at 4°C until you are ready to proceed with the next amplification. The virus should be protected from light; thus, it is beneficial

Table 2
Assessing Efficacy of Baculoviral Infection of Sf9 Cells

Signs of infection	Phenotype	Description
Early (first 24 h)	Increased cell diameter; increased size of cell nuclei (must view at 40× or greater).	In Sf9 cells, the increase in cell diameter is only 25–50%; nuclei may fill the entire cell.
Late (24–72 h)	Cessation of cell growth; granular appearance; detachment.	Cells appear to stop growing when compared to untransfected cells; vesicular appearance in cells; cells will release from the plate (i.e., float).
Very late (>72 h)	Cell lysis	Cells are lysed, cell debris is present, and clearing from the monolayer is observed.

Note: This table describes the appearance of Sf9 cells at various times after infection with baculovirus. A particular phenotype of infected Sf9 cells indicates the appropriate time to collect amplified virus and the efficiency of infection during protein expression. This information is presented as described by the manufacturer's protocol: Bac-to-Bac Baculovirus Expression System Version C December 19, 2002, 10359; also available online at (http://www.invitrogen.com/content/sfs/manuals/bactobac_man.pdf).

to wrap the tube in foil. The virus is stable at 4°C for quite some time, but it can be stored for longer periods at –80°C (*see* **Note 22**).

14. If you wish to scale up for high-throughput dicing, *see* **Subheading 3.9.1**.

3.2.5. Producing a P2 Viral Stock

The P2 viral stock is produced by amplifying the P1 viral stock (i.e., infecting Sf9 cells with the P1 viral stock). For amplification, the titer of the P1 viral stock must be known. Titer, in PFU/mL (plaque-forming units/mL) or number of virions per milliliter, is determined by a viral plaque assay; however, we find that this procedure is very time-consuming and not reliable. Thus, we make assumptions about the viral titer, and in most cases, this is just as reliable as the plaque assay. Generally, the titer of the P1 viral stock is approx 1×10^6 PFU/mL. The titer could be as high as 5×10^6 PFU/mL, but it is usually not the case for the Dicer P1 viral stock. Keep in mind that the titer of the P1 viral stock is dependent on the length of time the cells are incubated after transfection (*see* **Subheading 3.2.4.**).

1. Amplification is accomplished by infecting Sf9 cells at an MOI (multiplicity of infection) of 0.05–0.1 (i.e., 1 viral particle per 10–20 cells). If you use 1 mL of

virus at 1×10^6 PFU/mL, then $(1–2) \times 10^7$ Sf9 cells can be infected. If you transfected several wells in a six-well dish in **Subheading 3.2.4.,** you will have more than 1 mL of virus and can, therefore, use several 10-cm dishes to amplify the virus, as may be necessary for producing a large amount of virus for high-throughput dicing (*see* **Subheading 3.9.1.** and **Note 23**).

2. Grow a suspension culture of Sf9 cells to mid-log phase ($[2–3] \times 10^6$ viable cells per milliliter with a viability of ≥97%). Do not use if the viability is < 97% because the quality of the P2 viral stock will be compromised. In a sterile tissue culture hood, dilute cells to 1×10^6 viable cells per milliliter in 10 mL of Sf900 II SFM medium. Plate the entire 10 mL in a 10-cm tissue culture dish. If required, add cells to additional 10-cm dishes.

3. Allow cells to attach for 30 min to 1 h in a nonhumidified incubator at $28 \pm 0.5°C$. If cells are healthy, they will attach in the first 15 min, but wait at least 30 min. In the hood, add 1 mL of P1 viral stock to the 10-cm plate and incubate at $28 \pm 0.5°C$ for 48 h.

4. Look for signs of infection and viral production (*see* **Table 2**). Isolate the P2 viral stock by transferring the medium (approx 10 mL) from the 10-cm plate to a sterile 15-mL conical tube and centrifuge for 5 min at room temperature and 500*g*.

5. Return to the hood and transfer the supernatant to a new sterile 15-mL conical tube, taking care to avoid the cells and cell debris in the pellet. It is better to leave some of the supernatant in the tube then to risk carryover of cells. Add FBS (**Subheading 2.2., item 13**) to 2% (v/v) and store at 4°C protected from light. If several dishes were infected, combine into a sterile 50-mL conical tube.

6. The titer of the P2 viral stock should be about $(0.5–1) \times 10^7$ PFU/mL.

3.2.6. Producing a High-Titer P3 Viral Stock

The P3 viral stock, like the P2 viral stock, is produced by amplification (that is, Sf9 cells are infected with the P2 viral stock). However, now a larger volume of viral stock can be produced because there is a larger volume of P2 viral stock than there is of P1 viral stock. Because the empirical titer of the P3 stock must be determined (*see* **Subheading 3.3.1.**), it is worthwhile to make as much P3 stock as possible (*see* **Note 24**).

1. Grow 100 mL of suspension culture of Sf9 cells to $(2–3) \times 10^6$ viable cells per milliliter with a viability of ≥97%. If the viability is <97%, the quality of the virus will be compromised.

2. If cell density is greater than 2×10^6 viable cells per milliliter, dilute to 2×10^6 viable cells per milliliter with fresh Sf900 II SFM; the final volume should be 100 mL.

3. In a sterile tissue culture hood, add 2 mL of the P2 viral stock to the 100 mL culture of Sf9 cells at 2×10^6 viable cells per milliliter. The MOI is again 0.05–0.1 because the cell number is 2×10^8 and the titer of the P2 stock is $(0.5–1) \times 10^7$ PFU/mL.

4. Incubate at $28 \pm 0.5°C$ and 135 rpm for 48 h.

5. To isolate the virus, portion the 100 mL of culture into two 50-mL conical tube. Spin at 500*g* for 5 min at room temperature. Return to the hood and carefully transfer the

virus-containing supernatant to fresh 50-mL conical tubes. It is essential to avoid carryover of cells and cell debris even if it means leaving some of the supernatant behind.

6. Add FBS to 2% (v/v) and store at 4°C protected from light. The titer of this P3 viral stock should be $(0.5–1) \times 10^8$ PFU/mL although for protein production, an "empirical titer" is used (*see* **Subheading 3.3.1.**). One hundred milliliters of virus should be enough for about 10 preparations of r-Dicer, enough for silencing about 500 genes. However, we suggest that you make 500 mL of P3 viral stock using all of the 10 mL of P2 viral stock (*see* **Note 24**). If an even larger quantity of r-Dicer is desired, enough for high-throughput dicing (*see* **Subheading 3.9.1.**), the amplification is easily scalable: Simply infect a larger volume of cells in multiple cultures or in a larger flask, keeping in mind that you will need more than 10 mL of P2 viral stock.

3.3. Expression and Purification of r-Dicer

For a gene silencing experiment, large quantities of r-Dicer can be produced by infecting a suspension culture of Sf9 cells with the recombinant Dicer baculovirus. This subsection describes how to (1) determine the empirical titer of the P3 viral stock, (2) express r-Dicer in Sf9 cells, and (3) purify r-Dicer.

3.3.1. Determine the Empirical Titer of the P3 Viral Stock

Each time a new P3 viral stock is generated (*see* **Subheading 3.2.6.**) the MOI for optimal protein production in Sf9 cells must be determined (*see* **Note 25**). A range of MOIs, 1, 2, 5, 10, 20, are tested to determine the MOI that results in optimal protein production or activity. The MOI, however, cannot be calculated unless the titer of the virus is known. The problem is that viral plaque assays, the method used to determine the titer in PFU/mL of virus, are unreliable and time-consuming. It is simpler to use an empirical MOI, because, essentially, you will be doing the same thing. Different volumes of the P3 viral stock are added to a fixed number of cells, incubated for 36–40 h, and then r-Dicer abundance is analyzed by Western blotting with the monoclonal antibody raised against the T7 epitope. The empirical titer is thus a volume of P3 viral stock, which produces a particular amount of r-Dicer. This procedure is described in the Bac-to-Bac Expression System manual and was adapted from Luckow and Summers *(51)*.

1. Grow a 50-mL culture of Sf9 cells to a density of $(2–3) \times 10^6$ viable cells per milliliter with a viability of >95%. In a sterile tissue culture hood, plate 6×10^5 viable cells per well in a 24-well dish. It works best to dilute the cell stock from $(2–3) \times 10^6$ to 6×10^5 cells per milliliter with Sf900 II SFM and then plate 1 mL per well. Place the plate in an incubator at 28 ± 0.5°C. Remember that CO_2 and humidity are not required. Cells will adhere in 30–60 min. In fact, if the cells are very healthy, they will attach very rapidly, as soon as 15 min. Nonetheless, allow the cells to attach for 1 h.

2. During the 1-h incubation, prepare a range of viral amounts. The titer of the P3 viral stock could vary by one to two orders of magnitude; thus, it is important to use a wide range of virus to ensure that protein can be detected. A good range is 6, 15, 30, 60, 120, 300, and 600 μL of virus. For the lower volumes, add the virus to a sterile microcentrifuge tube containing the appropriate amount of fresh SF900 II SFM medium such that the total volume is equal to 600 μL (*see* **Note 26**). Be sure to include 600 μL of fresh medium lacking virus as the uninfected control.

3. After the 1-h incubation, move the cells back to the tissue culture hood. Aspirate the medium, wash with 1 mL of fresh Sf900 II SFM, aspirate again, and add 300 μL of fresh Sf900 II SFM. When adding medium to the cells, run it down the side of the well so that the cell monolayer is not disrupted. Also, it is important that the cells do not dry out, so you must work quickly with only 12 wells.

4. Add the previously prepared viral dilutions or medium alone, each at 600 μL. After all additions, gently rock the plate in a circle and return to the incubator.

5. Look for signs of infection 36–38 h after adding the virus. In comparison to the uninfected well, the other wells should contain fewer cells and cells that are 25–50% larger than uninfected cells. The infected cells should not be dividing. The shape of a dividing Sf9 cells is unique, it becomes long and narrow with thin extensions from each end whereas nondividing cells are very round. Although it can be difficult to observe, the density of cells infected with virus is less than uninfected cells.

6. After visualizing the cells, aspirate the medium, wash the cells with 1 mL of D-PBS and add 200 μL of 1X SDS Laemmli sample buffer. The cells will lyse immediately and the DNA will make the lysate very viscous. To reduce the viscosity, shear the DNA either by passing the lysate through an 18- or 21-gauge needle or by completing a freeze–thaw cycle. Both methods are equally effective. To freeze the lysates, place the entire plate in a –20°C or –80°C freezer; –80°C is appropriate for long-term storage. Thaw the samples, transfer to a microcentrifuge tube, and heat the samples at 95°C for 3 min. If the samples were not frozen, simply transfer to a microcentrifuge tube and heat samples at 95°C for 3 min. Load 80 μL of the lysate on a 6% 100:1 SDS–polyacrylamide gel (*see* **Note 27**).

7. Transfer proteins to Immobilon-P in 1X transfer buffer containing 10% methanol as suggested by the manufacturer; we transfer at approx 1.5 A and 4°C for approx 1 h. Complete the Western blot per your normal protocol or the suggested protocol in **Note 28**, using the monoclonal antibody against the T7 epitope at a 1:5000–1:10,000 dilution. The full-length r-Dicer will migrate at approx 225 kDa (*see* **Fig. 2A**). There will be several immunoreactive bands, but the largest band at approx 225 kDa is the one to evaluate.

8. Determine which volume of P3 virus results in optimal protein production. This volume will be scaled proportionally to the cell number to infect a much larger culture for producing large quantities of r-Dicer (*see* **Subheading 3.3.2.**). In general, expression will peak at a particular volume of virus, usually between 15 and 60 μL (*see* **Fig. 2A**). The distribution is explained by the fact that, if the MOI is too low,

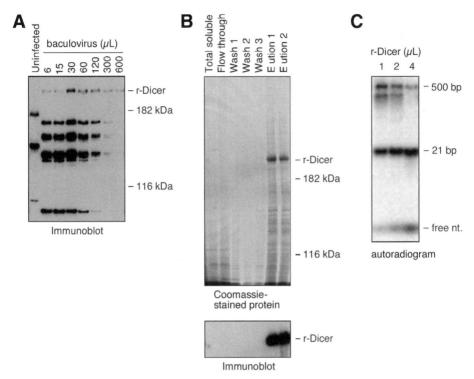

Fig. 2. **(A)** Expression of r-Dicer in Sf9 cells. Total cellular lysates from uninfected Sf9 cells or cells infected with various volumes of recombinant baculovirus encoding Dicer were electrophoresed in a 6% 100 : 1 SDS gel, transferred to a PVDF membrane, and subjected to immunoblotting with the anti-T7-epitope antibody. **(B)** Partial purification of r-Dicer. Using cobalt Sepharose, r-Dicer was partially purified from the total soluble fraction. An aliquot from the total soluble fraction (Total soluble), the supernatant containing unbound protein (Flow through), the first (Wash 1) and second (Wash 2) supernatant fractions of the β-wash, the second supernatant fraction of the δ-wash (Wash 3), and the first (Elution 1) and second (Elution 2) supernatant fractions of the γ-elution was electrophoresed in a 6% 100:1 SDS gel and stained with Coomassie blue (**upper panel**) or subjected to immunoblotting with the anti-T7-epitope antibody (**lower panel**). **(C)** Activity of r-Dicer. For 16 h at 37°C, various volumes of partially purified r-Dicer were incubated with 1 μg of an approx 500-bp double-stranded RNA (dsRNA) that was internally labeled with [α-^{32}P]-UTP. Reactions were electrophoresed in a 15% native polyacrylamide gel and relative amounts of radioactivity corresponding to undigested dsRNA (500 bp) and d-siRNA (21 bp) were evaluated with a PhosphorImager.

then the r-Dicer yield will also be minimal because cells are not infected simultaneously, and, if the MOI is too high, expression is also compromised because of to cell death. Expression levels may demonstrate other patterns. For example, the highest level of expression is seen at the lowest volume of virus, and expression decreases as MOI increases, usually the result of a very high-titer virus. The converse is also possible; expression increases linearly with increasing MOI, the result of a very low-titer virus. If protein is not detectable, make sure the T7 antibody is functional and Western blotting conditions are optimal (*see* **Note 29**). Likewise make sure that baculoviral production and infection were done correctly (*see* **Note 30**).

9. The P3 viral stock is useful only if peak production occurs at an empirical titer of ≤60 μL because of the practical limitations imposed by culture size. For example, if r-Dicer is expressed at 60 μL, then 20 mL will be required to infect a 100-mL culture at a density of 2×10^6 cells per milliliter. Any volume of virus above 20 mL is impractical. If expression is not optimal at empirical titers ≤60 μL, a new P3 stock should be made. Begin with transfection ensuring that the P1 stock has a titer ≤1 × 10⁶ PFU/mL and that subsequent amplifications produce the appropriate titer. This may include allowing P1 production to occur for 5 d (*see* **Subheading 3.2.4.**). Before making a new P3 stock, you can try expressing r-Dicer for various lengths of time (e.g., 24, 48, 72, and 96 h because peak protein production can vary with time (*see* **Note 29**), and although we attain the highest yields of r-Dicer 36–38 h postinfection, the P3 stock you made may provide optimal expression at a time-point other than 36–38 h.

10. The empirical titer should also be determined for Sf9 cells grown in suspension because suspension cultures are used to produce large quantities of r-Dicer. Evaluating protein expression in a 24-well plate simply provides a rough estimate of the virus required for optimal expression. Grow three flasks, 50 mL each, to a density of 2×10^6 viable cells per milliliter; make sure that the viability is >95%. Choose the viral volume that provided the best expression in the 24-well plate and scale up proportionally to infect 50 mL of cells at 2×10^6 viable cells per milliliter. Use this volume as the center point to infect one of the cultures and then use one-half the volume and twice the volume of virus to infect two additional cultures. For example, if 30 μL of P3 viral stock yielded the most r-Dicer, use 2.5, 5, or 10 mL to infect the 50-mL cultures.

11. Transfer 300 μL to a microcentrifuge tube 36–40 h postinfection and spin at 500*g* for 5 min at room temperature. Carefully remove the supernatant and add 1 mL of D-PBS. Spin again, remove supernatant, and lyse in 200 μL of 1X SDS Laemmli sample buffer. Load 80 μL per well and proceed with Western blot as described in **steps 6** and **7**. Determine which volume of P3 virus is optimal for production of r-Dicer (*see* **steps 8** and **9**). This procedure is ideal for determining the optimal postinfection time at which r-Dicer expression is maximal because it is simple to remove a small aliquot from the culture at various time-points. First test a broad range of times, 24, 48, 72, and 96 h, at three different MOIs and then narrow in on a specific time by measuring expression at 2-h intervals (*see* **Note 31**).

3.3.2. Expression of r-Dicer

This subsection describes how to express Dicer for subsequent purification. The purification protocol, presented in **Subheading 3.3.3.**, was designed to isolate r-Dicer from a set number of cells. If more r-Dicer is required, both expression and purification can easily be scaled up (*see* **Subheading 3.9.1.** and **Note 32**). Expression is the key to consistently producing active r-Dicer. From time to time, there will be some variation in cell growth after infection; thus, to normalize the purification protocol, the cell density at the time of infection is used. The purification strategy described in **Subheading 3.3.3.** is based on having infected 4×10^8 cells. Cell pellets can be used immediately for purification of r-Dicer or stored at –80°C for at least 1 yr.

1. From a mid-log phase culture with a density of $(2–3) \times 10^6$ viable cells per milliliter and a viability >95%, seed two 100-mL cultures at a density of 5×10^5 viable cells per milliliter. Use a fresh 250-mL flask and the cell culture techniques learned in **Subheading 3.2.2**. This will be enough r-Dicer to silence about 50–100 genes, but if a larger amount is desired, *see* **Note 32**; for high-throughput dicing, *see* **Subheading 3.9.1**.
2. Count the cells 48 h after seeding the cultures and determine viability. The cell density should be $(2–3) \times 10^6$ viable cells per milliliter with a viability > 95%. Combine the appropriate volume of cells, virus, and, if needed, fresh medium such that the final volume is 100 mL and the density is 2×10^6 viable cells per milliliter. For example, if the cell density is 3×10^6 viable cells per milliliter, then you would use 67 mL of cells, 28 mL of fresh Sf900 II SFM medium, and 5 mL of virus. The same 250-mL flask (**step 1**) can be used; simply remove 33 mL from the 100 mL of cells and add the fresh medium and virus. Note that the optimal volume of virus was determined as described in **Subheading 3.3.1.** and may vary from 2 to 10 mL. The amount of fresh medium can be increased or reduced, depending on the amount of virus that is used.
3. Harvest the cells 36–40 h after infection. Portion the cultures into 50-mL conical tubes and spin at $500g$ for 5 min. Centrifugation can be at 4°C or room temperature.
4. Pour off medium and wash the cell pellets with 50 mL D-PBS. Ideally, all four pellets should be combined in one tube. Add 50 mL D-PBS to one tube, recap, and resuspend the cell pellet by tapping the bottom of the tube. Add to the next tube and resuspend. Repeat until all cell pellets are resuspended in one tube.
5. Pellet cells again at $500g$ for 5 min.
6. Pour off D-PBS, place cell pellet on ice, and proceed with purification of r-Dicer (*see* **Subheading 3.3.3.**) or snap-freeze in liquid nitrogen and store at –80°C. This pellet containing 4×10^8 cells is sufficient for a single purification and for silencing 50–100 genes.

3.3.3. Purification of r-Dicer

The r-Dicer is only partially purified and this is sufficient for high activity. Cobalt affinity resin is used to bind the N-terminal 6xHis-tag on r-Dicer. Contaminating proteins are removed by washing, and r-Dicer is eluted with imidazole. The r-Dicer is then dialyzed and the resulting activity is assessed. If more r-Dicer is required, the purification can easily be scaled up (*see* **Subheading 3.9.1.** and **Note 33**).

1. If the pellet has been stored at −80°C, it should thaw on ice. If the pellet is fresh, then be sure that it is placed on ice directly after centrifugation (*see* **Subheading 3.3.2., step 6**). Add protease inhibitors to ice-cold α-extraction buffer at a final concentration of 1 mM PMSF, 0.55 μg/mL leupeptin, 0.55 μg/mL pepstatin, 1 μg/mL aprotinin, and make 10 mL of α-extraction buffer containing protease inhibitors for each pellet of 4×10^8 cells. Resuspend the cell pellet, 4×10^8 cells, in 9 mL of ice-cold α-extraction buffer containing protease inhibitors by pipeting up and down; do not vortex.

2. Sonicate cell suspension with 5×10 s pulses at approx 20–25 W in an ice-water slurry. Allow samples to chill for 30 s between the 10-s pulses to prevent heating of the sample. Complete lysis of the Sf9 cells is very important for obtaining a high yield of r-Dicer. The efficacy of sonication may vary; thus, the efficiency of the lysis can be measured by mixing 10 μL of the sonicated lysate with 10 μL of trypan blue; sonication is complete when cells do not exclude the dye.

3. Remove insoluble fraction and cell debris by centrifugation at 10,000g for 20 min at 4°C. Glass Corex tubes work well for centrifugation. About 50% of r-Dicer is present in the soluble fraction.

4. During the centrifugation, equilibrate a 1-mL bed volume of cobalt Sepharose with 25 vol of α-extraction buffer; it is not necessary to add protease inhibitors to the α-extraction buffer used for resin equilibration. Resuspend the cobalt Sepharose slurry and transfer 1.7 mL to a 15-mL conical tube (*see* **Note 34**). Spin at 700g and 4°C for 2 min. The resin packs tightly, so the supernatant can be poured off. Add 12 mL of α-extraction buffer and resuspend the resin by tapping and inverting the tube. Centrifuge again, pour off supernatant, and add another 12 mL of α-extraction buffer. Centrifuge one last time and pour off supernatant and place resin on ice.

5. Mix the 10,000g supernatant (**step 3**) with the equilibrated 1-mL bed volume of cobalt Sepharose (**step 4**) for 1 h at 4°C using a conventional rotator.

6. Collect resin by centrifugation at 700g and 4°C for 5 min.

7. Wash the resin twice with 12 vol of β-wash buffer, 12 mL of buffer per 1 mL of resin, by mixing on a conventional rotator for 10 min at 4°C. Resin is collected between washes by centrifugation at 700g and 4°C for 2 min.

8. Two additional washes with δ-wash buffer (**Subheading 2.3., item 26**) are used to remove the detergent, NP-40. Washes are performed as in **step 7**.

9. r-Dicer is eluted by six successive elutions with 1 mL of γ-elution buffer. Resuspend resin in 1 mL of γ-elution buffer by pipetting up and down several times. Collect resin by centrifugation at 700g and 4°C for 2 min and then transfer the supernatant to a fresh 15-mL conical tube. All six elutions can be combined in one 15-mL tube. Avoid carryover of the resin (*see* **Note 35**).

10. To ensure the r-Dicer is free of contaminating resin, spin the 6 mL of elution at 700g and 4°C for 2 min and then transfer the supernatant to a fresh 15-mL conical tube. Avoid carryover of resin.

11. Using a syringe attached to an 18- or 21-gauge needle, load the 6 mL of elution in a Slide-A-Lyzer as described by the manufacturer. Be sure to presoak the cassette. Be

careful with the membrane; do not pierce with the needle and avoid contacting it with your fingers. Dialyze at 4°C against 3 L of dialysis buffer. Three successive 1-L dialyses are performed—the first two for 2 h and the third overnight. The order in which the timed dialyses are performed is irrelevant (*see* **Note 36**). Remove the dialyzed solution with a fresh syringe and needle. Transfer to several RNase-free microcentrifuge tubes. Keep in mind that a very large amount of r-Dicer may be produced by scaling up the expression and purification (*see* **Subheading 3.9.1**).

12. r-Dicer is stored at 4°C and is stable for up to 1 yr. The 225-kDa r-Dicer can be detected by Coomassie staining (5 µL), silver staining (1 µL), or Western blotting (1–3 µL) after electrophoresis in a 6% 100:1 SDS–polyacrylamide gel (*see* **Fig. 2B**) or 4–20% gradient gel. The monoclonal antibody raised against the T7 epitope is used for the Western blot at a dilution of 1:10,000 (*see* **Note 28**). However, the tag on the r-Dicer may be lost over time, greater than 1 mo.

13. The r-Dicer activity must be assessed and this is more important than measuring the quantity of the approx 225-kDa band by SDS-PAGE. Select a dsRNA to dice (*see* **Subheading 3.4.3.**); we generally use our control dsRNA from firefly luciferase, but any dsRNA will do. Mix 1 µg of dsRNA with 50 mM $MgCl_2$ (**Subheading 2.3., item 30**), reaction dilution buffer, RNase-free H_2O (**Subheading 2.1., item 8**), and 1, 2, or 4 µL of r-Dicer as described below. Pipet up and down gently to mix and centrifuge briefly to collect the liquid at the bottom of the tube. Incubate at 37°C overnight (approx 16 h) (*see* **Note 37**).

Reaction Conditions	Final concentrations
x̲ µL dsRNA (1 µg)	100 ng/µL dsRNA
0.5 µL of 50 mM $MgCl_2$	2.5 mM $MgCl_2$
4, 3, 1 µL reaction dilution buffer[a,c]	250 mM NaCl
1, 2, 4 µL r-Dicer	30 mM HEPES
x̲ µL H_2O[b]	0.05 mM EDTA
10.0 µL	16–64 ng/µL r-Dicer

[a]pH 8.0
[b]Volume depends on concentration of dsRNA.
[c]Volume depends on the volume of r-Dicer used; the volume of r-Dicer and reaction dilution buffer must always equal 5 µL so that the final concentration of NaCl, HEPES, and EDTA is correct.

14. Pour a native, 1X TBE buffered, 15% polyacrylamide gel (*see* **Note 38**). The gel must remain cool so that the duplexed d-siRNAs remain intact. Prechill for 15 min and prerun the gel without sample at 10 W for at least 10 min at 4°C. We generally use 0.5X TBE for the running buffer because buffering capacity is sufficient and electrophoresis is similar to the use of 1X TBE.

15. Wash the unpolymerized acrylamide from the wells before loading the samples. An 18-gage needle attached to a 25-mL syringe works well.

16. Add 1.2 µL of 10X RNA loading buffer to the 10-µL reaction and mix. Centrifuge briefly to collect contents. Load half of the reaction (5.6 µL) on the native 15% polyacrylamide gel (*see* **Note 39**). Run at 10 W for about 1.5 h at 4°C. The

difference in mobility between the large dsRNA and d-siRNA is easily detectable; thus, little resolution is required. Stain the RNA by submerging the gel in H_2O containing 0.5 µg/mL ethidium bromide for 10 min. Discard the staining solution and rinse briefly with H_2O three times. Photograph the gel with exposure to UV light.

17. Determine which volume of r-Dicer results in cleavage of 70–80% of the large dsRNA (*see* **Fig. 2C**). Also look for the highest yield of d-siRNA. Generally, we find that 2 µL of r-Dicer will dice 70–80% of the large dsRNA and results in the highest yield of d-siRNA (*see* **Note 40**).

18. For reproducibility, it is best to set a unit value for r-Dicer activity. Our unit value is 1 unit of r-Dicer will dice 1 µg of dsRNA to 80% completion in 16 h at 37°C. Usually, our r-Dicer is 0.5 units microliter. The concentration of the 225-kDa band is generally 160–200 ng/µL. If the protein yield or activity is suboptimal, *see* **Note 41**.

3.4. Selection and Production of In Vitro Transcription Templates

To generate d-siRNAs, you will need an approx 550-bp dsRNA, which is transcribed in vitro from a DNA template. We use an approx 550-bp region because in vitro dicing has been optimized for this size. Because most genes are larger than 550 bp, you will need to select a particular region of the gene, referred to as the target region. It is most efficient to make the in vitro transcription template by PCR amplification. This subsection describes (1) general guidelines for selecting a target region of a gene, (2) how to design primers for PCR amplification of the target region, and (3) how to use PCR to generate an in vitro transcription template corresponding to the target region.

3.4.1. Selecting a Target Region

1. For any gene silencing experiment, you must have control d-siRNAs. Typically, d-siRNAs generated from exogenous genes should be a good control because there is very little nucleotide identity to any endogenous gene. However, it might not really be a control because it will not have a target mRNA. In order to provide a target, control d-siRNAs that target an endogenous gene irrelevant to the gene that is being studied could be used. The problem is that the genes might be adventitiously linked. It is difficult to choose a control and thus could be considered to be a personal preference, but it might matter which is used (*see* **Note 42**).

2. If you choose to use control d-siRNAs derived from an exogenous gene, we have found particular regions of the EGFP (enhanced green fluorescent protein), Firefly luciferase (GL3), or Renilla Luciferase (RL) genes that provide a sufficient control. Primers used to amplify the target region are shown in **Table 3**. Sequence corresponding to the genes can be found with the following accession numbers: EGFP, 675-1394 in U57609.1, pEGFP-N3 cloning vector (Clontech, Palo Alto, CA); GL3, 280-1932 in U47296.2, pGL3 control cloning vector (Promega, Madison, WI); RL, 694-1629 in AF025845, pRL-SV40 coreporter vector (Promega, Madison, WI).

3. First, find the nucleotide sequence of the gene you wish to silence. If you do not already know the sequence of the gene, you should use a public database; Homologene and Unigene are good resources http://www.ncbi.nlm.nih.gov/HomoloGene/ and http://www.ncbi.nlm.nih.gov/UniGene/. These sites also provide information on calculated orthologs and gene clusters. Furthermore, the LocusLink and OMIM (Mendelian inheritance site) sites can provide more information, including a history of the gene, accession numbers for the mRNA and protein, and, finally, a gene RIF (reference into function). It is much easier to find the desired sequence on these sites than Entrez-nucleotide searches because they do not contain partial and patented sequences.

4. A potential reason that gene silencing with d-siRNAs might not work is that the wrong gene is diced. Take special care to find the gene you wish to silence. The nomenclature can be confusing. Often, the name of the gene in the database will be different from the name by which you know the gene. The gene RIF as well as cluster and ortholog information can be used to ensure that you have found the correct nucleotide sequence. Be sure that the sequence you find is from the organism in which you will be doing gene silencing. This seems to be an impossible mistake, but it has happened. Also, take into account splice variants and whether or not all variants should be silenced. It should be simple to find a target region that will silence all variants. In some cases, it will be impossible to find a 550-bp region that will silence only one splice variant, but in some cases, it might be possible to select a region that might silence only a portion of the splice variants. Finally, make sure that you discover all potential isoforms and be sure that you are targeting the correct isoform.

5. An approx 550-bp target region needs to be selected from the gene. The potency of gene silencing does not change much when different target regions are selected. However, we have found that approx 550-bp segments located at or near the 3' end of the coding sequence work slightly better than segments in the 5' end. In addition, if the template is to be amplified from cDNA, it is more likely to amplify a segment from the 3' end than one from the 5' end. Thus, we choose the last approx 550 bp of the coding region. The exogenous control templates generated with the primers listed in **Table 3** are located at the 5' end of the coding region because they were designed before recognition that the 3' end of the coding region provides the best target. For other target options, see **Note 43**.

6. Although we do not worry about it very much because it has not been a problem, there is one last concern for selection of a target region—nucleotide homology with other genes. Check the region you select for homology with other genes using a BLAST (blastn) search (http://www.ncbi.nlm.nih.gov/BLAST/) or a ClustaW alignment. When compared to homologous genes, the selected target region should not contain stretches of consecutive nucleotides longer than 15. Several single mismatches (e.g., 4 within 21 nucleotides) make longer stretches acceptable. It should be very easy to find at least one target region, approx 550 bp in length, that is near the 3' end of the coding region and has minimal homology to any other gene in the genome (*see* **Note 44**).

Table 3
Primer Sequence for Generating Exogenous In Vitro Transcription Templates

Exogenous gene products		Forward and reverse primer sequence	G.S. T_m / F.L. T_m	PCR conditions
Green fluorescent protein (GFP)	Forward (69)	5'-GCGTAATACGACTCACTATAGGAAACGGCC ACAAGTTCAGCG-3'	59.6°C/ 66.9°C	As described in **Subheading 3.4.3.**
Acc. no. U57609.1	Reverse (563)	5'-GCGTAATACGACTCACTATAGGGTGTTCTG CTGGTAGTGG-3'	53.6°C/ 65.7°C	
Firefly Luciferase (GL3)	Forward (114)	5'-GCGTAATACGACTCACTATAGGAACAATTG CTTTTACAGATGC-3'	50.4°C/ 63.1°C	As described in **Subheading 3.4.3.**
Acc. no. U47296.2	Reverse (615)	5'-GCGTAATACGACTCACTATAGGAGGCAGAC CAGTAGATCC-3'	54.0°C/ 65.1°C	
Renilla Luciferase (RL)	Forward (119)	5'-GCGTAATACGACTCACTATAGGAAAAACAT GCAGAAAATGCTG-3'	51.2°C/ 63.4°C	As described in **Subheading 3.4.3.**
Acc. no. AF025845	Reverse (620)	5'-GCGTAATACGACTCACTATAGGTTGAATGG TTCAAGATAT-3'	43.3°C/ 61.4°C	

Note: Control d-siRNAs can be generated from exogenous genes. This table provides the location of the target region in respect to the start codon (in parentheses next to primer orientation), the respective accession number (Acc. No.), forward and reverse primer sequence including the T7 phage polymerase promoter 5' to the gene specific sequence, the melting temperature (T_m) of the gene-specific region (G.S.) and of the full-length primer (F.L.), and recommended PCR conditions for producing the in vitro transcription template.

A

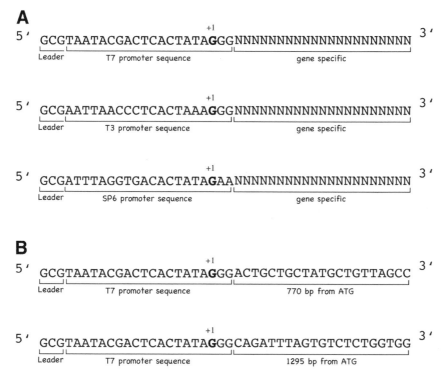

B

Fig. 3. (**A**) Schematic of PCR primers used to amplify in vitro transcription templates. The primer should contain, from the 5′-end, a short leader sequence, then a phage polymerase promoter (T7, T3, or SP6) and 18–25 nucleotides of a gene-specific sequence. (**B**) Schematic of PCR primers used to amplify an in vitro transcription template for production of cyclin A2 dsRNA. The primer contains the essential elements as described in (A).

3.4.2. Primer Design

The primers must contain a gene-specific region and the T7 polymerase promoter for in vitro transcription (*see* **Fig. 3**). Primer design software will make the process more efficient.

1. First select 18–25 nucleotides of gene-specific sequence that will prime the sense strand of your target sequence. Use basic primer design rules ensuring that the gene-specific region does not form an extended hairpin or extensive homodimers or heterodimers. Then, make sure that the primer is still free of hairpins and homodimers or heterodimers when the T7 polymerase promoter is added to the primer. Avoid GC-rich regions within the gene-specific portion of the primer because the T7 polymerase promoter contains GC-rich regions. Repeat the process for the antisense primer. Remember to design primers for the control (*see*

Subheading 3.4.1. and **Note 42**). Other phage polymerase promoters and polymerases could be used (*see* **Note 45**).

2. The melting temperature of the primers, the T_m, in the gene-specific region should be as close to 57°C as possible, under standard PCR conditions (50 mM NaCl, 2 mM MgSO$_4$, and [primer] = 0.4 μM). If the T_m is ≤55°C or ≥60°C, yields might be decreased. The T_m will increase when the T7 polymerase promoter sequence is added, so that is why the T_m of the gene-specific region should be as close to 57°C as possible. Note that the T_ms of the control primers (*see* **Table 3**) are not as suggested here because the primers were designed before these rules were established. Nonetheless, the primers work well mostly because the template is amplified from a sublconed gene.

3. Order the smallest scale of primers available. Standard purity (e.g., desalted) is sufficient.

3.4.3. PCR Amplification of the In Vitro Transcription Template

There are many types of in vitro transcription template, each with its own purpose. For making dsRNA that is to be diced, the best template is a PCR product (*see* **Note 46**). In order to simplify the process, we have designed a two-step PCR strategy that allows the PCR product to be used without any gel purification or other column-based purification systems (*see* **Note 47**). This makes the process time and cost efficient as well as amenable to 96-well format (*see* **Subheading 3.9.2.**). The first PCR is used to amplify the template from cDNA. A fraction of this PCR is then used as the template in a second PCR. This strategy enriches for the template and dilutes the cDNA component. Using nested primers in the second step of a two-step PCR strategy will increase the odds of amplifying the correct gene but is generally not necessary (*see* **Note 48**).

1. Using the primers designed in **Subheading 3.4.2.**, set up a PCR reaction that will amplify your in vitro transcription template. The source template for your PCR is cDNA (*see* **Subheading 3.1.2.1.**). If you will be silencing many genes, it is a good idea to have cDNA from the species of interest on hand. If you wish to use a subcloned gene as your template, you will need to get rid of the template after **step 3** (*see* **Note 49**). Remember to make a template for your control (*see* **Subheading 3.4.1.** and **Note 41**). If the control is an exogenous gene, it will likely be amplified from a vector (*see* **Note 49**). Even if the in vitro transcription template is amplified from a vector, two rounds of PCR as described below should be used.

2. The cycling parameters are very important; the PCR must enrich for the correct template. Heat reaction at 95°C for 2.5min to denature template and antibody. For the first three cycles denature at 94°C for 30 s, anneal at 55°C for 30 s, and extend at 68°C for 45 s. For the next 27 cycles, denature at 94°C for 30 s, anneal at 60°C for 30 s, and extend at 68°C for 45 s. Changing the annealing from 55°C to 60°C increases the specificity of the primers for the correct template and works because

after the first round, the entire primer, gene-specific sequence plus T7 promoter sequence, is represented in the desired product.

3. We use Platinum Taq High Fidelity because of the large yield. Mix the following in a PCR tube and overlay with oil if necessary (*see* **Note 50**):

   5 μL of 10X High Fidelity Buffer; 2 μL of 50 mM MgSO$_4$; 1 μL of 10 mM dNTPs (10 mM each of dATP, dCTP, dGTP, dTTP); 1 μL of 20 μM forward primer; 1 μL of 20 μM reverse primer; 1 μL cDNA; 0.2 μL Platinum Taq High Fidelity; 38.8 μL H$_2$O.

4. There is no need to check the results of the first PCR on an agarose gel, unless you are curious. Remember that if the target mRNA is not abundant, it might be difficult to see the product after staining with ethidium bromide (*see* **Note 51**).

5. Now, you will use a fraction of the first PCR as the template in the second PCR. Mix reaction components as described below and overlay with oil if necessary (*see* **Note 50**). Heat reaction at 95°C for 2.5min for denature template and antibody. For 30 cycles, denature at 94°C for 30 s, anneal at 60°C for 30 s, and extend at 68°C for 45 s (or 1 min per kilobase).

   5 μL of 10X High Fidelity Buffer; 2 μL of 50 mM MgSO$_4$, 1 μL of 10 mM dNTPs (10 mM each of dATP, dCTP, dGTP, dTTP), 1 μL of 20 μM forward primer, 1 μL of 20 μM reverse primer, 2 μL first PCR, 0.2 μL Platinum Taq High Fidelity, 37.8 μL H$_2$O.

6. Run 5 μL of the second PCR on a 1% agarose gel as previously described. The Low Mass Ladder is a good standard for estimating the concentration of the template. The band should be approx 550 bp and the concentration is generally 20–40 ng/μL (*see* **Fig. 4A**). If the yield is low or multiple products are apparent, see **Note 52.**

7. If the band is the appropriate size, the odds of it being the incorrect product are low; however, you can confirm that you have the correct band by digesting the template with a restriction enzyme that will give a noticeable size change. Alternatively, sequence the product with either the sense or antisense primer; generic T7 primers cannot be used because the sequence is present on both ends of the product (*see* **Note 53**). Sequencing will also confirm that the product is not contaminated with other pieces of DNA that are too dilute to detect by ethidium bromide staining.

8. This amplification strategy should enrich for the correct band, but if more than one band is present, the correct band must be isolated by gel purification (*see* **Note 47**). Alternatively, PCR conditions can be adjusted to produce just one band (*see* **Note 52**). There should not be a concern for mutations in the template or contaminating products that are undetectable by ethidium bromide staining (*see* **Note 54**).

9. Fifty microliters of template is enough for five transcription reactions or production of 50 μg of d-siRNA probably more than you will ever need. However, for some templates, such as control templates you will need many transcription reactions. To generate more template set up another PCR as described in **step 5**. Because you will be using only 2 μL of the first PCR, you will be able to generate as much template as you need—125 in vitro transcriptions.

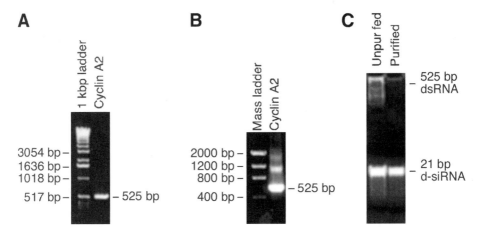

Fig. 4. (**A**) PCR amplification of an in vitro transcription template. The in vitro transcription template corresponding to a region of the human cyclin A2 gene was amplified from HeLa cell cDNA with reverse transcription (RT)-PCR. After two rounds of amplification, a fraction of the second PCR reaction was then electrophoresed in a 1% agarose gel and detected with ethidium bromide staining. (**B**) In vitro transcription. The cyclin A2 target region was transcribed from a PCR template using the MEGAscript in vitro transcription kit. A fraction of the reaction was then electrophoresed in a 1% agarose gel and detected with ethidium bromide staining. (**C**) In vitro dicing and purification of d-siRNAs. Fifty micrograms of cyclin A2 dsRNA was incubated with 100 µL of r-Dicer at 37°C for 16 h. One percent of the reaction was removed and saved for electrophoresis. The d-siRNAs were then purified from the remainder of the reaction with the Micro-to-Midi. Total RNA Purification System. One percent of the reaction before purification (unpurified) and 200 ng of purified d-siRNA (purified) were electrophoresed in a 15% native gel and stained with ethidium bromide. The undigested substrate dsRNA (unpurified) can be detected at approx 525 bp and the d-siRNA product (unpurified and purified) can be detected at approx 21 bp.

3.5. In Vitro Transcription and Dicing

Double-stranded RNA (dsRNA) is made with the MEGAscript in vitro transcription kit as directed by the manufacturer. The dsRNA is then diced with r-Dicer (*see* **Subheading 3.3.**). This subsection describes (1) how we use the MEGAscript kit and (2) how to set up an in vitro dicing reaction.

3.5.1. In Vitro Transcription

Both the sense and antisense strands are transcribed in a single tube and the transcripts anneal as they are made. Other protocols suggest transcribing each strand in a separate tube and then annealing; however, this requires extra work and does not seem to be necessary because strands anneal as well or better during transcription than in a subsequent annealing step. Also, avoiding a separate annealing reaction is important if in vitro transcription is scaled up to 96-well format. Furthermore, RNA is more stable as a duplex; therefore, if both strands are present, RNase activity is less of a problem.

We follow the general guidelines suggested by Ambion when making dsRNA with the MEGAscript kit and have incorporated additional technical information that it has kindly provided. We have tested our approach to ensure that we can avoid purifying the in vitro transcription template away from PCR components as well as to ensure that we can avoid purifying the dsRNA before dicing. This was done so that dicing would be amenable to the 96-well format, but because it works so well, we use the same approach when making d-siRNAs for a few genes.

1. Remember to make a control dsRNA for generating control d-siRNAs. We generally use a region of the firefly Luciferase gene (*see* **Table 3**). Set up the reactions as follows in a RNase-free microcentrifuge tube (*see* **Note 55**) 8 μL template (from the second 50-μL PCR; about 160–300 μg); 8 μL NTPs (2 μL each at 75 m*M*); 2 μL of 10X reaction buffer; 2 μL T7 enzyme mix.

2. Mix the reaction by gently pipetting up and down, centrifuge briefly, and incubate at 37°C for 2.5 h. Use an incubator that will prevent condensation so that the salt concentration in the reaction will remain the same, ensuring maximum yield. Incubators used for bacterial growth or hybridization ovens work well; be sure to verify temperature with a thermometer. Also, thermal cyclers with heated lids (set to 37°C) also work well; use a tube that will fit in the thermal cycler.

3. After 2.5 h at 37°C, add 1 μL RNase-free DNase I (provided in the kit), pipet up and down to mix, centrifuge briefly, and return to 37°C for 30 min. This will eliminate the DNA template. If the reaction volume was scaled up or down, add the appropriate amount of RNase-free DNase I.

4. Heat at 75°C for 5 min and then at room temperature for 5 min. This step facilitates annealing of "branched" or single-stranded regions within the dsRNA (*see* **Note 56**). If a thermal cycler is used for this step, be sure that the lid temperature is set no higher than 80°C.

5. Dilute the reaction to 50 μL with RNase-free H_2O (included in kit). If the transcription reaction volume was scaled up or down, dilute appropriately; for example, if the reaction volume was 10 μL, dilute to 25 μL. Use native agarose gel electrophoresis to estimate the concentration and quality of the dsRNA. Usually, the concentration would be estimated by measuring the absorbance at 260 nm, but the dsRNA is not purified away from the NTPs or protein, both of which will contribute to the absorbance at 260 nm thereby preventing an accurate measurement. This is not a problem because the concentration of the dsRNA can be estimated from agarose gel electrophoresis and a gel must be run anyway to evaluate the quality of the dsRNA. The dsRNA concentration will be estimated by comparing the intensity to the intensity of a dsRNA of known concentration (*see* **Note 57**).

6. Mix 0.5–1 μL of each dsRNA sample with 1 μL of 10X RNA loading buffer and 8.5–9 μL RNase free H_2O (provided in MEGAscript kit. Prepare three different

amounts of a dsRNA for which you know the concentration (i.e., the standard); 0.5 μg, 1 μg, and 2 μg are appropriate (*see* **Note 57**). The standards should also be formulated in a final volume of 10 μL by mixing the dsRNA with RNase-free H_2O and 10X RNA loading buffer. Also, it is a good idea to run the Low DNA Mass Ladder as a size marker (**step 10**).

7. Load on a 1% agarose gel and run as described previously, ensuring that reagents and apparatus are RNase-free. Ethidium bromide should be added to the agarose prior to casting. dsRNA should **not be heated** prior to electrophoresis and should **not be loaded in a denaturing buffer** (the 2X buffer provided in the MEGAscript kit or any other buffer that contains formamide or urea) because you want to ensure that the sense and antisense strands are annealed.

8. The bromophenol blue marker should be run to the center of an 8-cm gel because minimal resolution is needed. To estimate the concentration, compare the intensity of the dsRNA of unknown concentration to the intensity of each mass of the standard dsRNA; we routinely use a GelDoc and the Quantity One® software (Bio-Rad, Hercules, CA) for quantification. Determine which mass of standard dsRNA most closely matches the mass of the dsRNA of unknown concentration or estimate the mass if the intensity falls in the middle of two standards. Calculate the concentration by dividing the mass of the unknown dsRNA by the volume loaded (*see* **Note 58**). The concentration should be between 0.6 and 2.5 μg/μL, with typical yields for a 20-μL reaction between 30 and 125 μg, depending on the template (*see* **Note 59**).

9. Furthermore, the quality of the dsRNA should be evaluated. The dsRNA should run as a sharp band. There might be some faint bands or a smear above the desired product; this is caused by secondary structure within the dsRNA or branched regions within the dsRNA (*see* **Fig. 4B**). However, smearing below the correctly sized band indicates degradation of the RNA (*see* **step 11**).

10. First, determine if the RNA is annealed or what fraction is annealed. The size of the dsRNA is a good determinant of annealing because if the sense and antisense strands are annealed, then the dsRNA will migrate at nearly the same rate as a DNA fragment of the same size (*see* **Fig. 4B**). For example, if the length of the dsRNA is approx 550 bp, it should migrate at nearly the same rate as an approx 550-bp DNA fragment. On the other hand, if the RNA migrates at about the same rate as a 300-bp DNA fragment, then the RNA is single stranded (*see* **Note 60**). Also, if the RNA annealed well, it will form a sharp band. For most of the dsRNAs tested, annealing occurs in the transcription reaction. Thus, annealing should not be a concern, but it might be the case that a fraction of the RNA is single stranded, if this is the case, then see **Note 61**.

11. Second, evaluate the integrity of the dsRNA. The dsRNA should be free of degradation products, which will migrate below the correctly sized band and will appear as a smear. Degradation could happen during electrophoresis. Thus, if degradation is seen, it is worthwhile to run another gel after ensuring that the electrophoresis apparatus is RNase-free.

12. After estimating the concentration and verifying the quality of the dsRNA, proceed to in vitro dicing.

3.5.2. In Vitro Dicing

In order to dice large dsRNA, you will need to know the activity of your r-Dicer (*see* **Subheading 3.3.**). In general 50–100 µg of dsRNA is sufficient starting material to generate enough d-siRNAs for a large number of experiments. In fact 10–20 µg of d-siRNA will be generated. This is enough for 25–100 transfections in a six-well dish. However, if a larger mass of d-siRNA is required larger masses of dsRNA will be required (*see* **Subheading 3.5.1., step 1** and **Note 55**).

1. The following reaction conditions have been established for approx 50 µg of dsRNA and 100 µL of r-Dicer at a concentration of 160–200 ng/µL (*see* **Subheading 3.3.**). Of course, the amount of r-Dicer to use depends on the activity; more or less might be required (*see* **Note 62**). As described below, mix appropriate amounts of the dsRNA (*see* **Subheading 3.5.1.**), 50 m*M* MgCl$_2$ reaction dilution buffer, r-Dicer (*see* **Subheading 3.3.**), and RNase-free H$_2$O in a RNase-free microcentrifuge tube. Remember to dice a control dsRNA.

Reaction Conditions	Final concentrations
x µL dsRNA (45–50 µg)	150–167 ng/µL dsRNA
15 µl of 50 m*M* MgCl$_2$	2.5 m*M* MgCl$_2$
50 µL Rxn Dil. Bfr.[a,b]	250 m*M* NaCl
100 µL r-Dicer[b]	30 m*M* HEPES
x µL H$_2$O[c]	0.05 m*M* EDTA
300.0 µL	50–70 ng/µL r-Dicer

[a]pH 8.0.
[b]Volume depends on the volume of r-Dicer used; the volume of r-Dicer and Rxn. Dil. Bfr. must always equal 5 µL so that the final concentration of NaCl, HEPES, and EDTA is correct.
[c]Volume depends on concentration of dsRNA.

2. Pipet up and down to mix the reaction components and centrifuge briefly and incubate at 37°C for 16–18 h (*see* **Note 63**). Do not vortex.

3.6. Purification of d-siRNAs

Two methods can be used to obtain pure d-siRNAs: the original method we devised or a method that was created by Invitrogen. Both methods purify the d-siRNAs away from the r-Dicer, salts, buffer, and uncleaved large dsRNAs. Removing the undigested dsRNA is critical because it is toxic to most cells in which siRNAs are used. Both purification schemes are capable of removing the undigested dsRNA, but the method developed by Invitrogen is more reliable, time efficient, and simpler than our original method; thus, it is preferred and is presented here. See **Note 64** for other protocol.

The purification scheme utilizes spin cartridges to separate undigested dsRNA and d-siRNAs as well as to remove the unwanted reaction components. Altering the isopropanol concentration allows differential binding of the undigested, large dsRNA, and d-siRNAs. After dicing, the reaction is mixed with a denaturing buffer and isopropanol so that only the undigested, large dsRNA is bound to the first spin cartridge. The d-siRNAs and reaction components are in the flowthrough fraction. To purify the d-siRNAs away from the rest of the unwanted components, they are bound to a second spin cartridge. The d-siRNAs will bind the spin cartridges when the isopropanol concentration is increased, but the proteins, salts, and NTPs are removed because they do not bind the spin cartridge. To ensure high purity, the d-siRNA is washed with an ethanol-based salt solution before it is eluted from the spin cartridge with H_2O.

The spin cartridges are available from Invitrogen in either the Block-it Dicer RNAi kit or in the Micro-to-Midi Total RNA Purification System. However, if you make your own r-Dicer and dsRNA, you will not need the complete Block-it Dicer RNAi kit; instead, use the spin cartridges provided in the Micro-to-Midi Total RNA Purification System. Because the Micro-to-Midi Total RNA Purification System is used to isolate total RNA from cells, the components are named as such; note that these names do not make complete sense when described below. The protocol presented below is different from the method described in either the Block-it Dicer RNAi kit or Micro-to-Midi Total RNA Purification System manual. These changes ensure complete removal of unwanted larger dsRNA and maximize the d-siRNA yield; thus, our protocol is presented here. This subsection describes (1) how to obtain pure d-siRNAs and (2) how to estimate the concentration and quality of the d-siRNAs.

3.6.1. d-siRNA Purification

1. Remove 5 µL from the dicing reaction before purification and save at 4°C (short term) or –20°C (long term) for analysis on a 15% native polyacrylamide gel (*see* **Subheading 3.6.2.**). This sample is important for analysis of dicing efficiency. If the d-siRNA yield is low, you will be able to determine if it is a purification problem or a dicing problem.
2. First, add RNase-free β-mercaptoethanol to the lysis buffer at a concentration of 1% (v/v); this must be prepared fresh. Prepare enough lysis buffer plus 1% β-mercaptoethanol for all dicing reactions—300 µL per reaction.
3. To the remaining dicing reaction, approx 295 µL, add 300 µL of lysis buffer containing 1% β-mercaptoethanol. Pipet up and down at least five times. Purification will not work well if the two solutions are not mixed to homogeneity.
4. Now, add 400 µL of RNase-free isopropanol to the approx 600 µL (**step 3**). The isopropanol concentration is 40% (*see* **Note 65**). Set pipet to 1000 µL and pipet up and down at least five times to mix. Again, purification will be compromised if the solutions are not mixed well; in fact, your d-siRNA will likely be contaminated with large dsRNA.
5. Apply half of the sample, approx 500 µL, to a spin cartridge and centrifuge at 14,000g for 15 s. The flowthrough is collected in the microcentrifuge tube. **Save**

the flowthrough because it contains the d-siRNA. Also, **save** the spin cartridge for the remaining approx 500 μL.

6. Transfer the spin cartridge to a fresh, RNase-free microcentrifuge tube and add the remaining 500 μL sample (**step 3**). Spin as in **step 4** and again **save** the flowthrough. Now, the spin cartridge can be discarded.

7. Combine the two flowthrough fractions from **steps 5** and **6**, approx 1000 μL, in an RNase-free, 2-mL microcentrifuge tube. Add 1000 μL of RNase-free isopropanol to the 2-mL tube, raising the concentration of isopropanol to 70%. Cap the tube and mix very well by vortexing for at least 10 s and inverting several times. You will lose a substantial amount of d-siRNA if the two solutions are not mixed to homogeneity (*see* **Note 66**).

8. Now, you will use a *new* spin cartridge to bind the d-siRNA. All of the 2 mL of solution (**step 6**) must be passed through this same cartridge, but the maximum volume that can be loaded on the spin cartridge is 500 μL, so the 2 mL will have to be added to the spin cartridge in 500-μL aliquots, four times.

9. Apply 500 μL to the *new* spin cartridge and spin at 14,000*g* for 15 s. Collect the flowthrough in the microcentrifuge tube.

10. Discard the flowthrough and again apply 500 μL to the same spin cartridge. Spin as in **step 8**.

11. Repeat **step 9** two more times.

12. Now, the d-siRNA bound to the spin cartridge will be washed with 1X wash buffer II. The wash buffer II is supplied as 5X and should be prepared fresh. To make 1X wash buffer II, you must ***dilute*** the 5X solution ***with*** 100% ***ethanol.*** If you dilute the 5X wash buffer II with H_2O, the d-siRNAs will be lost. You will need 1 mL of 1X wash buffer II per column.

13. Add 500 μL of 1X wash buffer II to the spin cartridge and spin at 14,000*g* for 15 s. Discard the flowthrough.

14. Repeat **step 12**.

15. Transfer the spin cartridge to a fresh, RNase-free microcentrifuge tube and spin at 14,000*g* for 30 s. This step is essential to remove all traces of the wash buffer.

16. Transfer the spin cartridge to another fresh, RNase-free microcentrifuge tube and add 40 μL of RNase-free H_2O to the center of the filter. Make sure that you do not disrupt the membrane with the pipet tip. Let the H_2O stand for 3 min. Centrifuge at 14,000*g* for 30 s.

17. Using the same collection tube, add another 40 μL of RNase-free H_2O. Again, let the H_2O stand for 3 min. Centrifuge at 14,000*g* for 30 s.

18. Store the d-siRNA at 4°C or on ice until you complete the tasks in **Subheading 3.6.2**.

3.6.2. d-siRNA Analysis

The concentration of d-siRNAs (ng/μL) is estimated by measuring the absorbance at 260 nm. The purity of the d-siRNAs is estimated by calculating the 260/280 ratio; thus, a reading at 280 nm is also required. This purity measurement assesses the relative amount of protein contamination. The concentration estimate, or at least the relative concentrations of the d-siRNAs, is confirmed by visualizing the d-siRNA after is has been electrophoresed through a 15% native polyacrylamide gel. The gel is also used to determine if the d-siRNA is double stranded and free of undigested large dsRNA (*see* **Fig. 4C**).

1. Dilute the d-siRNA 50-fold; add 2 μL to 98 μL RNase-free TE, pH 8.0. A_{260}/A_{280} ratios are not reliable when measuring the purity of RNA if the measurement is taken in an unbuffered solution like H_2O; thus, TE is used *(52)*.

2. Mix the diluted d-siRNA by vortexing and measure the absorbance at 260 nm and 280 nm. Use the same TE (**Subheading 2.1., item 2**) as the blank. The optimal absorbance range is 0.05–0.3. Readings outside this range are not reliable. If you diced 50 μg of dsRNA and eluted in 80 μL, the A_{260} readings should be between approx 0.06 and approx 0.12 and the A_{280} readings between approx 0.03 and approx 0.06. The concentrations of d-siRNA can vary as much as 2.5-fold, but should be approx 100–250 ng/μL.

3. Estimate the concentration of the d-siRNA (in ng/μL):

$$[\text{d-siRNA}] = (A_{260})(32 \text{ ng/μL})(50)$$

4. Estimate the purity of the d-siRNA:

$$\text{Purity} = A_{260}/A_{280} \quad \text{should be} \geq 2$$

5. Prepare a 15% native polyacrylamide gel (*see* **Subheading 3.3.3., step 13** and **Note 38**).

6. Prepare the d-siRNAs and the samples of dicing reactions for electrophoresis. Transfer 1 μL of the purified d-siRNA to a tube containing 8 μL RNase-free H_2O and 1 μL of 10X RNA loading buffer. To the 5-μL sample of the dicing reaction (*see* **Subheading 3.6.1., step 1**) add 4 μL of H_2O and 1 μL 10x RNA loading buffer. Vortex each sample to mix and centrifuge briefly to collect the contents of the tube. A d-siRNA of known concentration can also be loaded on the gel (*see* **Note 67**).

7. Load the entire 10 μL on the 15% native gel and run at 10 W and 4°C for approx 1.5 h. Be sure the gel was prechilled and prerun (*see* **Subheading 3.3.3., step 13** and **Note 38**). The approx 21-bp d-siRNA will migrate in the middle of the xylene cyanol and bromophenol blue dyes.

8. Stain the nucleic acid by submerging the gel in H_2O or 1X TBE containing 0.5 μg/mL ethidium bromide for about 10 min. Rinse three times with H_2O and visualize and photograph with UV light (*see* **Fig. 4C**).

9. Look for the following: (1) analyze the dsRNA species in the samples taken before purification, ideally dicing should be at least 50% but not greater than 80% complete (*see* **Note 68**); (2) d-siRNAs must be free of uncleaved large dsRNA; (3) d-siRNAs should be double stranded; single-stranded RNA species migrate faster than double-stranded species (*see* **Note 69**); (4) the relative concentration of the control d-siRNA and the experimental d-siRNA should appear equal; (5) the concentration estimate determined by spectrophotometry should be accurate.

10. If d-siRNAs are contaminated with large dsRNA, d-siRNA must be repurified or remade; usually, repurification will work (*see* **Note 70**). The d-siRNAs should be

double stranded, but sometimes a portion, up to 10%, will be single-stranded RNA and will migrate faster than the double-stranded d-siRNA. This is not a concern. The control d-siRNA should be as concentrated as the experimental so that it is really a control and so that the knockdown can be attributed to specific gene silencing and not to merely a presence of a particular concentration of d-siRNA. Ensure that the concentration of all d-siRNAs correlates well with the A_{260} estimate and previous preparations. If you prepared several d-siRNAs, then the concentration might vary substantially; therefore, you should see different intensities. This helps in verifying concentration estimates.

11. If the A_{260} concentration estimates look accurate, dilute each d-siRNA to 100 ng/μL. If some of the concentrations are <100 ng/μL, then sets of d-siRNAs (e.g., experimental and control) can both be diluted to 50 ng/μL.

12. Aliquot the d-siRNAs, generally into 20 μL, and store at −80°C.

13. If the yield of d-siRNAs is nil or minimal, check the following: (1) the efficiency of dicing from the unpurified aliquot (70–80% of the input dsRNA should be diced); (2) ethanol and not water was added to wash buffer II; (3) d-siRNAs were not efficiently eluted (add more water to the columns and elute).

3.7. Silencing Gene Expression with d-siRNAs

To silence a gene, d-siRNAs must be introduced into the cytosol. Many delivery methods have been successful including liposomal-based transfection *(7,12)*, electroporation *(13)*, or microinjection (Firestone et al. unpublished data). Other methods such as pinosome-mediated endocytosis or PEI (polyethylinimine- or calcium phosphate-mediated transfection have been reported to work. Peptide conjugation *(53,54)* has been used to deliver siRNAs and might be useful for delivering d-siRNAs to cell lines that are difficult to transfect. Using either lipid transfection or electroporation, d-siRNAs have been used to silence gene expression in a variety of cell types, including HEK 293, HeLa, NIH/3T3 *(11)*, RBL (rat basophilic leukemia) *(13)*, rat hippocampal neurons *(12)*, A431 human fibroblasts (Heo and Meyer, unpublished data), and zebrafish embryos (Firestone et al., unpublished data).

Growth and transfection conditions for HeLa cells or other cell lines vary to some degree between individuals, so the information provided here might not be important to you. On the other hand, transfection can be very fastidious. Slight changes in a transfection protocol or materials can affect the efficiency. Thus, a detailed procedure is outlined below including all the materials that we use. The conditions, at least with one particular transfection reagent, provide efficient delivery of d-siRNAs and DNA, thereby inducing efficacious and specific gene silencing with minimal toxicity. We present the following: (1) general guidelines to silence a gene for the first time; (2) instructions for plating HeLa cells; (3) instructions for transfecting HeLa cells; (4) a guideline for establishing transfection in any cell type.

3.7.1. Silencing Gene Expression with d-siRNAs

It is very simple to suppress gene expression with d-siRNAs. Two parameters need to be determined: the amount of d-siRNA to use and the posttransfection time-point at which silencing is optimal. The silencing must be specific (i.e., reduce the abundance of the target protein but not other proteins.) Also, the desired levels of silencing should not occur at the cost of optimal cell health. The following approach has allowed us to attain gene silencing in HeLa cells on the first try for 20 genes. This approach should be amenable to other cell types so long as transfection conditions have been optimized (*see* **Subheading 3.7.4.**).

1. It is most efficient to first determine the amount of d-siRNA to use to attain the desired level of protein reduction. In general, a standard concentration of d-siRNA can be used to silence any given gene. However, some targets can be silenced at a lower concentration, which is useful to know if you desire to silence multiple genes simultaneously because the total concentration of d-siRNA that can be used is limiting (*see* **step 7**). Silencing of other targets might require a higher d-siRNA concentration; thus, it is best to do a titration.

2. If you are cotransfecting d-siRNAs and DNA (e.g., a fluorescent reporter construct), use a final d-siRNA concentration of 1, 3, or 10 nM (*see* **Table 4** and **Note 71**). If you are transfecting only d-siRNAs, use a final concentration of 5, 12.5, or 25 nM (*see* **Table 4**). These concentrations are often lower than suggested for chemically synthesized siRNAs, but are appropriate. The final d-siRNA concentration is determined as femtomoles of d-siRNA per microliter of total volume covering cell monolayer where the total volume is the volume of the growth medium used during transfection (OPTI-MEM I) plus the volume of transfection mix (*see* **Table 4**). To convert the mass of a d-siRNA to moles, use 640 g per mole per basepair (*see* **Note 72**).

3. Forty-eight hours posttransfection, evaluate cell health and determine the efficacy of gene silencing by measuring the remaining abundance of the target protein or mRNA (*see* **Subheading 3.8.1.**). If the d-siRNA concentration does not exceed the recommended amount, cell health should be similar to untransfected cells. The level of silencing will vary depending on the target, but one, or all, of these d-siRNA concentrations should specifically reduce protein levels by 50–95% (*see* **Note 73**). Ensure that silencing is specific and not the result of toxicity.

4. After determining an ideal concentration of d-siRNA to use, it is worthwhile to monitor silencing at 24, 48, and 72 h posttransfection. One concentration of d-siRNA is sufficient; use the concentration of d-siRNA that provided desired level of silencing (**steps 2** and **3**). Although most of our targets are reduced by >70% at 48 h posttransfection, silencing can be detected as early as 12 h or as late as 72 h. For subsequent phenotypical analysis, it can be beneficial to determine the half-life of the protein or the posttransfection time at which desired silencing is attained. A time-course of 12- to 24-h intervals for as long as 72 h is sufficient. For example, assess the level of gene silencing at 12, 24, 36, and 48 h. Silencing will

Table 4
Concentration of d-siRNA Recommended for Transfection

Plate format	Total TFC. volume	Transfection without DNA		Transfection with DNA		
		d-siRNA mass (ng)	Final [d-siRNA] (n*M*)	d-siRNA mass (ng)	Final [d-siRNA] (n*M*)	DNA mass (ng)
96-Well	~50 µL	3.5–17.5	5–25	0.7–7	1–10	1–5
24-Well	~300 µL	20–100	5–25	4–40	1–10	5–10
12-Well	~575 µL	40–200	5–25	8–80	1–10	12.5–25
6-Well	~1.1 mL	75–375	5–25	14.5–145	1–10	25–50

Note: d-siRNAs are effective gene silencers at a range of concentrations. This table suggests the range of concentrations in which d-siRNAs trigger specific knockdown when transfected alone or when cotransfected with DNA into HeLa cells and other easily transfectable cell lines. The final d-siRNA concentration (n*M*) is defined as nanomoles of d-siRNA per liter of total transfection (TFC.) volume, which is equal to the volume of medium and transfection mix added to the cells during transfection. The suggested mass in nanograms (ng) of d-siRNA and DNA is provided for various plate formats.

not usually benefit from a 96-h incubation and, in fact, some targets might begin to accumulate between 72 and 96 h (*see* **step 9** and **Note 74**). If measuring the half-life of the protein, do not plate different amounts of cells as suggested in **Table 5**; use the cell density for latest time-point.

5. If desired, the evaluation of silencing at various d-siRNA concentrations (**step 2**) and posttransfection time-points (**step 4**) can be monitored in a single experiment.

6. Keep in mind that different transfection reagents or methods, or transfection in different cells lines, might be more or less efficient. Thus, the optimal d-siRNA concentration might need to be determined (*see* **Subheading 3.7.4.**). If the desired amount of silencing is not seen, refer to the troubleshooting information in **Subheading 3.8.2.**

7. It can be very useful to silence multiple genes. The limiting factor is the total concentration of d-siRNA. In most cases, the maximum concentration of d-siRNA used is either 10 or 25 n*M*, depending on whether or not the d-siRNA is cotransfected with DNA. However, silencing is possible at a 5- to 10-fold lower concentration (*see* **Table 4**). Therefore, in theory, 5–10 genes could be silenced simultaneously; simply cotransfect 5–10 different pools of d-siRNAs that target 5–10 different genes such that the suggested maximum concentration of d-siRNAs is not exceeded. For example, mix five different d-siRNAs such that the final concentration of each will be 2 n*M* and the total concentration will be 10 n*M*; then, mix with DNA and transfect. To date, we have simultaneously silenced three genes. Silencing more than three genes might be impossible because of limited quantities of the RNAi machinery.

8. It can be very useful to reconstitute or "add back" the protein that was silenced. First, it is an easy way to validate the specificity of gene silencing because

Table 5
Recommended HeLa Cell Density

Plate format	Plating volume	Cell density: 12 to 24-h experiment	Cell density: 36 to 48-h experiment	Cell density: 72 to 96-h experiment
96-Well	100 µL	$(4–5) \times 10^3$ cells/well	$(2.5–3) \times 10^3$ cells/well	$(1.5–2) \times 10^3$ cells/well
24-Well	1 mL	$(2–2.5) \times 10^4$ cells/well	$(1.5–2) \times 10^4$ cells/well	$(1.2–1.6) \times 10^4$ cells/well
12-Well	1 mL	$(4–5) \times 10^4$ cells/well	$(3–4) \times 10^4$ cells/well	$(2.5–3) \times 10^4$ cells/well
6-Well	2 mL	$(1–1.2) \times 10^5$ cells/well	$(0.8–1) \times 10^5$ cells/well	$(6–8) \times 10^4$ cells/well

Note: Cell density is important for analysis subsequent to d-siRNA-mediated gene suppression. Suggested here are a range of cell densities each appropriate for a particular plate format and analysis at a specific time-point. Suggested plating volume ensures proper dispersal of cells within the well.

reconstitution will alleviate the phenotype if the aberration is the result of the lack of a particular gene. In fact, reconstitution is a requirement for publication in particular journals *(55)*. Second, it provides a method for mutational analysis in a nearly null background, making structure function experiments more efficient. The easiest way to perform a "knockdown reconstitution" is to target the endogenous mRNA using d-siRNAs made against a noncoding region like the 3′ UTR (untranslated region) or 5′ UTR *(see* **Subheading 3.4.1.***)* and cotransfect an expression construct containing only the coding region of the gene. The endogenous gene but not the exogenous gene will therefore be silenced. This "gene swap" allows mutated versions of the gene, both truncations and point mutants, to be analyzed. Keep in mind that some genes will not have large 3′ UTRs or 5′ UTRs such that in vitro transcription and dicing will not be efficient *(see* **Note 43***)*.

It is a good idea to tag the exogenous gene with a fluorescent reporter (e.g., YFP) for visualization and the fact that the exogenously expressed protein will be larger than the endogenous gene so that when evaluating knockdown by Western blotting, it will be simple to distinguish between the two forms. Transfection should be performed as described for the cotransfection of DNA and d-siRNAs. Be sure to include d-siRNAs cotransfected with an empty vector and an irrelevant d-siRNA as controls.

9. In some cases, prolonged silencing will be desired, (e.g., phenotypical assays after long-term silencing, like 5–14 d). Although d-siRNAs are not capable of sustained gene silencing, the cells can be continually transfected every 3–4 d. Simply trypsinize the cells, replate and transfect 12 h later. It is essential to ensure that the protein does not accumulate between transfections, otherwise the phenotype might not be correct.

Thus, have enough cells for measuring protein abundance (*see* **Subheading 3.8.1.**) at each replating time-point. Exogenous d-siRNAs, used for a control, might be problematic for multiple transfections. We have noted that d-siRNAs lacking a target might have a longer half-life than d-siRNAs with a target. The persistence of d-siRNAs is problematic if the levels accumulate above the concentration at which silencing is accompanied by off-target and nonspecific effects.

3.7.2. Plating HeLa Cells for Transfection

Cell plating is often overlooked as an important factor governing transfection efficiency. However, the density and especially the distribution of the cells within a well are very important for attaining reproducible and efficient transfection and gene silencing. Furthermore, the health and growth rate of the cells are extremely important for reproducible and efficient transfection. If the cells do not divide, then very stable proteins will not be diluted by cell division (*see* **Note 73**). Passage number is also very important. The growth characteristics of HeLa cells are relatively constant between passage 5 and 25; therefore, this is the range in which we do experiments (*see* **Note 75**). If using a cell line other than HeLa cells, plate cells at an appropriate density (*see* **Subheading 3.7.4.**) and follow recommended culture conditions.

1. HeLa cells are maintained as suggested by ATCC with minor modifications. HeLa cells are cultured in complete growth medium (i.e., DMEM containing 10% FBS and 1X penicillin–streptomycin–glutamine in a 37°C, humidified incubator with 5% CO_2). Cells are routinely maintained in a 10-cm dish and split once the confluency reaches 50% (2.4×10^6 cells in total). ATCC recommends subculturing two to three times per week at 1:2 to 1:6, but lower densities such as 1:10 to 1:20 are acceptable. Cells are split by detaching with trypsin and replating in a fresh 10-cm dish as described below. Cell health should be optimal and is monitored by careful analysis of cell morphology via inverted light microscopy at ×10 magnification (*see* **Note 76**). Unhealthy cultures should be discarded. Frozen cell stocks should be generated (*see* **Note 77**).

2. HeLa cells are plated in a 24-well, 12-well, or 6-well dish 12–18 h before transfection (*see* **Note 78**). The assays subsequent to gene silencing will determine which plates to use. If more cells are required, then 6-well plates should be used; if fewer cells are needed, then 24-well plates will be sufficient.

3. Prewarm complete growth medium at 37°C. A warm medium facilitates timely cell attachment after plating. It is better to warm only the amount of medium required for cell plating because glutamine and other components are sensitive to extended heating. This approach will ensure that the large stock of complete growth medium maintains integrity over time.

4. Move the 10-cm dish of HeLa cells to a tissue culture hood and aspirate the growth medium. Add 10 mL D-PBS, aspirate, then add 3 mL trypsin–EDTA, and return to the 37°C incubator. The trypsin is inhibited by serum; thus, washing the cells with D-PBS or trypsin-EDTA is essential.

5. Cell detachment is monitored via microscopy. Cells should detach in 3–5 min. Longer incubations will damage cell integrity (*see* **Note 79**). Although agitation during trypsin treatment might result in clumping, it is better to detach the cells in a shorter period of time; thus, it might be necessary to swirl the trypsin.

6. When approx 80% of the cells are detached, move the plate back to the hood and add 7 mL of prewarmed complete growth medium (**step 3**). The medium should be repeatedly dispersed over the plate to collect the attached cells. Clumps should be disseminated by pressing a 10-mL pipet tip against the bottom of the 10-cm dish while dispensing the cells. Transfer all 10 mL of cells to a sterile 15-mL conical tube.

7. Using a hemacytometer determine the cell density. Ten microliters of the cell suspension should be used and the cell number should be determined for four quadrants (*see* **Note 80**). Typical densities are approx 2×10^5 cells per milliliter, which corresponds to approx 50% confluency. It is acceptable for the confluency to be between 30% and 70%. Trypan blue staining is not essential.

8. Using the suggested cell densities and plating volumes in **Table 5**, plate cells in the appropriate number of wells and dishes; remember to include wells for controls. To eliminate variation in cell number between each well, make a large stock of cell suspension at the appropriate cell density by diluting the cells in fresh, prewarmed (**step 3**) complete growth medium. Dilute the cells such that adding 1 or 2 mL to a well will result in the correct cell number. For example, in a 24-well plate, transfer 1 mL of cells, at a density of $(2–2.5) \times 10^4$ cells per milliliter, to each well. Prepare a large stock of diluted cells (e.g., 30 mL in a 50-mL conical tube), so that cell number is similar between wells and plates. Remember that the cells will settle to the bottom of the conical tube; thus, the suspension must be mixed well before cells are aliquoted into the plates. Cells can also settle in the pipet during transfer to the wells; therefore, work with 12 mL.

9. After dispensing cells into a well, gently rock in a circular pattern to distribute the cells and place in the incubator. While cells are attaching, they tend to clump in the center of the well, especially in 12- and 6-well plates. Clumping will decrease transfection and gene silencing efficiency; thus, it must be avoided. Ten to fifteen minutes after the cells are plated or when some of the cells have attached, gently rock the plate in a circle to redistribute the unattached cells; monitor by microscopy. Repeat two more times. This will ensure that the cells are evenly dispersed.

3.7.3. Transfection of HeLa Cells

We have tested many commercially available lipids and found that Genesilencer provides very efficient transfection and gene silencing in HeLa cells. Very small amounts of d-siRNAs are required to attain high levels of knockdown. Thus, smaller amounts of d-siRNA can be prepared. Also, there is little concern that the apparent knockdown is the result of a general toxicity. This lipid also permits the cotransfection of d-siRNAs and DNA, an important capability for analyzing gene silencing and the resulting phenotype. The protocols for transfecting d-siRNAs or cotransfecting d-siRNAs and DNA vectors are presented here. Because transfection efficiency can be affected by the materials used (e.g., composition of the plastic), all materials are listed. Note that RNase-free barrier tips are used for all manipulations.

1. Transfection efficiency of HeLa cells is improved twofold to threefold when transfection is performed in the absence of serum. Thus, complete growth medium must be replaced with serum-free medium. Three to four hours after transfection, serum levels are restored to 10% by adding an equal volume of 2X medium.

2. Transfection in the smallest possible medium volume improves efficiency. This requires that more medium be added 12–18 h posttransfection, which does take extra effort, but is worth it because less d-siRNA will have to be used and transfection is much more efficient.

3. Using an inverted light microscope, assess the quality of plating (*see* **Subheading 3.7.1.**). The cells should be distributed evenly throughout the well and free of clumps. The cell density should be optimal for subsequent assays but is not a concern for transfection efficiency because Genesilencer provides efficient transfection at low and high cell densities.

4. If cell density and distribution are suitable, then warm the appropriate amount of OPTI-MEM I to 37°C; enough to wash each well once as well as enough to overlay cells during transfection (*see* **Table 6**).

5. While OPTI-MEM I is warming, transfer the appropriate amount of d-siRNAs or mixture of d-siRNAs and DNA into 1.5-mL microcentrifuge tubes. This can be done in a nonsterile environment (e.g., a benchtop centrifuge) because it is quick, the growth medium contains antibiotics, and risk of fungal or mycoplasm infection is low. Plan the experiment according to **Subheading 3.7.1.** using suggested amounts of d-siRNA and DNA (*see* **Table 4**). If you are using a range of d-siRNA concentrations, then you can add RNase-free H_2O to equalize the volumes. For the lowest d-siRNA concentrations, it might be necessary to dilute with H_2O to ensure accurate pipetting. Be sure to have a Genesilencer alone or Genesilencer plus H_2O control. If you use H_2O, use the same volume as used for the d-siRNAs.

6. Move culture dish to hood, aspirate complete growth medium, and add appropriate volume of prewarmed OPTI-MEM I to wash the cells (*see* **Table 6**). To avoid dessicating the cell monolayer, work with a maximum of 12 wells in a 24-well plate, one 12-well plate, or one 6-well plate.

7. Aspirate OPTI-MEM I and add the correct volume of prewarmed OPTI-MEM I suggested for transfection (*see* **Table 6**). Place the plate back in the incubator.

8. Place the vial of siRNA diluent and Genesilencer siRNA transfection reagent in the hood. The reagents do not need to be kept at 4°C during transfection. Also, place an aliquot of OPTI-MEM (not the same aliquot that was prewarmed in **step 4**) in the hood; usually aliquoted in a 50-mL conical tube and should be at either 4°C or room temperature. Prepare a master mix of the diluent and OPTI-MEM I using volumes suggested in **Table 6**. Prepare the master mix for one more transfection than you will need. For example, if you are transfecting 9 wells in a 12-well dish, make a master mix to accommodate 10 transfections; mix 250 µL diluent with 150 µL OPTI-MEM

Table 6
Transfection Specifications

Plate format	OPTI-MEM I volume		Diluent/OPTI-MEM I mix		Genesilencer/ OPTI-MEM I mix	
	Wash volume	Volume for transfection	Diluent volume (µL)	OPTI-MEM I volume (µL)	Genesilencer volume (µL)	OPTI-MEM I volume (µL)
96-Well	N/A	N/A	2.5	15	0.25	25
24-Well	500 µL	250 µL	10	15	3.5	25
12-Well	1 mL	500 µL	25	15	5	25
6-Well	2 mL	1 mL	25	15	5	25

Note: The Genesilencer transfection agent is routinely used to transfect d-siRNAs or cotransfect d-siRNAs and DNA into HeLa cells. This table lists volume of each reagent used for each plate format to attain optimal transfection efficiency.

      I in a 1.5-mL microcentrifuge tube. Pipet up and down to ensure complete mixing.

9. Move the tubes containing d-siRNAs or a d-siRNA/DNA mixture to the hood and transfer the appropriate amount of the diluent/ OPTI-MEM I mixture (e.g., 40 µL for a 12-well plate) to each tube and then pipet up and down three times to mix. Incubate for 5 min with the tubes uncapped. It is easier and more time efficient to make the 5-min incubation continual (i.e., start the incubation after you transfer the diluent/ OPTI-MEM I mixture to the first tube). It is possible to handle about 18–24 transfections at once.

10. During the remainder of the first 5-min incubation, prepare the Genesilencer/ OPTI-MEM I master mix using the volumes suggested in **Table 6**. Again, prepare the master mix for one more transfection than you will need. For example, if you are transfecting 9 wells in a 12-well dish, make a master mix to accommodate 10 transfections; add 50 µL of Genesilencer reagent to 250 µL OPTI-MEM I in a 1.5-mL microcentrifuge tube and mix by pipetting up and down about five times. It is important to add the Genesilencer reagent to the OPTI-MEM I. Transfer the appropriate amount of the Genesilencer/ OPTI-MEM I mixture (e.g., 30 µL for a 12-well plate) to each tube; do not pipet up and down. Again, incubate for 5 min in a continual fashion (**step 9**). The tubes remain uncapped.

11. At the end of the 5-min incubation, move the tissue culture dish to the hood and add the transfection mix (diluent/Genesilencer/OPTI-MEM I/nucleic acid) in a dropwise fashion to the OPTI-MEM I that is covering the cells. Swirl several times to mix and put the plate back in the incubator. Be sure to swirl to mix after every sixth well to avoid accumulation of transfection mix in the center of the well.

12. Add an equal volume of 2X medium 3.5–4 h after adding the transfection mix. Account for the extra volume imposed by addition of the transfection mix (*see* **Table 4**). For example, add approx 600 µL to each well in a 12-well plate. Be sure to prewarm the 2X medium at 37°C before adding it to the cells.

13. Additional medium can be added to each well 12–18 h after transfection. Because the cells are adjusted to a 50 : 50 mixture of 2X medium and OPTI-MEM I, it is beneficial to add more of the 50:50 mixture and not just complete growth medium. Prepare enough of the 50:50 mixture for all of the wells and prewarm to 37°C prior to addition.

14. Transfection can be accomplished with other reagents (*see* **Note 81**). If you are using a cell line other than HeLa, transfection can potentially performed in the presence of serum (*see* **Note 82**). Be sure to plate the appropriate density if a cell line other than HeLa is used.

3.7.4. Optimizing Transfection

Using the aforementioned Genesilencer transfection protocol, it should be very simple to silence HeLa gene expression with d-siRNAs. However, if you are using a different cell line or transfection reagent, it is worthwhile to optimize transfection conditions because poor transfection is the most prominent cause of a failed gene silencing experiment. To determine the optimal transfection conditions, we have used d-siRNAs to silence both transiently and stably expressed YFP. Although other reporters could be used, such as Luciferase, fluorescent reporters are advantageous because silencing and transfection efficiency can be evaluated in single cells. This subsection describes how to optimize transfection conditions. To begin, we describe how to optimize transfection of d-siRNAs using a cell line that stably expresses a fluorescent reporter. Subsequently, we describe how to optimize transfection conditions for cotransfection of a d-siRNA and an expression construct.

1. Although transient expression is simpler than generating a stable cell line, cotransfection of a fluorescent reporter and d-siRNAs is not optimal. At high concentrations of d-siRNA, very little vector gets delivered to the cells, preventing evaluation of high d-siRNA concentrations. Furthermore, some transfection reagents cannot efficiently cotransfect d-siRNAs and DNA. If DNA and d-siRNAs are cotransfected, you will be optimizing only for cotransfection, which might not be appropriate for future experiments. Finally, cotransfection only measures target cleavage, like reverse transcription (RT)-PCR, and cannot be used to measure the half-life of the protein, or accurately evaluate the silencing conditions required to knock down an endogenous protein. Thus, it is more informative to optimize transfection of d-siRNAs by silencing a stably expressed reporter, such as GFP (*see* **Note 83**). Other fluorescent reporters can be used, including CFP and YFP; d-siRNAs targeting GFP will silence CFP and YFP because their nucleic acid sequence is nearly identical. An in vitro transcription template encoding a fragment of GFP can be made with the primers shown in **Table 3**. A control, such as firefly or Renilla luciferase, is required (*see* **Table 3**). Prepare dsRNA and d-siRNAs as described in **Subheadings 3.5.** and **3.6.**

2. To optimize transfection, cell number, amount of transfection reagent and concentration of d-siRNA must be evaluated. All three variables can be tested in a single

Table 7
Optimizing Transfection

Plate format	Amount of lipid (µL)	Final [d-siRNA] (nM)	Amount of DNA (ng)
96-Well	0.25, 0.5, 1, 2	3.2, 6.25, 12.5, 25, 50, 100	2.5, 5, 10, 20, 40, 80
24-Well	1, 1.75, 3.5, 5	3.2, 6.25, 12.5, 25, 50, 100	5, 10, 20, 40, 80, 160
12-Well	1.25, 2.5, 5, 10	3.2, 6.25, 12.5, 25, 50, 100	10, 20, 40, 80, 160, 320
6-Well	1.25, 2.5, 5, 10	3.2, 6.25, 12.5, 25, 50, 100	25, 50, 100, 200, 400, 800

Note: Poor transfection efficiency is the most likely cause of a failed gene silencing experiment. Although easily transfectable cells can be used for many experiments, additional cell lines might be important for analysis of a phenotype. The Genesilencer transfection reagent provides efficient transfection in many cell types, but transfection of d-siRNAs or cotransfection of d-siRNAs and DNA for a particular plate format should be optimized using combinations of the suggested amount of lipid, DNA, and d-siRNA concentration ([d-siRNA]).

experiment, but if you want to simplify the experiment, evaluate the amount of lipid and mass of d-siRNA first. It is best to do the experiment in duplicate.

3. If you have or are willing to make a stable cell line expressing a fluorescent reporter, follow the directions here; otherwise, skip to **step 9**. Using either a 96-well plate or several 24-well plates, set up a range of transfection conditions that test the cell number, amount of transfection reagent, and mass of d-siRNA (*see* **Table 7**). Test three to four amounts of transfection reagent versus five to six different concentrations of d-siRNA such that each volume of transfection reagent will be tested for each d-siRNA concentration. This is easiest to think about in a grid format. For each ratio of transfection reagent to d-siRNA concentration, evaluate three or four different cell densities, 20%, 40%, 60%, or 80% confluent. Be sure to replicate the transfection conditions with a control d-siRNA; we routinely use d-siRNAs generated from Firefly Luciferase. Also, include transfection reagent alone so that cytotoxicity, relative to untransfected cells, can be attributed to either the nucleic acid or transfection reagent.

4. If the transfection reagent is Genesilencer follow the protocol in **Subheading 3.7., step 3**, but if not, follow the manufacturer's protocol. Evaluate a range of transfection reagent volumes, d-siRNA concentrations, and cell densities, each of which is centered on values suggested by the manufacturer.

5. Evaluate cell health 24–28 h posttransfection and determine relative fluorescence with an inverted fluorescent microscope; we routinely use the ImageXpress (Molecular Devices). Ensure that fluorescence intensity is not saturating so that quantification will be accurate. Cell number should be determined by staining nuclear DNA with Hoechst 33342 as directed by the manufacturer (*see* **Note 84**).

6. Compare fluorescence intensity of cells transfected with GFP d-siRNA to cells transfected, under the same conditions, with GL3 or other control d-siRNAs.

Determine which concentration of d-siRNA, volume of transfection reagent, and cell number resulted in optimal gene silencing without compromising cell health. It is important to determine the cause of cytotoxicity, either transfection reagent or nucleic acid, so that the appropriate adjustments can be made.

7. Generally, a range of d-siRNA concentrations will be effective, such as 1–30 n*M*. Optimal transfection conditions will result in a 40–50% reduction in fluorescent protein at 24 h and 70–80% at 48 h. Under the conditions where there is gene silencing, there will be some cells that are expressing the fluorescent reporter at very high levels, similar to the control, but in the majority of the cells there should be minimal or a complete lack of expression. To ensure that the reporter is truly silenced, a second reporter can be cotransfected with the d-siRNAs (*see* **step 8**).

8. If desired, a fluorescent reporter can be cotransfected with d-siRNAs (*see* **Table 7**). Once you have narrowed in on d-siRNA-only transfection conditions, try various amounts of DNA to optimize cotransfection. For example, keep cell number, volume of transfection reagent, and d-siRNA concentration constant while titrating DNA mass. Include a control d-siRNA for all conditions and evaluate silencing efficacy and potency as described in **steps 6** and **7**. Expression levels of the untargeted reporter should increase with increasing mass of DNA, but at higher concentrations, it is likely that the transfection efficiency (*see* **Note 85**) will decrease as a result of cytotoxicity. Make sure that the excitation spectra of the stably expressed fluorescent reporter and the cotransfected reporter can easily be separated (e.g., use GFP and RFP or YFP and RFP). CFP, GFP, and YFP cannot be used together because GFP d-siRNAs will silence CFP and YFP expression. We have found that silencing of GFP and presumably delivery of GFP d-siRNAs correlates well with expression of RFP such that cotransfection of an expression construct can be used to mark cells that have been transfected with d-siRNAs and ensure that the apparent gene silencing is not the result of the lack of transfection.

9. If a stable cell line is not available, silencing efficacy can be determined by cotransfecting an expression vector and d-siRNAs. Set up an array of d-siRNA concentration, volume of transfection reagent, and cell number with both GFP d-siRNAs and a control d-siRNA as described in **step 3**. To avoid adding another variable (mass of DNA), it is worthwhile to first determine how much of the expression construct is required to attain 50–70% transfection efficiency. Determine optimal mass of DNA to use by again evaluating expression (*see* **step 5**) at a variety of volumes of transfection reagent and masses of DNA (*see* **Table 7**). Keep in mind that when cotransfecting d-siRNA and DNA, less DNA will be required than when DNA is transfected alone. As described in **step 8**, GFP and RFP expression constructs can be cotransfected with d-siRNAs such that the RFP can be used to normalize expression.

10. Determine optimal transfection conditions based on silencing efficacy as described in **steps 6** and **7**. It is very likely that higher concentrations of both the control d-siRNA and the GFP d-siRNA will result in a fewer number of expressing cells. We believe this is the result of a competition between the d-siRNAs and DNA and not the result of nonspecific silencing of the control d-siRNA because expression of the untargeted reporter is also decreased. Note that unlike

cotransfection of d-siRNAs and a single expression construct, the correlation between expression of two reporters and d-siRNAs is rather low. However, the transfection efficiency of the untargeted reporter can be used to assess the general levels of transfection and is usually constant between samples.

11. If transfection efficiency cannot be optimized by altering the volume of transfection reagent, concentration of d-siRNAs, and cell number then try transfection in the absence of serum. This, however, should be a last resort because many cell lines detach or initiate apoptosis in the absence of serum, even during incubations as short as 30 min.

3.8. Assaying and Troubleshooting Gene Silencing

For most gene silencing experiments, it is important to know that protein levels have been reduced. To evaluate protein abundance, we most commonly use Western blotting, but immunofluorescence is equally effective and has the advantage of analyzing knock down in single cells even though it is more difficult to quantify than Western blotting. If an antibody for the protein of interest is not available, silencing can be evaluated by measuring mRNA abundance with RPA (RNase protection assay), Northern blotting, or quantitative RT-PCR, but keep in mind that this might not be sufficient for publication, especially if the phenotype is questionable and if reconstitution experiments are not included. This subsection presents (1) general information about measuring protein abundance and evaluating knockdown and (2) information for troubleshooting a failed gene silencing experiment.

3.8.1. Assaying Gene Silencing

We have found some reagents that make evaluating knockdown much easier, thus they are described here.

1. To make cellular lysates for Western blotting we use either the PARIS kit or the M-PER lysis reagent as directed by the manufacturer. We use the PARIS kit to extract cellular protein and isolate total RNA. It can also be used to fractionate the nuclear and cytoplasmic extract but we have not done this. Both protease inhibitors and phosphatase inhibitors can be included in the M-PER lysis reagent and both of the lysis buffers in the PARIS kit. It is important to remove residual medium before lysis by washing the cells with D-PBS after aspirating the growth medium. Passing the lysate through an 18- or 21-gage needle attached to a syringe ensures complete lysis and shears DNA, thereby decreasing viscosity. Also, a "snap-freeze" in liquid nitrogen and a thaw will effectively shear DNA and allows for a stopping point at which the lysates can be stored at –80°C. Traditional lysis buffers, like RIPA, are also appropriate but are not as convenient or efficient. When isolating RNA for DNA-sensitive application such at RT-PCR or DNA microarray analysis, contaminating DNA should be removed as suggested by the protocol (a good reagent is DNA-free; Ambion, Austin, TX; cat. no. 1906).

2. Before electrophoresis, it is essential to normalize the total amount of protein loaded in each well. We have had the best success with the BCA protein assay kit because it is compatible with a wide variety of solutes present in lysis buffers. We use the microplate procedure as described by the manufacturer. Note that we use the 1 : 20 ratio by diluting 5 μL of lysate in 100 μL of "working reagent," but it is also acceptable to use 2.5 μL of lysate in 50 μL of "working reagent" so long as it completely covers the surface of the well. We use 96-well plates and a Bio-Rad plate reader at 595 nm. The recommended wavelength is 562 nm, but 540- to 595-nm measurements are equally accurate. We use BSA for a standard curve at 250, 500, 750, 1000, 1500, and 2000 μg/mL. Both the samples and the standard curve are measured in duplicate. Avoid bubbles because they interfere with measurement. Be sure to measure the background of the "working reagent" (i.e., prepare "working reagent" without any protein). Calculate the net absorbance of each sample by subtracting the background value of the "working reagent." Determine which lysate is the least concentrated and dilute all other lysates to this concentration with the appropriate lysis buffer. Using the standard curve, calculate the total protein concentration.

3. We generally load 15–20 μg of total protein with 1X sample buffer in each lane of an SDS-PAGE gel (16 × 16 cm); for reagents and instructions, *see* **Note 27**. To detect scarce proteins, more protein should be loaded, up to 30–40 μg, or for detection of abundant proteins load less, as little as 10 μg. Note that smaller gels require less protein.

4. We often test Western blotting conditions before evaluating knockdown. Simply load various amounts of total protein in four wells (e.g., 10, 20, 30, or 40 μg; repeat loading two or three times so that the membrane can be cut into strips and detection can be tested at various primary antibody dilutions (**step 6**).

5. We transfer protein to Immobilon-P in 1X transfer buffer containing 10% methanol as directed by manufacturer. After transfer, the membrane is dried by submersion in 100% methanol and "air-drying" or purely "air-drying" as suggested by manufacturer. Complete drying is essential for the "rapid detection protocol" (**step 6**). After testing various membranes, we found that Immobilon-P provides the most consistent results; it is therefore strongly recommended.

6. For blotting, we use the "Rapid immunodetection method on Immobilon-P using chemiluminescence" protocol provided by Millipore (tech note TN051; http://www. millipore.com/publications.nsf/docs/TN051) (*see* **Note 28**). Most primary antibodies work well at 0.1–0.2 μg/mL in antibody diluent (**Subheading 2.3., item 15**); incubations are generally for 1 h at room temperature. Biosource International (Carmillo, CA) is a very good source for antibodies. About two out three antibodies that Santa Cruz Biotechnology (Santa Cruz, CA) produces are good. Secondary antibodies are used at 0.3–0.6 μg/mL in antibody diluent, incubations are generally for 30 min to 1 h at room temperature. The Zymax HRP conjugate antibodies work very well.

7. For detection, follow suggestions in **Note 28**. Make sure that the intensity is within the linear range such that knockdown can be accurately quantified. Protein

abundance is quantified with Quantity One software (Bio-Rad, Hercules, CA) directly after collection with the GelDoc (Bio-Rad, Hercules, CA), or if film is used, abundance is quantified with Adobe Photoshop after the film is scanned. If the intensity is out of the linear range, remove the substrate by washing the membrane in D-PBS; then, as previously described, reapply the substrate for only 30–60 s and wait until luminescence intensity is in the linear range.

8.   An antibody against Actin can be used to control for equal loading of protein in each lane. The goat and mouse antibodies are equally effective, but, in general, anti-goat HRP-conjugated secondary antibodies have a higher level of background, so the mouse anti-Actin antibody is used more frequently. Having both antibodies is advantageous because the membrane will not have to be stripped of the antibody used to detect the target protein before the abundance of Actin is measured; stripping of antibodies often removes protein as well. We use the goat antibody when the primary antibody against the target protein is produced in mice, and we use the mouse antibody when the primary antibody is raised in goat or rabbit. Other abundant, stable proteins such as GAPDH or Lamin A/C can also be used as a loading control.

9.   Determine the remaining protein abundance and normalize to the levels of the loading control, such as Actin. Protein levels of the target gene should be reduced by 50–95% relative to the controls. The most important aspect of a gene silencing experiment is specificity; that is, the abundance of the target should be similar in untransfected cells, mock transfected cells, and cells transfected with the control d-siRNA (*see* **Note 86**).

10. If an antibody is available, ensure that silencing is specific by measuring the abundance of an isoform of the target. If an isoform cannot be evaluated, measure the abundance of a protein that is most closely related to the target at the nucleotide level. The levels of the isoform should not change, but if they do, it could be that the protein stabilities are linked. Thus, further experiments such as measuring mRNA abundance and reconstitution should be performed. Also, the half-life of the target should be determined to evaluate whether changes in the closely related protein occur after loss of the target.

11. The phospho-specific eIF2α antibody can be used to determine if nonspecific inhibition of translation contributed to the knockdown. The translation initiation factor eIF2α is phosphorylated and inhibited in response to large dsRNA (*16*) or high levels of siRNA (*7*).

3.8.2. Troubleshooting Gene Silencing

d-siRNA-Mediated gene silencing is very efficient in HeLa and other cells. However, there might be instances in which gene silencing is unsuccessful; thus, this subsection provides information to troubleshoot a failed gene silencing experiment.

1.   Poor transfection is the most likely cause of a failed gene silencing experiment. Transfection should be optimized as described in **Subheading 3.7.4.**

Transfection efficiency can be monitored by cotransfecting a fluorescent reporter with d-siRNAs. Although expression of a reporter gene often correlates with gene silencing, this approach might not be appropriate for all cell types or transfection reagents. d-siRNAs can, therefore, be directly labeled with Cy3 or FAM and detected by fluorescent light microscopy to ensure delivery of the d-siRNAs into the cells. Confocal microscopy should be used to distinguish between d-siRNAs in the cell or those stuck to the cell membrane. In theory, d-siRNAs can be labeled by incorporating FAM-labeled UTP, or other labeled nucleotides, into the large dsRNA during in vitro transcription. However, the yield of dsRNA and d-siRNAs are reduced; thus, it is much easier to label the d-siRNA after they are purified.

2. If transfection efficiency is low, gene silencing will be masked by the large population of untransfected cells; thus, biochemical assays will not be feasible unless the transfected cells can be separated from untransfected cells. FACS (flow-activated cell sorting) can be utilized to isolate the transfected cells from the untransfected cells. Transfected cells can be marked with fluorescently labeled d-siRNAs or by cotransfection of d-siRNAs and a fluorescent reporter. Alternatively, silencing and phenotypes can be monitored with a single cell-based assay. Transfected cells can be detected by cotransfection of d-siRNAs and a fluorescent reporter or fluorescently labeled d-siRNAs and then silencing can be monitored by immunofluorescent staining. Keep in mind that the fluorescence used to detect transfected cells and knockdown must have discernible excitation and emission spectra. Cotransfection of a fluorescent reporter and d-siRNAs has worked well for gene silencing and subsequent phenotypical analysis in rat hippocampal neurons, primary cells that are very difficult to transfect *(12)*.

3. Others have used multiple transfections, as frequent as every 24 h, to increase the amount of gene silencing. However, we have found that the increase in gene silencing is usually the result of increased toxicity and employ multiple transfections only if the experiment requires that the gene be silenced for a prolonged period (see **step 9** in **Subheading 3.7.1.**). Multiple transfections do not result in transfecting additional cells; usually, the same cells are transfected. Also, control d-siRNAs do not have an mRNA target and might persist longer than d-siRNAs that do have a target. Thus, the control d-siRNAs is not really a control any more because they might be present at a higher cellular concentration.

4. It is possible that silencing does occur, but is not specific; for example, the control d-siRNAs also silence the target. If this is the case, reduce the concentration of d-siRNAs or increase the cell density. It is a good idea to use a range of concentrations, starting as low as 0.5 n*M*. If off-target silencing is encountered, simply reduce d-siRNA concentration.

5. It is also possible that the nonspecific gene silencing is the result of transfection reagent toxicity. If this is the case, simply reduce the amount of transfection reagent used or increase the cell density; use a different transfection reagent. Ensure that the nuclease-free H_2O is free of toxins. For example, DEPC (diethylpyrocarbonate) is often used to prepare RNase-free H_2O, but if it is not completely removed,

it will be toxic to cells. If you suspect the RNase-free H_2O is toxic, you can simply add the H_2O to the cells or add d-siRNAs that have been dissolved in the H_2O to the cells in the absence of transfection reagent. Similarly, you can systematically test each component used in transfection for toxicity.

6. Silencing might not be apparent because the protein has a very long half-life and does not turn over because the cells do not divide. Make sure the cells are healthy after transfection so that they can divide. In HeLa cells undergoing efficient d-siRNA-mediated gene silencing, protein half-life should never be greater than 20 h because HeLa cells divide about every 18–20 h and translation should be completely inhibited. Cytotoxicity will prevent cell division; to ensure optimal cell health, make sure that the appropriate cell density, d-siRNA concentration, and transfection reagent amount are used. Some d-siRNAs will silence essential genes, resulting in poor cell health. This is fine so long as controls, lipid alone and control d-siRNAs, do not affect cell health. Essential genes can be studied at early time-points posttransfection, before cells become unhealthy.

7. Gene silencing might fail because d-siRNAs do not actually target the gene of interest. Make sure that the in vitro transcription template encodes the gene of interest and not a closely related gene. Also, make sure that d-siRNAs have been made from the species in which silencing is being attempted. Finally, make sure that d-siRNAs target all splice variants.

8. It is possible that gene silencing only appears to fail. Detection could fail; for example, the primary antibody might not be specific for isoforms, detects closely related proteins, or an incorrect protein. Likewise, nonspecific binding of the secondary antibody would mask silencing. This problem is more likely to occur when using immunofluorescence; thus, make sure that all appropriate controls are included, such as primary and secondary antibodies alone. If the protein level remains unchanged after treatment with d-siRNAs, then it is worthwhile to measure the level of the targeted mRNA.

9. If silencing does not appear to be specific (i.e., the control d-siRNA reduces target protein abundance), first try decreasing the amount of d-siRNA used. If this does not help, then make sure the exogenous control is not homologous to the target gene or, for that matter, any gene in the genome of the species of interest, because protein stability might be adventitiously linked. Even single mismatches should abolish silencing; thus, the control d-siRNA can have some complementary fragments, but, overall, it should not contain long stretches of identical matches to any gene in the genome of the species of interest. A potential problem with using exogenous d-siRNAs as a control is if multiple transfections are to be performed. The exogenous control d-siRNAs will not be consumed inside the cell, and if multiple transfections are performed, the control d-siRNAs will accumulate and lead to nonspecific gene silencing.

3.9. High-Throughput Dicing

In vitro dicing was developed so that large numbers of genes could be silenced. This subsection describes the protocols necessary for silencing genes in a 96-well format. The high-throughput dicing protocol presented here is an extension of all the previously described methods in **Subheadings 3.1.–3.8.**, where each method has been scaled up to accommodate a 96-well format. Thus, each method is briefly described and focuses on the additional information required to produce a large collection of d-siRNAs. Furthermore, many of the materials are identical and will be cross-referenced to the relevant section, or if the reagent has not yet been listed, it will be described in **Subheading 2.9.** Multichannel pipets (**Subheading 2.9., items 1** and **2**) and related equipment (**Subheading 2.9., item 3**) are required.

3.9.1. Preparation of r-Dicer

r-Dicer is prepared exactly as described in **Subheading 3.3.**; however, a larger amount must be prepared. In order to make r-Dicer, you must complete the tasks outlined in **Subheadings 3.1.–3.3.**, but each process must be scaled up. See **Subheading 3.9.3.** to decide how much r-Dicer will be required.

1. Generally one bacmid preparation, 10–20 µg (*see* **Subheading 3.1.**) is sufficient to make 5–10 L of P3 viral stock, which is sufficient to make enough r-Dicer to dice about 25,000 genes. However, if more is required, the dicer bacmid can be amplified from the DH10Bac-dicer bacmid glycerol stock (*see* **step 6** in **Subheading 3.1.4.3.**). Using a pipet tip or inoculation loop, inoculate a 3-mL LB culture with a portion of the glycerol stock. LB should contain 50 µg/mL kanamycin (**Subheading 2.1., item 33**), 10 µg/mL tetracycline, 7 µg/mL gentamycin. Grow liquid cultures overnight to stationary phase at 37°C and 250 rpm and isolate bacmid DNA as described in **Subheading 3.1.4.2.**

2. In general, 1 µg of bacmid will provide 1 mL of P1 viral stock, then 10 mL of P2 viral stock, and then 500 mL of P3 viral stock, which is sufficient to prepare enough r-Dicer for silencing 2500 genes although the yields might vary by twofold to threefold. With these quantities in mind, first determine how much r-Dicer you will need. This depends on the number of genes to be silenced and the quantity of d-siRNA desired (*see* **Subheading 3.9.3.**). Next generate a large amount of P1, P2, and P3 viral stock as described in **Subheading 3.2.** To generate more P1 viral stock, simply transfect multiple wells in a six-well dish. Likewise, simply scale up viral amplification to generate more P2 and P3 viral stocks; just prepare enough P3 so that the empirical titer only has to be determined once (*see* **Subheading 3.3.1., Notes 24–28**). If P3 stocks must be prepared in batches, mix a small aliquot from each batch, 1–5 mL, and use the mixture to determine empirical titer.

3. Expression of r-Dicer is also easily scalable. It is reasonable to infect 20–30 cultures at once, 100-mL each in a 250-mL flask. Alternatively, even larger flasks can be used (*see* **Note 32**) although we have not tested expression in larger culture volumes. Flasks fabricated from glass can be used instead of the more expensive disposable flasks; however, care must be taken to ensure that the virus is completely removed. Glass flasks should be washed with mild detergent, rinsed very well with deionized H_2O, and then autoclaved twice for 30 min at $\geq 121°C$. Then, again, flasks could be used either only for infection of the Sf9 culture or only for growth of the culture such that small amounts of contaminating virus should not be a problem (*see* **Note 87**).

4. Batch purification is easily scalable, simply use a 50-mL conical tube instead of a 15-mL conical tube and scale up each component by a factor of 4 (*see* **Note 33** and **Subheading 3.3.3.**). It is reasonable to work with four purifications at once, each batch in a 50-mL tube. You will need 32, 100-mL cultures and appropriate amount of recombinant baculovirus, 16 mL of resin (**Subheading 2.3., item 24**), 16 times the volume of buffers (**Subheading 2.3., items 19–23, 25–27,** and **29**), and 8 dialysis cassettes (**Subheading 2.3., item 28**) but will produce enough r-Dicer to silence between 800 and 1600 genes. Be sure to assess r-Dicer activity before setting up 1600 dicing reactions (*see* **Subheading 3.3.3.**). The activity of the r-Dicer does need to be high because of reaction volume constraints (*see* **step 2** in **Subheading 3.9.3.**).

3.9.2. Selection and Production of In Vitro Transcription Templates

1. Target regions should be selected as described in **Subheading 3.4.1**. Ensure that target regions have minimal nucleotide identity to other genes. It is better to choose regions closer to the 3′-end of the gene, especially if templates will be made with RT-PCR (*see* **Note 88**). The 3′-UTR can be used as the target region.
2. In vitro transcription templates should be PCR products. Design primers as described in **Subheading 3.4.2**. Primer design software should be used to make the process more efficient.
3. Generate in vitro transcription templates via PCR in 96-well format (**Subheading 2.9., items 5** and **6**) as described in **Subheading 3.4.3.** using a PCR supermix (**Subheading 2.9., item 4**) and evaluate with agarose gel electrophoresis; a 96-well format is available. PCR reaction size can be decreased to 20 μL. The best approach is to amplify templates from a validated set of genes such as an expression library, but it is still a good idea to use two amplification steps. Using a second round of PCR, amplify the in vitro transcription template from a small portion (1–2 μL) of the first PCR reaction. This will allow simple regeneration of the in vitro transcription template, increase the yield of the template, and dilute the source template (vector DNA) so that the risk of contaminating transcription products is reduced. Be sure to remove the vector DNA after the first round of PCR (*see* **Note 49**) prior to the second round of PCR amplification. Nested primers can be used for the second round of amplification (*see* **Note 48**) if desired, but they are not necessary because PCR conditions (*see* **Subheading 3.4.3.**) will enrich for the correct template. There should not be a concern about generating mutations even though *Taq* polymerase is being used (*see* **Note 54**). PCR templates do not need to be purified prior to in vitro transcription. If a library is not available as a source template for PCR amplification, see **Note 88** for alternative approaches.

3.9.3. In Vitro Transcription and Dicing

About 3 μg of d-siRNA for each target can be generated very easily with high-throughput dicing. This is enough material for 500–1000 transfections in a 96-well plate (*see* **Subheading 3.9.5.**). On the average, 3 μg of d-siRNA can be produced from approx 15 μg of dsRNA and 30 μL of r-Dicer. Thus,

100 genes could be silenced with 3 mL of r-Dicer. If more d-siRNA is required, simply scale up.

1. In vitro transcription, using the MEGAscript kit as described in **Subheading 3.5.1.**, can be performed in 96-well format (**Subheading 2.9., item 5**). Use the maximum volume of template (*see* **Subheading 3.9.2.**) allowed per reaction size (e.g., 4 µL in a 10-µL transcription or 8 µL in a 20-µL transcription). It is not necessary to normalize the amount of template because transcription yields will be variant regardless of template mass. Ensure that the plate is well sealed to prevent evaporation during incubation. "Branched" regions should be eliminated as described in **step 4** in **Subheading 3.5.1.** (a thermal cycler works well for this; do not allow the lid to heat to >80°C) after removing the template with RNase-free DNase I. The MEGAscript kit can be purchased in bulk to reduce the cost. Also, to decrease the expense of in vitro transcription, the reactions can be scaled down by 50%. The yield of dsRNA will be at the least 15 µg and will provide approx 2–3 µg of d-siRNA, enough for hundreds of transfections in a 96-well format (*see* **Subheading 3.9.5.**).

2. The yields of dsRNA will vary by twofold to threefold (*see* **Note 89**). Concentration of the dsRNAs should be estimated by comparison to a dsRNA of known concentration after agarose gel electrophoresis; a 96-well format is available. Determine the concentration of the least concentrated sample, likely to be approx 1.3 µg/µL (15 µg in total from a 10-µL transcription reaction), and determine the relative concentrations of the more concentrated dsRNAs. The quality of the dsRNA can also be assessed after electrophoresis.

3. Set up the in vitro dicing reaction as described here and in **Subheading 3.5.2.** The reaction can be scaled up as needed. It is not necessary to adjust the concentrations of the dsRNAs such that equal amounts of dsRNA would be present in each well (*see* **Note 89**); it is better to dice the entire transcription reaction. Be sure that the ratio of dsRNA to r-Dicer is appropriate for the least abundant dsRNAs so that "overdicing" can be avoided (*see* **Notes 40** and **89**). For example, if the minimum mass of dsRNA used is 10 µg, then reduce the amount of r-Dicer to 20 µL, or if the r-Dicer is more active than normal, reduce the volume of r-Dicer accordingly.

Reaction conditions	Final concentrations
11 µL dsRNA (15–50 µg)	200–670 ng/µL dsRNA
3.75 µL 50 m*M* MgCl$_2$	2.5 m*M* MgCl$_2$
7.5 µL Rxn Dil. Bfr.[a,b]	250 m*M* NaCl
30 µL r-Dicer[b]	30 m*M* HEPES
22.75 µL H$_2$O[c]	0.05 m*M* EDTA
75.0 µL	64–80 ng/µL r-Dicer

[a]pH 8.0.

[b]Volume depends on the activity of r-Dicer; the volume of r-Dicer and Rxn. Dil. Bfr. must always equal 37.5 µL so that the final concentration of NaCl, HEPES, and EDTA is correct.

[c]Volume depends concentration of dsRNA.

A master mix of in vitro dicing components should be made and added to the dsRNAs already present in a 96-well plate; be sure to mix well by pipetting up and down. Also, make sure the foil plate seal is well sealed to prevent evaporation of samples. Incubate at 37°C for 16 h and proceed with purification (*see* **Subheading 3.9.4.**).

3.9.4. d-siRNA Purification in 96-Well Format

The purification scheme, like that in **Subheading 3.6.1.**, utilizes differential isopropanol concentrations to separate d-siRNAs from undigested large dsRNA, but instead of spin cartridges, the RNA is bound to 96-well filter binding plates. This procedure requires a centrifuge and rotor that can accommodate 96-well plates and 1400*g*.

1. It is not critical that the efficiency of the dicing reaction be analyzed; however, if desired, 2 µL of the reaction can be removed prior to purification. This aliquot can be stored at 4°C (short term) or –20°C (long term) prior to electrophoresis; 0.5 µL of 10X RNA loading buffer and 2.5 µL of RNase-free H₂O should be added as a master mix prior to storage.
2. For high-throughput dicing, the typical reaction size is 75 or 150 µL, depending on how much dsRNA is generated and on the quantity of d-siRNA desired (*see* **Subheading 3.9.3.**).
3. Working with two or four 96-well plates, purify d-siRNAs. First, transfer the dicing reaction to a 1-mL, 96-well plate to accommodate the increase in volume because of the addition of lysis buffer and isopropanol. To the dicing reaction, add an equal volume of lysis buffer that contains 1% β-mercaptoethanol. For example if the dicing reaction was 75 µL add 75 µL lysis buffer. Pipet up and down at least five times. Purification will not work well if the two solutions are not mixed to homogeneity. Make a large volume of lysis buffer containing 1% β-mercaptoethanol.
4. Now, add isopropanol to 40% (v/v) (e.g., add 100 µL of isopropanol to the 150 µL of lysis buffer and dicing reaction). Mix very well by pipetting up and down at least five times. Again, purification will not work well if the solutions are not mixed well; in fact, your d-siRNA will likely be contaminated with large dsRNA.
5. Apply the mixture to the first filter plate and centrifuge at 1400*g* for 1 min. The flowthrough is collected in a 1-mL, 96-well plate. **Save** the flowthrough; it contains the d-siRNA.
6. Discard the filter plate and increase the isopropanol concentration in the flowthrough to 70% (v/v) and mix well by piptetting up and down at least five times (e.g., add 250 µL isopropanol to the 250 µL of flowthrough). Solutions must be mixed to homogeneity to prevent loss of d-siRNA.
7. Transfer the sample to a second filter plate resting on a *New* 1-mL, 96-well plate and spin again at 1400*g* for 1 min. The d-siRNAs will now bind the filter plate.
8. Discard the flowthrough and add 500 µL of 1X wash buffer II. Spin again at 1400*g* for 1 min. Use the same 1-mL, 96-well plate as in **step 7** to collect the wash buffer.
9. Repeat **step 8**.

10. Discard flowthrough and centrifuge again at 1400*g* for 2 min to remove residual ethanol. The same 1-mL, 96-well plate can be used.
11. Place the filter plate with the bound d-siRNAs on a 96-well collection plate and add 100 μL of RNase-free H_2O. Let H_2O stand for 3 min and then centrifuge at 1400*g* for 1 min.
12. Repeat **step 11**.
13. The d-siRNA concentration should range from 15 to 45 ng/μL. The concentration can be estimated with several methods. First, use agarose gel electrophoresis and compare d-siRNAs to d-siRNAs of known concentration. A 4% agarose gel should be used and will allow the quality of the d-siRNAs to be evaluated as well; a 96-well format is available. Two microliters of the sample is sufficient for detection with ethidium bromide. Second, d-siRNA concentration can be estimated by measuring the absorbance at 260 nm. However, the concentration is at the lower limit of detection, so the sample cannot be diluted; therefore, use RNase-free disposable cuvets or disposable 96-well plates such that the sample can be returned to the stock plate or use a spectrophotometer that requires only 1–2 μL of sample. Absorbance at 280 nm can also be measured for assessing the purity of d-siRNAs. Generally, d-siRNAs are extremely pure. Third, use RiboGreen as directed by the manufacturer.
14. We find that it is not always necessary to normalize d-siRNA concentration, but if desired, use of an automated liquid handler prevents error. Aliquot d-siRNAs, into appropriate volumes, depending on downstream application. Material should not go through more than 5–10 freeze–thaw cycles and should be stable for greater than 1 yr at –80°C.
15. Note that the same protocol can be used if the dicing reaction is 150 μL; just double the amount of lysis buffer (**step 3**), isopropanol (**Subheading 2.6., item 4,** and **steps 4–6**), and H_2O (**steps 11** and **12**) used at each step; it is not necessary to double the volume of wash buffer (**steps 8** and **9**).

3.9.5. Transfection of d-siRNAs

The Genesilencer transfection reagent is used to transfect d-siRNAs into HeLa cells as described in **Subheading 3.7.3.** with a few modifications to accommodate the 96-well format. If you wish to use a different cell line or transfection reagent, see **Subheadings 3.7.3.** and **3.7.4.**

1. HeLa cells are plated in complete growth medium 12–18 h before transfection as described in **Subheading 3.7.2.** We use a multichannel pipet to plate cells at the densities suggested in **Table 5**. Depending on downstream application, cells can be plated in different 96-well plates; use **item 17** or **18** in **Subheading 2.9.** for fluorescent imaging or **item 10** in **Subheading 2.8.** for other applications. Many other plate formats, including 384- and 1536-well, are available, but we have not used these routinely.
2. In the 96-well format, HeLa cells are still transfected in the absence of serum. Three to four hours after transfection, serum levels are restored to 10% by adding an equal

volume of 2X medium (**Subheading 2.7., item 8**). However, the cells do not need to be washed with OPTI-MEM I prior to transfection.

3. Using an inverted light microscope, assess the quality of plating. The cells should be distributed evenly throughout the well and free of clumps. The cell density should be optimal for subsequent assays, but it is not a concern for transfection efficiency because Genesilencer (**Subheading 2.7., item 6**) provides efficient transfection at low and high cell densities.

4. Transfer the appropriate amount of d-siRNAs or mixture of d-siRNAs and DNA into a 96-well plate. This can be done in a nonsterile environment i.e., the bench-top because it is quick, the growth medium contains antibiotics, and risk of fungal or mycoplasm infection is low. Plan the experiment according to **Subheading 3.7.1.** using suggested amounts of d-siRNA and DNA (*see* **Table 4**). If all 96 d-siRNAs are at a variety of concentrations (*see* **steps 13** and **14** in **Subheading 3.9.4.**), use an average concentration, which falls within the effective range (*see* **Table 4**) such that the lowest concentration will be >1 nM and the highest concentration will be ≤10 nM.

5. Place the vial of siRNA diluent (**Subheading 2.7., item 6**) and Genesilencer siRNA transfection reagent (**Subheading 2.7., item 6**) in the hood. The reagents do not need to be kept at 4°C during transfection. Also, place an aliquot of OPTI-MEM in the hood, usually aliquoted in a 50-mL conical tube and can be at either 4°C or room temperature for transfection. Prepare a master mix of the diluent and OPTI-MEM I using volumes suggested in **Table 6**. Prepare the master mix for one more row and one more column than you will need (e.g., 9 rows and 13 columns equals 117 times the volumes). Invert tube several times to ensure complete mixing. Portion into each well of a 12-well strip; in each column, there will be enough mixture for all eight rows.

6. Move the 96-well plate containing d-siRNAs or a d-siRNA/DNA mixture to the hood and transfer 17.5 µL of the diluent/OPTI-MEM I mixture to each well and then pipet up and down five times to mix. Incubate for 5 min; it is easier and more time efficient to make the 5-min incubation continual (i.e., start the incubation after you transfer the diluent/OPTI-MEM I mixture to the first row).

7. During the remainder of the first 5-min incubation, prepare the Genesilencer/OPTI-MEM I master mix using the volumes suggested in **Table 6**. Although the manufacturer suggests using 1 µL of Genesilencer per will, we routinely use only 0.25 µL combined with 25.75 µL of OPTI-MEM I. Again, prepare the master mix for one more row and one more column than you will need (e.g., 9 rows and 13 columns equals 117 times the volumes). It is important to add the Genesilencer reagent to the OPTI-MEM I. Invert tube several times to ensure complete mixing. Portion into each well of a 12-well strip; in each column, there will be enough mixture for all eight rows.

8. Transfer 26 µL of the Genesilencer/OPTI-MEM I mixture to each well; it is not necessary to pipet up and down. Again, incubate for 5 min in a continual fashion (**step 6**).

9. At the end of the 5-min incubation, move the 96-well plate of cells to the hood. Working with one row at a time, aspirate complete growth medium and overlay

with the nucleic acid/diluent/Genesilencer/OPTI-MEM I transfection mix. Ensure that all of the complete growth medium is removed. Multiwell aspirators or a multi-channel pipet can be used to remove the medium; to avoid removing cells, do not make contact with the bottom of the plate. Also, make sure that the exchange of medium is quick so that cells do not dry out.

10. Add an equal volume of 2X medium (**Subheading 2.7., item 8**). 3.5–4 h after adding the transfection mix, generally approx 50 µL. Be sure to prewarm the 2X medium at 37°C before adding it to the cells.

11. Additional medium must be added to each well 12–18 h after transfection. Because the cells are adjusted to a 50:50 mixture of 2X medium (**Subheading 2.7., item 8**) and OPTI-MEM I, it is beneficial to add more of the 50:50 mixture and not just complete growth medium. Prepare enough of the 50:50 mixture for all of the wells and prewarm to 37°C prior to addition.

12. Transfection can be accomplished with other reagents (*see* **Note 81**). If you are using a cell line other than HeLa, transfection can be potentially be performed in the presence of serum (*see* **Note 82**). However, we have not tested other transfection reagents or cell lines in the 96-well format. Be sure to plate the appropriate density if a cell line other than HeLa is used (**step 1**).

4. Notes

1. Sodium phosphate-buffered solutions are made by mixing dibasic Na_2HPO_4 and monobasic NaH_2PO_4 to provide the correct molarity. The ratios of the monobasic and dibasic solutions can be altered to provide the correct pH as described in table B.11 on page B.21 in *Molecular Cloning: A Laboratory Manual (36)*. Note that the correct pH will not be attained unless the solution is at the appropriate volume; that is, if you are making a 1-L solution, do not check or adjust pH until the solution volume is set to 1-L. For example, to make 500 mL of α-extraction buffer at pH 8.0, mix 23.3 mL of 1 M Na_2HPO_4 (dibasic) with 1.7 mL of 1 M NaH_2PO_4 (monobasic), 30 mL of 5 M NaCl, 5 mL of NP-40, and approx 540 mL of H_2O. The pH should be very close to 8.0; if not, adjust accordingly with concentrated HCl or 10 N NaOH. The other sodium-phosphate-buffered solutions required for r-Dicer purification are made in a similar fashion.

2. Genomic DNA should be removed with the QIAshredder as described by the manufacturer. Elute RNA twice, each time with 50 µL of RNase-free H_2O. Jurkat cells may also be used. Alternatively, human total RNA, mRNA, or cDNA can be purchased from a commercial source.

3. These PCR conditions should produce enough of the correctly sized Dicer insert for subcloning; however, listed here are potential solutions to problems encountered. If the second PCR (**Subheading 3.1.2.3., step 6**), at 50 µL, does not produce enough insert for cloning, then multiple reactions could be performed and combined. To concentrate the DNA, combine the reactions and load onto a single purification column from the QIAquick PCR Purification Kit (*see* **step 6** in **Subheading 3.1.2.3.**). The binding of the Dicer insert might have to be done in steps if more than two 50-µL reactions are combined because the capacity of each

column is 750 μL. The binding capacity of each column is 10 μg, but this is not a concern because it is unlikely that four PCR reactions would yield more than 10 μg. Do not simply reamplify a third time because this increases the odds of introducing a deleterious mutation.

Do not exceed 2 μL of cDNA per 50-μL volume of PCR. This will not increase the yield and actually might decrease the yield. One unit of polymerase might be used, but the use of 2.5 units is recommended for targets greater than 3 kb. Further optimization might require increasing the [MgSO$_4$] to 2 mM, increasing the [dNTP] to 0.6 mM, or titrating in the enhancer solution.

If many bands are present, the specificity needs improvement. To increase the specificity, the annealing temperature could be increased from 55°C in increments of 2.5°C. The optimal annealing temperature is 5–10°C below the T_m of the primers. Thus, if the specificity is poor, a two-step cycle can be performed with these primers: Denature at 94°C for 15 s and then extend at 68°C for 6 min. Two-step cycling is good for primers with high a T_m.

Other high-fidelity polymerases can be used; we suggest Platinum *Pfx* because we are familiar with it and the error rate is extremely low. Lower-fidelity polymerases such as Platinum Taq High Fidelity (Invitrogen), *Pfu* (Stratagene), or Vent (New England Biolabs) could be used, but keep in mind that these enzymes are not as accurate as *Pfx*.

4. We merely suggest an efficient method to digest the DNA. Restriction digests reactions can be modified as you wish (e.g., volume, time, and amount of enzyme can be changed as desired). Also, equivalent enzymes can be purchased from other companies (e.g., Invitrogen).

5. There are important steps in the gel extraction protocol that make a difference in the recovery of the DNA from the agarose and in the efficiency of the subsequent ligation: (1) add 1 gel volume of isopropanol to the agarose piece dissolved in QG buffer because fragments are greater than 4 kb and this will improve the yield; (2) the optional wash with 500 μL of the QG buffer greatly improves ligation efficiency; (3) the optional 2- to 5-min incubation with the PE buffer also greatly improves the ligation efficiency (**Subheading 3.1.2.5.**).

6. Calculate the moles of vector with the following:

Moles of vector = (moles)(bp)/660 g;
so for 4856 bp, 50 ng is 16 fmol.

Calculate the moles of insert with the following:

Moles of insert = (moles)(bp)/660 g;
so for 5775 bp, 60 ng is 16 fmol.

The T4 DNA ligase should be the last component added to the tube. Ligation conditions can be modified so that they are familiar to you; for example, volume, time, incubation temperature, and amount of enzyme can be changed as desired. This ligation might be difficult because the insert is larger than the vector.

If ligation is not successful, (i.e., no colonies are obtained after transformation [*see* **Subheading 3.1.2.6.**], positive controls should be included: (1) add a control to ensure that the ligase is working (e.g., use a digested insert and vector known to ligate); (2) add a control to ensure that the transformation was successful (e.g., transform in 1–5 ng of supercoiled plasmid); (3) test the *Not*I and *Xho*I restriction enzymes for activity by cutting the pFastbac-HTC with each enzyme in separate tubes; analyze by looking for a linearized vector on a 0.8% agarose gel; (4) ensure the insert is being digested by the *Not*I and *Xho*I enzymes in separate reactions; test ligation of the insert to itself in a typical ligation reaction; analyze by looking for a smear on a 0.8% agarose gel.

Ligation can be optimized by combining 0.5-fold to 10-fold molar excess of insert with 50 ng of digested vector. Using multiple molar ratios is helpful in case the estimation of DNA concentration is incorrect. Further optimization, such as increasing the amount of vector or ligation at 16°C for 4–16 h, might be required to produce a construct. If you prefer, the equivalent enzyme could be purchased from other companies (e.g., Invitrogen).

7. Ideally, there will be very few colonies on the plates from the vector-only control and a larger number of colonies on the plates from the vector plus insert transformations. The vector should have been digested to near completion with *Not*I and *Xho*I and these ends are not compatible for relegation; thus, there should be very few colonies.

If there are more colonies on the vector-only control plate than on vector plus insert plates, it does not rule out the possibility of having created the construct; two things can be tried. First, look for the concentration of insert, which resulted in greater than twofold fewer colonies relative to the vector-only control. If the insert was digested well with both enzymes and is in sufficient excess over the vector, then vector religation will be inhibited and the colonies seen could be the insert ligated into the vector. Fewer colonies on vector plus insert plates than on vector-only control plates could be seen at all concentrations of insert, but it is likely to be the case for the ligations containing threefold molar excess of insert or higher. Second, the 5′-monophosphates could be removed from the vector with phosphatase enzymatic activity to prevent religation. A good phosphatase is Antarctic Phosphatase (New England Biolabs; M0289S) because it is very active and is easily inactivated by incubating at 65°C for 5 min. The vector can be dephosphorylated directly following restriction digest without exchanging buffer as long as the reaction is supplemented with 1X Antarctic Phosphatase reaction buffer. The reaction is set up and performed as directed by the manufacturer. Incubate at 37°C for 15 min and inactivate by heating at 65°C for 5 min. Proceed with ligation.

If there are absolutely no colonies on any of these plates; make sure the plates contain only ampicillin and not another antibiotic, decrease the incubation time or units of restriction enzymes, or see troubleshooting information provided in **Note 6** or in **ref.** *36*.

8. The number of colonies to select depends on how well you think the ligation worked. If there were only a few colonies on the vector-alone control plate and

two to three times more on the vector plus insert plates, four colonies should be sufficient to find a positive subclone. It is a good idea to have more than one sub-clone in case deleterious mutations are present in the Dicer insert. More colonies (e.g., 10) can be selected if you think the odds of selecting a colony that contains the subclone are much lower.

If the odds of finding a subclone containing the Dicer insert are very low, 20–30 colonies might need to be screened. This is a large number of colonies to screen by DNA isolation and restriction digest analysis. It is easier to screen by PCR. Select each colony with a pipet tip and resuspend in 5 µL of LB. One microliter will be used in the PCR reaction and the remainder will be used to start a culture.

If the PCR and agarose gel analysis are to be completed in the same day, the colonies resuspended in the LB can be stored at 4°C. If this is not the case, the bacteria should be streaked on a LB plate containing 100 µg/mL ampicillin; six to eight colonies can be streaked on one 10-cm LB plate after dividing into sections delineated with a Sharpie.

The PCR can be set up with the following primers:

Forward: 5′-GGTCCACGAGTCACAATCAAC (T_m = 55.9°C; starts at nucleotide 2102 in NM_177438.1)

Reverse: 5′-CTCTTCTTCATCATGCAAATCAAGCTC (T_m = 56.4°C; ends at nucleotide 2437 in NM_177438.1)

Because most of the volumes are rather small, all of the PCR compo-nents can be combined in a master mix and added to a PCR tube containing 1 µL of the colony resuspended in LB; overlay with mineral oil if necessary. 2 µL of 10X High Fidelity reaction buffer; 0.4 µL of 10 m*M* dNTPs; 0.8 µL of 50 m*M* MgSO$_4$; 0.4 µL of 10 µ*M* forward primer; 0.4 µL of 10 µ*M* reverse primer; 1 µL of colony resuspended in LB; 0.1 µL Platinum Taq High Fidelity; H$_2$O to 20 µL.

Heat the mixture at 94°C for 5 min then cycle 30 times:

Denature: 94°C for 30 s;
Anneal: 55°C for 30 s;
Extend: 68°C for 30 s (1 minute per kilobase).

Load 10–20 µL of the reaction in a 1% agarose gel containing 0.5 µg/mL ethid-ium bromide. Untransformed bacteria can be used as a negative control and the previously made Dicer insert can be used as a positive control.

A positive clone will produce a band at 335 bp. Identify the positive clones and start a culture from either the 4 µL of LB or from a colony on the LB plate depending on the timeliness with which the procedure was completed. It is a good idea to pick two to four positive clones in case the sequence of one

particular clone is incorrect. Proceed with DNA isolation as described in
Subheading 3.1.2.7.

The primers described in **Subheading 3.1.2.3.** can also be used to screen for
positive clones, but the band produced will be very large. If they are used, the
extension phase, at 68°C, must be extended to 6 min.

9. Producing baculovirus requires a decent amount of work and time; therefore, it is
   very important to know that the Dicer insert is in frame with the 6xHis tag and
   that the Dicer coding region is free of deleterious mutations. This is important to
   know when troubleshooting protein production; that is, the lack of Dicer production can be attributed to factors other than errors within the subclone. If the
   selected clone contains numerous mutations, select another clone for sequencing.
   If the second clone also has numerous mutations, the subcloning process
   (**Subheadings 3.1.2.3.–3.1.2.7.**) must be repeated. Ensure that the PCR reaction is
   performed as described in **Subheading 3.1.2.3.** and per manufacturers' suggestions; excess dNTPs, a large number of cycles >40, and reamplifications can
   introduce mutations.

   If by some chance the clone contains only one or two deleterious point mutations, then they can be repaired using the QuikChange II XL Site-Directed
   Mutagenesis Kit as directed by the manufacturer (Stratagene, La Jolla, CA; 200521).
   Be sure to sequence the entire gene again to confirm that additional mutations
   were not introduced during the intended mutagenesis.

10. The use of one M13 primer, either forward or reverse, and a gene-specific primer
    close to either M13 primer site can reduce the length of the amplified region. Just
    be sure to choose the correct primer orientation. The M13 forward primer is at the
    5' end of the Dicer insert and the M13 reverse primer is near the 3' end. See the
    Bac-to-Bac instruction manual in order to calculate the length of the expected
    band with the gene-specific primer.

11. Starting in **Subheading 3.1.4.1.** a blue colony can be selected and carried through
    the process described in **Subheadings 3.1.4.1.–3.1.4.3.** as a control. This will help
    to confirm that what you see on the gel after the PCR is correct. It is possible that
    with some of the bacmids you selected, two bands will be present after the PCR:
    one at the correct size and one at 300 bp. This could be because the selected
    colony contained both a bacmid that contains the Dicer insert and one that does
    not. It is also possible that the PCR did not work well and smaller bands will be
    amplified. Thus, it is nice have the blue colony as a control to determine whether
    or not the contaminating small bands are from bacmids lacking the dicer insert or
    nonspecific amplification. However, in selecting a bacmid to use in preparation of
    baculovirus, it is preferred to use the bacmid, which produces only a single band
    at 8206 bp, and it must be confirmed that contaminating bands are the result of
    nonspecific amplification and not the result of contaminating bacmid that is lack-
    ing an insert.

12. If desired, cells can be cultured with antibiotics, 10 μg/mL gentamycin or
    100–200 U/mL or 100 μg/mL penicillin–streptomycin. Also, fungicidal agents
    can be used—amphotericin B at 0.25 μg/mL, but this agent will really slow cell

growth. It is very easy to maintain a sterile culture without antibiotics or fungicides and the cells double much faster if antibiotics are not used. Therefore, we prefer to avoid the use of antibiotics.

13. There are some problems with growing cells outside of an incubator or in a large room, the greatest being fluctuation in temperature and the second greatest being inability to set the room temperature near 28°C. Fluctuations in temperature make it very difficult to maintain an Sf9 culture. When the temperature is higher, the cells will divide more frequently, and when the temperature is lower, the rate of cell division decreases. Differences in rates of cell division make it much more difficult to maintain the culture at a high cell viability. Sf9 cells will grow at lower temperatures, a minimum of 26.5°C, but cells do not divide as frequently. The bottom line is that it is much more difficult to grow cells on a benchtop shaker.

14. The best way to clean an incubator is to autoclave any removable parts. Before heating, it is good to wipe all parts with 70% ethanol. For the parts that cannot be removed, clean well with a diluted Lysol solution. Concentrated Lysol solutions can be diluted 1:50. The walls and other interior can be cleaned in this manner. Avoid cleaning the interior of the incubator with ethanol because the vapors are toxic to cells.

15. Sf9 cells will also grow very well in any treated, sterile tissue culture vessel. The growth conditions are the same: $28 + 0.5$°C, no CO_2 or humidity required. It is difficult, however, to maintain cell viability in adherent cultures because cells must be sloughed off the plate in order to passage. This mechanical force decreases cell viability. In addition, more time and expense are required to grow adherent cultures because cell density per milliliter of medium is limiting. We plate the Sf9 cells only when necessary (e.g., for transfection, small-scale amplifications of virus, and empirical measurement of the viral titer required for protein production). Adherent cultures should be a last resort.

16. Sf9 cells are very sensitive to lack of glutamine; thus, care must be taken to preserve glutamine in the Sf900 II SFM. The entire bottle of media should never be prewarmed at 28°C. The media should be protected from long exposure to light (e.g., the bottle of media should be wrapped in aluminum foil). Also, entry of light into the incubator should be blocked; aluminum foil works well for this.

17. To calculate the cell density and viability, use the following:

$$\text{Density (viable cells/mL)} = \frac{(2)(1 \times 10^4)(\text{No. of cells in } n \text{ quadrants})}{n \text{ quadrants}}$$

where $n = 2$–4;

$$\% \text{ Viability} = \frac{(100)(\text{No. of viable cells})}{\text{Total number of cells}}$$

18. Cell health after beginning a new culture from a frozen stock varies greatly. There is a 50% chance that the cells will immediately grow as if they have been in

culture for quite some time and reach mid-log phase growth at a density of $(2–3) \times 10^6$ viable cells per milliliter and a viability of >90% after only 3 d. This is usually the case for the frozen stocks we have prepared. However, commercial cell stocks are not as reliable. The other 50% of the time, the cells will struggle to divide normally, reaching a density of only $(6–7) \times 10^5$ viable cells per milliliter and viability of 60–70% after 3 d. It really depends on the quality of the cells when the frozen stocks were made and on the accuracy with which the cells are thawed. Nonetheless, even if the cells do not grow well immediately, after three to five passages the cell viability and growth rate should improve. Expect viability to range from 80% to 90% and a density of $(1–2) \times 10^6$ viable cells per milliliter after 3–4 d in culture, and after 5–6 d cells should reach mid-log phase growth with a viability >90%. If this is not the case, try to rescue the cells (*see* **Note 20**) or thaw out another frozen stock. If the cells were purchased from Invitrogen, contact technical support for advice and potentially a replacement.

19. Because the disposable flasks are expensive, it is tempting to reuse the flasks. This is not a good idea because the viability and growth rate often decrease. However, glass flasks can be reused. Using glass requires more work because the flasks must be cleaned very well and then autoclaved. We do not use glass flasks because we are concerned that after infections with baculovirus, residual virus might be impossible to remove. If you wish to use glass flasks, create two subsets: one for cell growth and one for infection.

20. It is common for freshly initiated cultures to grow slowly with low viability. It is difficult to figure out what to do with a culture to improve the growth rate and viability. The key is to find a balance between improving the viability and growth rate. The growth rate will remain low if cells are split before they reach mid-log phase growth. However, to ensure that viability is increasing, cells must have fresh media. Thus, the dilemma is deciding when to split cells. With a freshly initiated culture, we first try to improve the viability in the first two or three passages by splitting the cells every 3 d. The viability will improve slowly. Generally, after the third passage, we try to improve the growth rate; thus the cells are not split until the density reaches mid-log phase, $(2–3) \times 10^6$ cells per milliliter. Attaining optimal growth is empirical, but the following suggestions might help. If the growth rate and viability have not improved in time for the third passages, there are two options. Note that the strategy will change after the third passage and an improvement in viability.

   First, if the density is approx 4×10^5 viable cells per milliliter, the cells must be spun down and resuspended in fresh media. After counting the cells, portion the culture into 50-mL conical tubes. Spin at $500g$ at room temperature for 5 min. Return to the hood and remove the media with a sterile pipet. Do not aspirate. Recap the tube and gently tap the bottom to disperse the cells along the walls of the tube. Dispense the fresh Sf900 II SFM down the side of the tube. Use the same volume that the culture was growing in, ensuring that the density is at least 3×10^5 viable cells per milliliter. It is alright if the density is as high as 5×10^5 viable cells per milliliter. Gently invert the tube several times to resuspend the cells, taking care to break up all cell clumps. It is alright to gently tap the tube. Dispense the culture into the appropriate flask (*see* **Table 1**).

Second, if the density is approx 7×10^5 viable cells per milliliter, the cells can be split to $(3–5) \times 10^5$ viable cells per milliliter and grown for 3 more days. At this point, the cells should be able to reach a density of $(2–3) \times 10^6$ viable cells per milliliter in 3–6 d with a viability of >95% when seeded at 3×10^5 viable cells per milliliter. If this is not the case, the cells must now be spun down and resuspended in fresh media.

21. If the viability of a well-growing culture or a freshly initiated culture drops below 95%, the culture can be rescued. First, try to revive the culture by watching the cell density very closely. As soon as the density is nearing $(1.5–2) \times 10^6$ viable cells per milliliter, split the cells to 5×10^5 viable cells per milliliter. This might require passaging the cells as often as every 2–3 d. Frequent cell splitting will help to improve the cell health. If the viability is still low, < 95%, the cells should be spun down and resuspended in fresh media as described in **Subheading 3.2.2., step 3**. Remember that cells should be spun down and resuspended in fresh media every 3 wk to maintain viability >95%. If viability does not improve after frequent cell splitting or resuspending in fresh media, a new culture should be started. Other variables, like temperature and shaking speed, can be tested and might improve cell health, but a new culture should be started and tested before changing anything because testing many variables is time-consuming and cultures will have to be maintained for several weeks at a low viability. See the instruction manuals provided by Invitrogen for additional help.

22. For long-term storage, a portion of the P1 virus can be placed at –80°C; however, the titer will decrease once the P1 stock is frozen. There is really no benefit to freezing an aliquot of the P1 stock at –80°C. One preparation of the dicer bacmid is sufficient for many transfections and more bacmid is readily accessible from a glycerol stock. Likewise, it is very easy to transfect Sf9 cells. Thus, we generally use most of the P1 virus to make a P2 stock. It is much easier to use most of the P1 and P2 stocks to make a P3 stock because the empirical titer of the P3 stock must be determined and it is more efficient to do this just once from a large volume of P3 stock rather than repeat the amplification process several times (*see* **Note 24**).

23. If you decide to make more than 10 mL of P2 viral stock, simply plate the appropriate number of 10-cm dishes. If several milliliters of P1 viral stock are available, it could be amplified in suspension culture, but we have found that amplification is more efficient in adherent cultures. As noted in **Note 24,** it is much easier to make as much P3 virus as possible; thus, if you need very large quantities of r-Dicer, you will need more than 10 mL of P2 viral stock.

24. It is a good idea to make as much P3 viral stock as possible (usually 500 mL) because the empirical titer must be determined (*see* **Subheading 3.3.1.**), and because this is a tedious process, it is better to do only once. In theory, P3 viral stocks could be made in batches, and if identical amplification conditions were used each time, the resulting titer of each batch would be similar. However, this is not the case; thus, the titer would have to be determined for each batch. It is

worthwhile to make as much P3 viral stock as possible for the following reasons. First, it is inexpensive to generate P2 and P3 viral stocks. Second, there is not a concern for the stability of the P3 viral stock because it is stable at 4°C for at least 1 yr so long as FBS is added to 2% (v/v). Also, after r-Dicer is expressed, the Sf9 cell pellets can be stored at –80°C for up to 1 yr; thus, the virus can be virtually stored as cell pellets. Finally, for high-throughput dicing, you will need a large amount of P3 viral stock (*see* **Subheading 3.9.**).

25. As a reminder, it is a good idea to have a large volume of P3 viral stock so that the empirical titer will only have to be determined once. You should already by familiar with viral titer and MOI from the description of viral amplification (*see* **Subheading 3.2.**). The viral titer is the number of virions per milliliter or plaque-forming units per milliliter. MOI is the number of virions used to infect each cell; for example, an MOI of 10 means that the virion number exceeds the cell number by an order of magnitude or there are enough virions such that each cell can be infected by 10 virions. Unlike viral amplification, protein expression is accomplished at a much higher MOI. The volume of virus to use for protein production is based on MOI and is calculated with the following formula so long as the titer is known or can be estimated:

mL of virus = (MOI)(Total cell number)/Titer of P3 viral stock

Although the titer of our P3 stocks is generally $(0.5–1) \times 10^8$ PFU/mL, we have found that establishing an empirical MOI is more efficient than calculating the titer.

26. Different volumes of virus can be used just make sure that the final volume of medium and virus is 600 µL. It is not necessary to use more than 600 µL of the P3 viral stock. In fact, 600 µL is even impractical because this translates to infecting a 100-mL culture with 200 mL of virus, an impractical feat. Six hundred microliters is suggested just to ensure that r-Dicer can be produced because if only lower amounts of virus were used and r-Dicer was not detectable, it would be impossible to determine if there was a problem with the clone or if there was a problem with viral production. Thus, the worst result of this experiment is that the highest yield of r-Dicer comes from 600 µL of P3 viral stock, meaning that the P1, P2, and P3 viral stocks will have to prepared again.

27. A short set of instructions for SDS-PAGE are provided here as adapted from **ref. 36**, but you can perform electrophoresis per your normal protocol. We generally detect r-Dicer after electrophoresis in a 6%, 100:1 (acrylamide:*bis*-acrylamide) gel because the resolution is better than that seen in a 6%, 29:1 (acrylamide:*bis*-acrylamide) SDS gel. However, if you prefer to run a 6%, 29:1 or a 4–12% gradient gel, it will work. Higher-percentage gels will not provide sufficient resolution to detect the full-length r-Dicer at 225 kDa. We generally use 16×16 cm gels, but if smaller gels with smaller wells are to be used, lyse the cells in 50 µL and load 20 µL. The lysate will be very viscous and this amount of protein is not optimal for the resolution provided by small gels, but it will work.

To prepare a 6%, 100:1 gel first make a 30%, 100:1 acrylamide:*bis*-acrylamide stock by mixing 371.4 mL of 40% acrylamide with 74.3 mL of 2%

bis-acrylamide and 54.3 mL of deionized H_2O; filter to remove particulates. Pour the 6%, 100:1 portion of the gel by mixing the following:

Resolving gel: 30 mL
6.5 mL H_2O;
6 mL of 30% 100:1 acrylamide:bis-acrylamide;
7.5 mL of 4X resolving buffer;
100 μL of 10% APS;
50 μL TEMED.

Overlay the gel solution with H_2O-saturated butanol or just H_2O to increase the rate of polymerization and produce a clean interface for the stacking gel. After the gel polymerizes, remove the H_2O-saturated butanol and wash well with deionized H_2O. If H_2O was used to overlay the gel, rinsing is not necessary. Pour the stacking portion of the gel and insert the comb.

Stacking gel: 10 mL
6.3 mL H_2O;
1.13 mL of 40% 29:1 acrylamide:*bis*-acrylamide;
2.5 mL of 4X stacking buffer;
30 μL of 10% APS;
30 μL TEMED.

After the gel polymerizes, remove the comb and wash the wells with 1X running buffer. Finally, load the sample and run the gel. Note that volumes required for resolving and stacking gels might need to be adjusted to accommodate the gel size.

28. Western blotting can be done many different ways. We use the protocol suggested by Millipore known as the "Rapid immunodetection method on Immobilon-P using chemiluminescence" (technical note TN051; http://www.millipore.com/publications.nsf/docs/TN051). This approach is slightly different than a general Western protocol. The membrane is dry when it is probed with the primary and secondary antibodies. The hydrophobic nature of the dry membrane prevents the antibodies from binding nonspecifically to the membrane; thus, the membrane does not have to be blocked with nonfat milk powder or BSA. The transferred proteins protrude from the surface of the membrane, improving interaction with the primary antibody.

This protocol is strongly recommended for several reasons. It is the most time-efficient method because time is not wasted blocking the membrane and incubation times with primary and secondary antibodies are reduced. It consistently provides a robust signal with very little background. Finally, very little primary or secondary antibody is used. If other blotting protocols are used, more T7 antibody might be required.

Ensure that the membrane is completely dry using one of the methods suggested by Millipore (Immobilon-P product insert); we either air-dry the membrane overnight on 3 mm Whatman paper or submerge in 100% methanol for 10 s and then air-dry for 15 min. Methods are equally effective but are used depending on the timeliness with which blotting is performed. If the membrane is not completely dry, the background will be very high.

When the membrane is dry, dilute the T7-HRP antibody 1:10,000 in antibody diluent and incubate with the membrane for 1.5 h at room temperature. Discard antibody, then rinse twice, and wash three times for 5 min each with D-PBS; do not add Tween 20 to the D-PBS used for washing. To produce the chemiluminescent signal, remove the membrane from the D-PBS, blot away excess D-PBS by holding the membrane upright and touching the bottom of it to a paper towel, and place it on a piece of Saran Wrap. Ensure that the Saran Wrap is flat against the benchtop and free of wrinkles and that all regions of the membrane are in complete contact with the Saran Wrap, free of air bubbles. Mix the Luminol/enhancer and peroxide buffer in a 1:1 ratio, 10 mL in total or 5 mL of each is sufficient for a 16 × 16 cm blot, and then in a dropwise fashion, overlay the membrane and incubate for 1.5–2 min. Note that incubations longer than 2 min might inhibit the HRP conjugate, resulting in absence of signal within the band, thus resulting in inaccurate quantification of the protein abundance.

Using forceps, lift the membrane up to drain off the substrate and blot away excess substrate by holding the membrane upright and touching the bottom of it to a paper towel. The membrane is placed inside of a heavy-duty sheet protector and air bubbles are removed. Use X-ray film or a detection device such as a GelDoc (Bio-Rad, Hercules, CA) to visualize proteins. Ensure that luminescence is in the linear range such that levels of r-Dicer production can be compared. If intensity is out of the linear range, remove the substrate by washing the membrane in D-PBS; then, as described earlier, reapply the substrate for only 30–60 s and/or wait until luminescence intensity is in the linear range.

r-Dicer can also be detected with the His monoclonal antibody (Clontech, Palo Alto, CA; 8916-1) diluted to 1:10,000 in antibody diluent. This antibody is not directly conjugated to HRP; thus, an HRP-conjugated secondary antibody is required. This antibody will bind to endogenous His epitopes, making detection of the full-length r-Dicer more difficult.

29. If you cannot detect protein at any MOI, make sure the T7 antibody is functional and that Western blotting conditions are optimal. A T7-tagged control protein is provided with the antibody so that the antibody and blotting conditions can be evaluated. However, the control protein is only 31 kDa; thus, in order to detect both proteins, samples must be electrophoresed in a 4–20% gradient gel. Alter blotting conditions such that the control protein can be detected and verify that the antibody is functional.

r-Dicer might be undetectable because of low yield or because of an error within the clone (e.g., a deleterious mutation) or with viral production and infection (*see* **Note 30**). First, determine if the yield is just low by repeating the titer experiment (*see* **steps 1–5** in **Subheading 3.3.1.**), but lyse the cells in a smaller volume (100 µL), and loading 80% of the lysate. Alternatively, the empirical titer can be evaluated by infecting cells in a six-well plates; simply scale up cell number and volume of virus by fivefold, such that after infection, each well will contain 3 mL of Sf900II SFM and virus. Lyse cells in 200–400 µL of 1X sample buffer and proceed as previously described.

The length of expression after expression affects the yield of r-Dicer. Initially, we found that r-Dicer expression peaked at 46 h postinfection; however, later we discovered that 36–38 h was even better. Thus, we routinely collect cells at 36–38 h postinfection. We have also found a correlation between the MOI and length of expression. Thus, if the yield of r-Dicer is suboptimal, it will be worthwhile to analyze expression at various MOIs and time-points postinfection; this analysis is best performed by infecting cells grown in a suspension culture (*see* **step 10** in **Subheading 3.3.1.**).

30. The r-Dicer might not be expressed because of an error in the subclone. Make sure the start codon of the Dicer coding region is in frame with the N-terminal 6xHis tag. Also, make sure that the Dicer coding region is free of deleterious point mutants or frame shifts. Finally, make sure that viral production was done correctly (*see* **Subheading 3.2.**).

31. Although we attain the highest yields of r-Dicer 36–38 h postinfection, other times could result in even greater expression. Thus, it is worthwhile to analyze expression at various MOIs and time-points postinfection. Because only a portion of the 50-mL culture will be used, the remainder can be saved for a small-scale purification; simply scale the purification appropriately (*see* **Subheading 3.3.3.**).

32. To make more r-Dicer, simply infect more cells. Either infect multiple 100-mL cultures, 20–30 is reasonable, or use larger flasks: 500 mL (Corning; distributed by Fisher Scientific, Suwanee, GA; 431145) or equivalent; 1 L (Corning, distributed by Fisher Scientific, Suwanee, GA; 431147) or equivalent; 2 L (Corning, distributed by Fisher Scientific, Suwanee, GA; nonbaffled 431255 or baffled 431256) or equivalent; 3 L (Corning; distributed by Fisher Scientific, Suwanee, GA; nonbaffled 431252 or baffled 431253) or equivalent. Do not exceed suggested culture volumes (*see* **Table 1**).

33. It is very simple to scale up the purification. Four pellets, each containing 4×10^8 cells, can be lysed and combined for purification. Use 9 mL of ice-cold α-extraction buffer, containing protease inhibitors, to extract protein from each of the four pellets. In order to sonicate efficiently, split sample into two 50-mL conical tubes or 50-mL Corex tubes. Likewise, the sample should be divided into equal portions for centrifugation in two 50-mL conical tubes. Proceed with the batch purification as described in **Subheading 3.3.3.**, but use a 50-mL conical tube instead of a 15-mL conical tube, scaling up each component by a factor of 4. For example, use four times the amount of resin—4 mL of bed volume. Wash with approximaterly four times the volume of β-wash buffer and δ-wash buffer; usually 12 mL of wash buffer is used for 1 mL of resin bed volume, but 48 mL of wash buffer will exceed the volume of the 50-mL conical tube, thus just use about 44 mL of each wash buffer. Elute with four times the volume of γ-elution buffer— 4 mL. Dialyze the 24 mL of the eluted protein in two dialysis cassettes, 12 mL each. It is reasonable to handle four, 50-mL purifications at once.

34. The resin bed volume has a capacity of approx 3 mg/mL. The resin is about 60% of the volume of the resin slurry; thus, 1.7 mL are used to obtain a 1-mL bed volume. For this particular bed volume of resin, 15-mL Falcon tubes are ideal. The volume is large enough to accommodate the following washes and only about 1% of the resin is lost over the numerous washing steps. In addition, use of a swinging buffer rotor and a 15-mL Falcon tube allows tighter packing of the resin. This prevents unnecessary loss of resin and bound protein. A portion of the r-Dicer will not bind; however, increasing the amount of resin does not improve binding and only dilutes the final amount of protein. Extended binding times do not improve the yield of r-Dicer, but merely increase the amount of contaminating proteins. Therefore, do not allow binding to proceed for longer than 1 h. In most cases, a 500-μL bed volume of resin does not have enough binding capacity for 4×10^8 cells (40 plates); thus, a 1-mL bed volume of resin is used.

35. It is better to lose some the r-Dicer than it is to carry over resin; thus, leave some of the elution fraction behind. Before dialyzing the elution fractions, it is essential that the r-Dicer be free of contaminating resin. The centrifugation suggested in **step 10** ensures complete removal of contaminating resin and is, therefore, not optional.

36. The order in which the timed dialyses are performed is irrelevant. Successive dialysis can be performed for 2 h, then overnight, and then for 2 h or overnight, then for 2 h, and then another 2 h.

37. The substrate dsRNA can be internally radiolabeled by adding 1 μL of α-P³²-UTP (Amersham, PB20383, or equivalent SP6/T7 grade UTP) to the transcription reaction (*see* **Subheading 3.5.1.** and **Note 55**). The amount of cold UTP does not need to be altered. Quantifying radiolabel is more accurate than ethidium bromide staining, but for most purposes, such as establishing an amount of r-Dicer to use for generating d-siRNAs, staining the gel with ethidium bromide is sufficient. One microgram of dsRNA might seem excessive, but it is important to mimic the in vitro dicing reaction so that the activity of r-Dicer can be determined under the conditions in which it will be used to prepare d-siRNAs for gene silencing.

38. Pour a native polyacrylamide gel by mixing components as described below. Denaturing PAGE also works, but you will not be able to determine if the d-siRNAs are double stranded after dicing. If a native gel cannot be used, pour a denaturing gel as described below. Dissolve urea in H_2O and TBE with heat (dissolving urea is endothermic) and then add the remaining components. Be sure to prerun gel and rinse the urea from the gels before loading the sample.

15% Native TBE gel: 50 mL
26.25 mL deionized H_2O;
18.75 mL 40%, 29 : 1 acrylamide;
5 mL of 10X TBE, pH 8.4±.1;
500 μL of 10% APS;
50 μL TEMED.

15% Denaturing TBE gel: 50 mL
10.5 mL deionized H_2O;
5 mL of 10X TBE, pH 8.4 ±.1;
21 g Urea
18.75 mL of 40%, 29 : 1 acrylamide;
500 µL of 10% APS;
50 µL TEMED.

39. RNA does not need to be purified away from proteins and reaction components prior to electrophoresis because the contaminants do not affect the mobility of the dsRNA or d-siRNAs. The reaction does not need to be quenched with EDTA because it is a component of the RNA loading buffer. Although Dicer will still bind the dsRNA in the absence of Mg^{2+}, loading buffer should contain EDTA to prevent binding of contaminating proteins to the RNA. Only one-half of the reaction needs to be loaded on the gel because 500 ng of RNA is easy to detect with ethidium bromide; if something happens to the gel, half of the reaction will be left for analysis.

40. We have found in vitro dicing to be more efficient if only 70–80% of the large dsRNA is diced. First, this increases the yield of d-siRNA. It seems counterintuitive that the yield is greater if dicing is not complete, but for some unknown reason, if r-Dicer consumes all of the substrate dsRNA, then either r-Dicer or a contaminating protein will consume the d-siRNA. This could be the results of unwinding and autocatalysis at 37°C. Second, if an enzyme is modifying or unwinding the d-siRNA once the substrate dsRNA is consumed, complete dicing might result in unfavorable selection of d-siRNAs, resulting in a less potent pool of gene silencers, because the most favorable siRNAs are easily unwound *(4,27)*; potent siRNAs could also be positively selected. Thus, we attempt to dice only 70–80%, so that if dicing is more efficient, there is room for error to prevent complete consumption of the substrate. Note that the maximal yield of d-siRNAs is approx 75% of input dsRNA. However, this is common for long incubations; a fraction of the input nucleic acid is always lost when treated with a modifying enzyme for an extended period of time.

41. First, do a Western blot to make sure that r-Dicer was expressed and purified. If the yield of r-Dicer is low, optimize either expression (*see* **Notes 29–31**) or purification. We have found that some purifications gain activity after storage at 4°C. It is likely that modification or even proteolysis or r-Dicer or some contaminating protein might occur at 4°C and hypothesize that this might be the cause of the improvement in activity. For example, something as simple as loss of the N-terminal tag might improve activity, but we have not yet tested removal of the 6x-His tag and the T7 tag is not removable. An even greater degree of proteolysis might be responsible for increasing activity and is consistent with previous findings that treatment with proteinase K can improve activity *(8)*. We have not yet substantiated our hypothesis or treatment with proteinase K, but it might be very useful. There are two important points to keep in mind. First, if activity is lower than

expected, test again in 2 wk. Second, if activity does improve after storage, plan ahead and make r-Dicer before you need it.

42. The presence of any siRNA in a cell must be controlled for, because high levels of siRNA increase the levels of eIF2α phosphorylation resulting in decreased levels of translation *(7)*. Because RNAi is so specific, any d-siRNA could potentially be used as a control so long as it does not affect the levels of the gene that is supposed to be silenced. On one hand, d-siRNAs that target exogenous genes such as Firefly Luciferase (GL3), Renilla Luciferase (RL), or GFP are good controls because when compared to the human genome, there is minimal nucleotide similarity. However, on the other hand, d-siRNAs that target exogenous genes are not a perfect control because they do not have a target. A target can be supplied by using a cell line that stably expresses an exogenous target or by cotransfecting an expression vector with the d-siRNAs. GFP, GL3, or RL coding sequences (see accession numbers in **step 2** of **Subheading 3.4.1.**) can be expressed in any mammalian expression vector; we generally use a CMV (cytomegalovirus) promoter. Targeting an endogenous gene as a control has the advantage that the control d-siRNAs have a target to silence, but the disadvantage of potentially affecting the levels of the gene that is intentionally being silenced.

43. Reconstituting the silenced gene can be very useful (*see* **step 8** in **Subheading 3.7.1.**). The simplest method to perform a reconstitution is to target the endogenous gene with d-siRNAs generated from the untranslated region (UTR), 3′ or 5′, and exogenously express the coding region. We have successfully silenced genes with d-siRNAs targeting the 3′-UTR but have not tried using d-siRNAs targeting the 5′-UTR. Generally, the 3′-UTR contains more nucleotides than the 5′-UTR from which to choose the target region and will be easier to obtain via PCR, but if the 3′-UTR is not of sufficient length, it is worthwhile to use the 5′-UTR. In some cases, there will not be enough UTR at either end to make a 550-bp dsRNA. We have not tried to make smaller dsRNAs, but it is worth to attempt it if a reconstitution experiment is essential.

Initially, the approx 550-bp target regions we selected started 100–150 nucleotides downstream of the start codon similar to a guideline used to select a siRNA target region. The logic here was to avoid competition between the d-siRNAs and translation initiation proteins that bind mRNA, but this guideline is not essential to potent gene silencing; any region will work. Furthermore, the full-length coding sequence can also be used Green Fluorescent Protein (GFP), Cdc25C, and B-raf were all silenced in HEK 293 cells when the full-length coding sequence was used. Furthermore, a full-length cDNA including the 5′- and 3′-UTRs can be used. The cyclin-dependent kinase, cdc2, was silenced in HEK 293 cells when the full-length cDNA including the 5′- and 3′-UTRs was used. However, for routine gene silencing, avoid using longer templates because the yield is lower and more single stranded or "branched" regions exist within the dsRNA; templates can be as long as 800 bp, but this length is not necessary. It is unclear if Dicer will process smaller dsRNAs, like 300 bp, and if 300 bp will really give a random pool of sufficient complexity, but it is likely it will; thus, if a small template is the only option, it is worth a try.

44. There has been some concern about silencing homologs or isoforms that have very similar nucleotide sequence to the target or even random genes that just so happen to have a small region of high nucleotide similarity to the target. This is an insightful point but is unlikely, as described in **Subheading 1.** Also, the degree of nucleotide homology between closely related genes is not always as high as commonly thought, especially within particular 550-bp regions. Even some conserved domains do not contain a high level of nucleotide identity. We have not determined the nucleotide homology for every gene in the genome, so there is a chance that some families have a higher degree of identity than those we have tested. One example is the high nucleotide identity between PKC-δ and PKC-ε in the 3′ end of the coding sequence. However, this is not a problem for us because there is little homology in the 5′ end; thus, this is the region we chose to target.

    For several genes, we have confirmed that genes closely related to the target are not silenced, including cyclin B1 and B2, Cdc25A and Cdc25C, and Cdk1 and Cdk2. For example, cyclin B1 and B2 are approx 50% identical at the nucleotide level and they contain 13–15 consecutive identical nucleotide stretches, but only the appropriate target is silenced. These results can be explained by the fact that even a single nucleotide mismatch between the d-siRNA and a mRNA will abolish silencing *(5,56)*. Although, in some cases, it might be advantageous to silence all isoforms and splice variants, this approach has not been attempted.

45. T7 phage polymerase is used more frequently than T3 or SP6 because the yields are generally larger; however, any polymerase will work.

46. In vitro transcription is most efficient from a PCR template because the runoff transcription can be done without modifying the template. When a vector is used as a template, it must be linearized with restriction endonucleases at each end of the template for the runoff transcription. The template must then be purified away from all of the reaction components, which is quite a bit of work. Although preparing a vector template is more work, there is an advantage. A single clone, free of any mutations, can be selected and easily amplified in bacteria, whereas a PCR template might contain some mutations and can only be amplified by performing another PCR. Nonetheless, a few mutations should not be a problem (*see* **Note 54**); thus, PCR templates are better. If a PCR template lacking T7 promoter sites is already available, an oligonucleotide linker containing the T7 promoter sequence can be added to the end using Topoisomerase (Invitrogen).

47. In vitro transcription yields from the MEGAscript kit are not affected if PCR reaction components are not removed from the template; thus, purification is unnecessary. If for some reason you wish to remove the reaction components from the template, use the QIAquick PCR Purification Kit as directed by the manufacturer. If PCR amplification results in multiple products, the correct template can be isolated via gel purification using the QIAquick Gel Extraction Kit as directed by the manufacturer. However, in vitro transcription is compromised after gel purification even if all suggested modifications to the protocol are employed. Thus, we find it beneficial to alter the PCR conditions such that only one band is amplified. This might require designing better primers (*see* **Subheading 3.4.2.**). Do not worry about contaminating bands that are not visible with ethidium bromide

staining. Note that all components in both of these kits are RNase-free when purchased even though it is not explicitly stated. Use proper technique to ensure components are not contaminated with RNases because the template will be added directly to the in vitro transcription after purification.

48. Nested PCR will increase the chance that a band that is amplified from cDNA is correct. Nested PCR requires two sets of primers. First, design a set of primers containing only gene-specific sequence and then design a second set of nested primers (primers within the amplified region) that contain the T7 phage polymerase promoter.

49. If a vector containing the gene of interest is used for the PCR template, then the vector DNA should be removed to avoid unwanted transcription products. If the vector was amplified in *dam*⁺ bacteria, and it is likely that it was, the DNA will be methylated. *Dpn*I will digest only methylated DNA; thus, the vector will be digested at many sites, but the PCR product will remain in tact. To digest, simply add 1 μL (20 units/ μL) of *Dpn*I (New England Biolabs, Beverly, MA; R0176S) directly to the PCR and incubate at 37°C for 1 h. It is important to heat-inactivate the *Dpn*I. If mineral oil was used to overlay the PCR, be sure that the *Dpn*I is added to the aqueous portion. It is a good idea to do a second round of PCR, using the first PCR reaction as the source template, to generate the in vitro transcription template. This will allow simple regeneration of the in vitro transcription template, increase the yield of the template, and dilute the source template (vector DNA) so that the risk of contaminating transcription products is reduced.

50. Because the templates will be added directly to the in vitro transcription reaction, it is better if mineral oil is not used. Newer PCR machines have heated lids, eliminating the need to use oil. If oil must be used, try to remove most of it. After PCR, place the reaction at –80°C for 10–20 min. When the reaction is returned to room temperature, the mineral oil will thaw before the aqueous portion. Use a pipet tip or aspirator to carefully remove the mineral oil.

51. If the target mRNA is thought to be scarce, 2 μL of cDNA can be used, but this is the maximum. Do not worry about yield because the second round of PCR ensures that low-abundance genes are sufficiently amplified. Do not add extra dNTP because this increases the chance of mutation.

52. If amplification was unsuccessful, either no product or multiple products, many variables can be altered. If there is no product, first try using 2 μL of cDNA in the first PCR reaction, but do not exceed 2 μL. Next, try altering PCR conditions: increase $MgSO_4$ up to 4 mM although this might decrease specificity, resulting in amplification of numerous genes; use a range of annealing temperatures 50–65°C. Use a gene-specific primer in the reverse transcription (*see* **Subheading 3.1.2.2.**). As a last resort, design new primers. If there are multiple products, try reducing the $MgSO_4$ to 0.5–1 mM, try raising the annealing temperature from 60°C to 65°C, try reducing the primer concentration to 0.2 μM, and, finally, as a last resort try designing new primers.

53. Prior to sequencing, the PCR product must be purified away from reaction components. Use the QIAquick PCR Purification Kit or the QIAquick Gel Extraction Kit as directed by the manufacturer. If gel extraction is used, be sure to follow the modified protocol so that the band can be sequenced.

54. In theory, the potency and accuracy of gene silencing could be affected if the template and, therefore, the d-siRNAs contain many point mutations, or if multiple products are amplified via PCR, the d-siRNA pool would be contaminated with undesired siRNAs. This, however, should not be a concern. The error rate of *Taq* high fidelity is quite low so long as the dNTP concentration does not exceed 0.2 m*M*. After 60 rounds of PCR, 2 separate amplifications, it is likely that the in vitro transcription template could have a single mutation, but because so many siRNAs are present in a d-siRNA pool, one or two mutations will not decrease silencing efficacy or increase off-target effects. It is very unlikely that contaminating PCR products are present, even at low concentrations. To ensure specific amplification, do not exceed the recommended concentration of primer or polymerase because it will facilitate nonspecific amplification of undesired products.

55. The yield of dsRNA is proportional to the amount of template used. Generally 8 µL results in highest yield and is, therefore, routinely used. The reaction size is scalable, the reaction can be doubled or cut in half, and the yields are proportional. Reaction size depends on the purpose. If the dsRNA will be diced frequently, such as a control dsRNA, it is worthwhile to double the reaction. The dsRNA is stable for at least 1 yr at –80°C so do not worry about producing too much. If a small amount of d-siRNA is required or for high-throughput gene silencing, the reaction can be cut in half. dsRNA can be radiolabeled by adding 1 µL of α-P^{32}-UTP (Amersham, PB20383, or equivalent SP6/T7 grade UTP) to the transcription reaction. The amount of cold UTP does not need to be altered.

56. If a particular approx 550-nucleotide region will not form dsRNA, the sense and antisense strands can be transcribed separately and then annealed. Also, the strands can still be transcribed simultaneously and then annealed. Be sure that the DNA template is removed before annealing. There are many methods to drive annealing, but we have routinely used the following. To the transcription reaction, add NaCl to 20 m*M* and HEPES to 10 m*M* (from a stock of 420 m*M* NaCl, 210 m*M* HEPES, pH 7.5–8.0), then heat at 95°C for 1 min 30 s, and then incubate at either 37°C for 1 h or at room temperature overnight. It is possible that RNA strand scission might occur when heated in the presence of divalent ions such as Mg^{2+} (which is present in the transcription reaction). However, noticeable strand scission has not been detected and, therefore, RNA is heated as described without adding EDTA.

57. Although dsRNA does not have to be routinely purified from reaction components, a purified dsRNA will be required to estimate the concentration. Thus, purification must be done at least once. We have modified the purification protocol suggested by Ambion to avoid the phenol/chloroform extraction; instead, the dsRNA is purified by two precipitations.

    After transcription treat with RNase-free DNase I and eliminate "branched" regions as described in **step 4** in **Subheading 3.5.1.** Protein and NTPs can be efficiently removed with a LiCl precipitation. Add 25 µL (nearly equal volume) of 7.5 *M* LiCl, 50 m*M* EDTA (provided in MEGAscript kit) to the 21-µL transcription reaction and vortex for about 10 s. Pellet dsRNA with in centrifugation at 14,000*g* and 4°C for 15 min. The pellet is very large and is visible. Carefully remove the supernatant and immediately resuspended dsRNA in 90 μL of RNase-free H_2O by pipetting up and down and vortexing for several seconds. It is not necessary to chill at –20°C prior to precipitation, but it is a good stopping point. Do not dilute with H_2O before precipitation because LiCl precipitation is more efficient when dsRNA is concentrated.

The dsRNA is then precipitated again with 5 *M* NH_4OAC and isopropanol to ensure complete removal of salts and NTPs. To the 90 μL, add 10 μL (1/10 vol) of 5 *M* NH_4OAC, 100 m*M* EDTA (provided in MEGAscript kit), and 100 μL of isopropanol (1 vol). Vortex for 10 s and pellet dsRNA with centrifugation, as described earlier.

After centrifugation, carefully remove the supernatant and wash the pellet with 1 mL of 70% EtOH (room temperature) during a 5- to 10-min centrifugation at 14,000*g* and 4°C. Remove the supernatant; it is important to remove all traces of ethanol. Remove the bulk of the ethanol, touch-spin to collect all traces of ethanol, and then use a fine pipet tip to remove the remaining ethanol. Dry the pellet for approx 10–15 min at room temperature with the cap open. Directly after removing the ethanol, the pellet will be visible; however, once it is dry, it will be difficult to see. Thus, remember where the pellet is located.

Resuspend the pellet in RNase-free H_2O (provided in the MEGAscript kit) so that the concentration of the dsRNA will be approx 1 μg/μL (typically 50 μL of H_2O is used to ensure a concentration at or above 1 μg/μL). Generally, for good templates, the yield is 80–100 μg. For templates that provide less efficient transcription and/or longer templates (>1000 bp), the yield is slightly lower (40–60 μg). If the reaction was scaled up or down, use the appropriate amount of; e.g., if the reaction was doubled, then resuspend in 100 μL of H_2O.

Estimate the concentration and purity of dsRNA by measuring the absorbance at 260 nm and 280 nm (*see* **step 2** in **Subheading 3.1.2.1.**). Dilute dsRNA in TE, pH 8.0 (**Subheading 2.1., item 2**) at 1 : 200–1 : 500 and use 40 μg/mL as a conversion factor; make sure that the A260 reading is between 0.05 and 0.3 absorbance units. The A_{260}/A_{280} ratio should be approx 2 or higher. Make sure that the concentration estimate is accurate because all future concentration estimates will be based on this dsRNA.

58. The concentration estimate is not 100% accurate, mostly because the dynamic range of ethidium bromide is small. However, for in vitro dicing, this is not important; only "ball-park" accuracy is required. It is just important to know that approx 50 μg of dsRNA is being added to the reaction (*see* **Subheading 3.5.2.**). If a GelDoc is not available, the histogram function in Adobe Photoshop is sufficient. For detection purposes, SYBRR gold nucleic acid gel stain can also be used and is advantageous for quantification because it has a larger dynamic range (Invitrogen, Carlsbad, CA; S-11494).

59. Most templates yield a large amount of dsRNA, but yield is variable. As suggested by Ambion, the length of the reaction can be altered to increase the yield. GC-Rich templates frequently result in a lower yield of dsRNA. If a larger yield is necessary, try designing a new template.

60. If there is a concern that the RNA is not annealed, a 2% agarose gel can be used to acess what portion of the RNA is annealed. Unlike a 1% gel a 2% agarose gel efficiently separates single-stranded and double-stranded RNA; the single-stranded RNA will migrate faster than the dsRNA. Note that the sense and anti-sense strands might migrate at different rates because of differential secondary structure. Alternatively, annealing can be confirmed by running an aliquot of the dsRNA in either native gel loading buffer or in denaturing loading buffer (2X loading buffer provided in MEGAscript kit). To denature the dsRNA, heat at 95°C for 5 min after adding an equal volume of the 2X loading buffer. Do not heat the RNA present in native gel loading buffer. Compare the mobility of each dsRNA on a 1% or 2% agarose gel. The denatured RNA should migrate faster.

61. There are many methods to anneal RNA strands; use the method described in **Note 56**. Longer dsRNAs do not anneal well; thus, it might be worthwhile to decrease the length if a larger dsRNA (>800 bp) is being used.

62. The volume of r-Dicer added is dependent on Director activity. The volume of the other components remains the same, except Rxn Dil Bfr, which changes in accordance with the volume of r-Dicer added so that the total volume of Rxn Dil Bfr and r-Dicer always equals 150 µL (this keeps the salt and buffer conditions optimal). Therefore, if the volume of r-Dicer is increased, the volume of Rxn Dil Bfr must be decreased and vice versa. If cleavage of the large dsRNA is inefficient, the amount of r-Dicer added to the reaction can be increased to a maximum of 150 µL (higher volumes result in inappropriate reaction conditions). If adding more r-Dicer does not help, try improving the purity of the prep or making a concentrated prep. On the other hand, recall that adding too much r-Dicer can be detrimental (*see* **Note 40**).

63. Large dsRNAs can be diced and stored at –80°C without purification. The d-siRNAs are stable for several weeks.

64. The original d-siRNA purification protocol we designed works and should be used as a last resort because it has several drawbacks. First, it is time-consuming. Second, it cannot completely remove the residual NTPs from the in vitro synthesis of the dsRNA, resulting in a slightly inaccurate absorbance at 260 nm and a chance for error in determining the concentration of the d-siRNAs. Finally, it is not suitable for high-throughput dicing. Nonetheless, the protocol is presented here.

    Remove large dsRNAs by spinning the 295-µL reaction (300-µL reaction minus 5 µL for electrophoresis) through a YM-100 microcon filter device (Millipore, Bedford, MA; 42413) at 500*g* and 4°C. Do not spin faster because the large dsRNA will then also pass through the membrane. Generally, it takes about 1 h 20 min to force the solution through the filter. Spin until the membrane is nearly dry (i.e. until approx 10 µL of the reaction remains on the top of the filter), at which point the liquid only covers about three-quarters of the entire filter. Spinning the membrane until it is dry or for longer than 1 h and 20 min will increase the chances of contaminating the d-siRNAs with larger dsRNA. It is better to have less d-siRNA than increase the chance that larger dsRNA will contaminate the d-siRNA; that is, it is better to have a smaller flow through volume.

The flowthrough contains the d-siRNA and is usually about 240–280 μL in volume. Precipitate d-siRNA with 0.5 *M* ammonium acetate and an equal volume of isopropanol after the addition of glycogen (Ambion, Austin, TX; 9510) to a final concentration of 150 ng/ μL and magnesium acetate to 10 mM. For example, to approx 240 μL of flowthrough, add the following:

---

15 μL of 10 *M* ammonium acetate (pH 7.0)
3 μL of 1 *M* magnesium acetate
9 μL of 5 mg/mL glycogen
<u>33 μL H$_2$O</u>
300 μL

Add 300 μL of isopropanol, vortex, and place at −20°C for 20 min. 3 *M* of sodium acetate (pH 5.2) can be substituted for ammonium acetate and should be used at a final concentration of 0.3 *M*. All solutions should be RNase-free.

Pellet d-siRNA by centrifugation at 14,000*g* and 4°C for 20 min. Remove the supernatant, add 1 mL of 70% ethanol, and centrifuge at 14,000*g* and 4°C for 10 min. Remove all traces of ethanol and air-dry the pellet for 10–15 min at room temperature; do not use a Speed-vac.

Resuspend the pellet in 40 μL of RNase-free H$_2$O. Estimate the concentration and evaluate the quality and quantity as described in **Subheading 3.6.2.** Occasionally, a small amount of large dsRNA will contaminate the d-siRNA. Dilute the d-siRNA to 300 μL and pass over a new YM-100 column. This will eliminate the remaining large dsRNA, but some of the d-siRNA will be lost on the column and during precipitation. Thus, it is often worthwhile to just start over. Gel purification can also be used for purification, but residual contaminants decrease transfection efficiency.

65. Double-stranded RNA fragments ≥50 bp bind the first spin cartridge when the isopropanol concentration is 40%. The original Invitrogen protocol suggests an isopropanol concentration of 33%. This works well if the dicing was perfect; that is, the only species in the reaction are the large undigested dsRNA and the d-siRNAs. However, dicing of some dsRNA produces intermediates, but the size is never <50 bp. Therefore, we have adjusted the isopropanol concentration to 40% to ensure elimination of any dsRNA species that will trigger a toxic response once introduced into the cell. Ethanol can be used instead of isopropanol, but we have not tried this.

66. The original Invitrogen protocol increases the isopropanol concentration to 60%, but we have found that increasing the concentration to 70% improves yield. Again, isopropanol can be substituted for with ethanol, but we have not tried this.

67. It is a good idea to prepare a large stock of d-siRNA to use as a standard measure of concentration. It can be used to verify absorbance but, more importantly, as a relative intensity for ethidium bromide staining. Aliquot the standard and load various amounts each time a 15% gel is run.

68. To ensure that d-siRNAs are functional, 70–80% of the substrate dsRNA should be diced (*see* **Note 40**).

69. Single-stranded RNAs 21 nucleotides in length can be made from chemical synthesis (Ambion, Dharmacon, Qiagen).

70. Dilute d-siRNA to 100 µL and repurify as described in **Subheading 3.6.1.** Make sure that the isopropanol concentrations are scaled appropriately: 40% for the first column and 70% for the second column.

71. It can be beneficial to cotransfect a DNA reporter with d-siRNAs. First, the transfection efficiency can be determined. At the DNA masses suggested in **Table 4,** the correlation between d-siRNA delivery and expression of the reporter gene is very high. Second, usually less d-siRNA is required; thus, material can be conserved. Finally, it is possible to reconstitute the silenced protein by expressing the coding region of the target protein from a plasmid (see **step 8** in **Subheading 3.7.1.**). We generally cotransfect d-siRNAs and a YFP (yellow fluorescent protein) expression vector, but other reporters such as CFP (cyan fluorescent protein), GFP (green fluorescent protein), RFP (red fluorescent protein) such as dsRed, mRFP, tDimer2, or HCred, β-galactosidase, β-lactamase, Firefly Luciferase, or Renilla Luciferase are acceptable. We favor the fluorescent reporters or the enzymes capable of generating fluorescent signal over Luciferase because the detection is noninvasive and can be evaluated at the single cell level.

    The mass of DNA to use for cotransfection is shown in **Table 4**. The amount used is dependent on the promoter driving expression. We commonly use CMV (cytomegalovirus promoters), so very little DNA is required to attain expression levels sufficient for detection. Using more DNA than suggested (thereby overexpressing the reporter gene even further) could result in poor cell health. If you cotransfect DNA that uses a different promoter—one that provides lower level expression like SV40—the optimal amount of DNA to use must be determined (*see* **Subheading 3.7.4.**).

    Table 4 lists a twofold range of DNA masses to use for cotransfection so that two different plasmids, each expressing a different gene, can be cotransfected, although, in some cases, the correlation between the expression of the two genes can be quite low. Using different promoters to drive expression of each gene might help. Nonetheless, there might be a case in which you want to transfect two genes, and if so, use equal amounts of each such that the total mass does not exceed the maximum of the mass listed in **Table 4**.

72. To determine how many nanograms of d-siRNA to use for any desired final [d-siRNA], use the following formula:

$$X \text{ nanograms of d-siRNA} = (\mu L \text{ Tfc. Vol.})$$

$$\times \left(X \frac{\text{nmol}}{\text{L}} \right) \left(\frac{L}{1 \times 10^6 \, \mu L} \right) \left(640 \text{ ng} \times \frac{21}{\text{nmol}} \right)$$

where microliters of Tfc. Vol. (transfection volume) is equal to the volume of the growth medium (optimum) during transfection plus the volume of the transfection mix (*see* **Table 4**). For example, to use 10 n*M* final [d-siRNA] in a six-well dish:

$$X \text{ nanograms of d-siRNA} = (\sim 1074 \ \mu L)\left(10 \ \frac{\text{nmol}}{\text{L}}\right)$$

$$\times \left(\frac{L}{1 \times 10^{6} \ \mu L}\right)\left(640 \ \text{ng} \times \frac{21}{\text{nmol}}\right)$$

which results in approx 150 ng of d-siRNA.

To convert nanograms of d-siRNA into femtomoles or vice versa, use the following formulas:

$$\text{Femtomoles of d-siRNA} = (X \ \text{ng})\left(\frac{1 \ \text{nmol}}{640 \ \text{ng}} \times 21 \ \text{bp}\right)\left(\frac{1 \times 10^{6} \ \text{fmol}}{1 \ \text{nmol}}\right)$$

$$\text{Nanograms of d-siRNA} = (X \ \text{fmol})\left(640 \ \text{ng} \times \frac{21 \ \text{bp}}{1 \ \text{fmol}}\right)\left(\frac{1 \ \text{ng}}{1 \times 10^{6} \ \text{fg}}\right)$$

73. The abundance, half-life, and transcription rate of mRNAs varies so the amount of d-siRNA to use also varies. d-siRNA-mediated gene silencing is limited by protein stability. If a protein is inherently stable, then the half-life will be equal to the length of the cell cycle so long as translation is completely abolished as a result of complete absence of the message. Thus, for very stable proteins, silencing >80% cannot be attained until three cell divisions have occurred.

74. Silencing could diminish between 72 and 96 h because the d-siRNAs are consumed or diluted below effective levels. Cells can be transfected multiple times if phenotypical analysis needs to be done after silencing a particular protein for several days.

75. Passage number here refers to the number of times cells have been split after thawing from a frozen cell stock. Because original cells stocks are usually made at or slightly before the fifth passage, the actual range of passage number is 10–30. Early passages of HeLa cells, shortly after cells have been thawed from a cell stock, grow and transfect pretty well, but occasionally and stochastically, cell growth will sputter. Thus, we usually wait for several passages before cells are used for experiments. Cells that have been passaged many times begin to grow rapidly and might affect phenotypic analysis and are, therefore, not used. To ensure that cells between passage 5 and 25 are always available, start a fresh culture once the passage number of the cells already in culture is nearing 18–20.

76. HeLa cells exist in a variety of shapes dependent on which stage of the cell cycle they are in: (1) G1-phase: cells are larger and flat, mostly round or triangular; (2) S- and G2-phase: cells are flat with protruding extensions, mostly oval; (3) M-phase: cells are smaller and rounded up, easily confused with a dying cell. Cells undergoing cytokinesis or that have recently divided can be seen as doublets. Cultures with small cells, detached cells, or nondividing cells are considered unhealthy and should be discarded.

77. To prepare frozen cell stocks, grow several 10-cm plates to approx 50% confluency. Trypsinize cells (*see* **steps 3–6** in **Subheading 3.7.2.**), determine cell number (*see* **step 7** in **Subheading 3.7.2.** and **Note 80**), and transfer to a sterile conical tube. Pellet cells at 500*g* at room temperature for 5 min. Remove supernatant and resuspend in 50% DMEM, 40% FBS,, and 10% DMSO such that cell density is

5×10^5 cells per milliliter. Transfer 1 mL of cell suspension to each cryotube and use a cell freezer to cool cells 1°C per minute at –80°C. If a cell cooler is not available, place cryotubes in a Styrofoam container and place at –80°C; the cooling rate will not be optimal, but the technique is sufficient. Transfer cells to liquid-nitrogen storage 12–18 h later; cells can be stored in vapor phase or submerged. Each 1-mL aliquot is sufficient to start a fresh culture in a 10-cm dish. To start a culture, thaw cells as quickly as possible at 37°C and transfer to a 10-cm dish containing 10 mL of prewarmed complete growth medium. Change medium 12 h later to remove DMSO.

78. We have not seen any improvements in transfection efficiency when cells are plated for longer than 18 h even though some transfection protocols recommended plating cells 24 h prior to transfection.

79. When cells are treated with trypsin for an extended period, cell plating is suboptimal because cells take a long time to attach to the plate and it is, therefore, difficult to attain optimal cell distribution.

80. At least four quadrants should be counted because cell number is low and stochastic variation could result in inaccurate cell counting. Determine cell density with the following formula (the cell number is specific to the hemacytometer:

$$\text{Density (cells/mL)} = \frac{\left(1 \times 10^4\right)\left(\text{No. of cells in 4 quadrants}\right)}{4 \text{ quadrants}}$$

81. Oligofectamine (Invitrogen, Carlsbad, CA; 12252-011) can be used to transfect HeLa cells as directed by manufacturer. Similar to delivery with Genesilencer, transfection is performed in the absence of serum. Transfection efficiencies of d-siRNAs are similar to Genesilencer, but Oligofectamine is not useful for cotransfecting DNA and d-siRNAs. Lipofectamine 2000 Transfection Reagent (Invitrogen, Carlsbad, CA; 11668-027) is capable of cotransfecting d-siRNAs and DNA but is not optimal for transfecting HeLa cells because of the cytotoxicity. Lipofectamine 2000 Transfection Reagent is, however, useful for other cell lines such as HEK 293 (American Tissue Type Collection, Manassas, VA; CRL-1573) or NIH/3T3 (American Tissue Type Collection, Manassas, VA; CRL-1658). Although fuGENE6 (Roche Applied Science, Indianapolis, IN; 11815091001), a nonliposomal reagent, provides efficient transfection of DNA into HeLa cells, it does not work well at all for d-siRNAs. Likewise, ExGen 500 (Fermentas, Hanover, MD; R051), a PEI (polyethylinimine) formulation does not provide efficient transfection of d-siRNAs in either HeLa or NIH/3T3 cells. Many other transfection reagents specially formulated for delivery of siRNAs are available; however, we have not yet tested these reagents.

82. If you wish to transfect a cell line other than HeLa cells with Genesilencer, it is likely that the cells can be transfected in the presence of serum. For example, we have transfected d-siRNAs and cotransfected d-siRNAs and DNA into HEK 293 and NIH/3T3 cells with Genesilencer in the presence of serum. Transfection efficiency, as evaluated by expression of a fluorescent reporter, is ≥85% for HEK 293 cells and approx 50% for NIH/3T3 cells, and efficiencies are not improved if

transfection is performed in the absence of serum. In fact, cell health and attachment is suboptimal if serum is removed.

To transfect cells in the presence of serum, follow the protocol described in **Subheading 3.7.3.** with the following changes. Prior to transfection, remove complete growth medium and replace with the appropriate volume of fresh growth medium as suggested in **Table 6**. Merely swapping the same media might seem ineffectual, but it serves two purposes. First, it allows unattached cells to be removed and, second, it permits the volume for transfection to be reduced, thereby conserving material and improving transfection efficiency. Suggested concentrations of d-siRNA and DNA (*see* **Subheading 3.7.1.** and **Table 4**) might have to be altered to attain optimal transfection efficiency and cell health (*see* **Subheading 3.7.4.**). However, the transfection mix is still formulated with diluent, Genesncer, OPTI-MEM I, and nucleic acid, as described in **Subheading 3.7.3**. Additional medium will not have to be added to the cells after transfection, but 12–18 h posttransfection, additional complete growth medium should be added because the volume of medium used during transfection is minimal. When plating cells for transfection (*see* **Subheading 3.7.2.**) cell numbers suggested in **Table 5** will not be optimal; thus, determine appropriate density for the cell line used (*see* **Subheading 3.7.4.**). Finally, if a transfection reagent other than Genesncer is used, follow manufacturer's protocol and recommendations regarding serum.

83. Cell lines stably expressing a protein fused to a fluorescent reporter can also be used because silencing is identical to a cell line expressing only the reporter. This is a useful alternative if a cell line expressing the fusion protein is already available.

84. Hoechst 33342 can be used at 1X (1:10,000 dilution of 10 mg/mL stock) to stain the nuclear DNA of living cells. Nuclear DNA will stain within 5 min when cells are placed at 37°C. Hoechst does not need to be removed prior to imaging. Cells will remain healthy for several hours in the presence of 1X Hoechst. However, if cells are going to be imaged both at 24 and 48 h, do not stain nuclear DNA until the 48-h time-point.

85. Transfection efficiency is defined as the percentage of cells expressing the reporter (i.e., the number of expressing cells divided by the total number of cells multiplied by 100). Hoechst staining is used to determine the total cell number. Under optimal transfection conditions with Genesilencer transfection efficiency should be ≥85%.

86. Transfection reagents and exogenous RNA can trigger cytotoxic responses, leading to a reduction in protein levels. Thus, for each experiment, it is imperative to have the proper controls: untransfected cells, cells treated only with transfection reagent, and cells treated with an irrelevant, "control" d-siRNA. Silencing is specific if the protein levels of the target gene are reduced by 50–95% relative to the controls. We have found that Genesilencer is not very toxic and control d-siRNAs usually have little affect on HeLa cell health; thus, the target protein abundance should be similar in all of the controls. However, in some cases, untransfected cells might be healthier than cells treated with only the transfection reagent, which might be healthier than cells treated with d-siRNAs, leading to some variance in target protein levels.

87. If glass flasks are used, they should be dedicated to production of r-Dicer. Trace amounts of other baculovirus might affect the yield of r-Dicer.

88. Templates can be amplified from cDNA (*see* **Subheadings 3.1.2.1., 3.1.2.2.,** and **3.4.3.** and materials within sections) by PCR. It is likely that many targets will not be represented in certain tissues; thus, if obscure genes or an entire genome is desired, use a diverse source of mRNA. The mRNA from different tissues can be isolated as described in **Subheading 3.1.2.1.** or can be purchased from commercial source. cDNA can be made by pooling mRNA from different tissues or cDNA can be made from each tissue and then pooled prior to PCR amplification.

    Because it would be very expensive to validate the accuracy of thousands of sequences, templates can be produced via nested PCR (*see* **Note 48**). This approach is not 100% accurate but will increase the odds that the correct gene has been amplified. It is also possible to create a verified library of in vitro transcription templates by subcloning each template after PCR so that the original template will be invariable and can be used in regeneration of the library. The TOPO cloning system (Invitrogen) is useful for subcloning PCR products.

    Although it is more efficient to amplify vector templates, preparation for in vitro transcription requires a substantial amount of work relative to in vitro transcription templates generated by PCR amplification (*see* **Note 46**). Vector templates could be used, but we recommend PCR products.

89. The yield of the dsRNAs will vary by twofold to threefold because transcription from some templates is more efficient. It is difficult to attempt to maximize yield of each template in a high-throughput format because yield is dependent on template quality and concentration and on length of transcription. The concentration of each dsRNA could be normalized, but if this is the case, be sure to dilute the most concentrated samples with the in vitro transcription buffer so that components present in the dicing reaction will be equivalent. For dilution, use an automated liquid handler to avoid human error and save time. The relative concentration of the dsRNAs can be determined with a plate reader capable of reading at 260 nm. Special, low-absorbance, 96-well plates (**Subheading 2.9., item 15**) or 96-well glass plates are required. Also, nucleic acid stains such as RiboGreen (**Subheading 2.9., item 16**) can be used. If the dsRNAs are normalized to the lowest concentration, just dice 15 µg.

    It is acceptable to use different masses of dsRNA, but the amount of r-Dicer used should be sufficient to dice 70–80% of the lowest mass of dsRNA (probably 15 µg), even though some of the reactions will contain larger masses (up to 50 µg). If dicing is performed such that 70–80% of the 50 µg is diced, then the yield of d-siRNAs from less abundant dsRNAs would be compromised because of "overdicing" (*see* **Note 40**). Therefore, perform the reaction such that the r-Dicer activity is appropriate for dicing 70–80% of the minimal dsRNA substrate mass. The percentage of large dsRNA that is diced will be lower for abundant dsRNAs, but this is not a concern because the complexity is similar to that attained when a large percentage of the substrate dsRNA is diced. However, the yields of the d-siRNAs will vary by twofold to threefold.

Acknowledgments

We thank Josh Jones, Delquin Gong, David Hendrickson, Ari Firestone, and members of the Ferrell lab for insightful discussion about and validation of the techniques presented here. Also, we thank Knut Madden, Adam Harris, Michaeline Bunting, Kerry Lowrie, Joe O'Connor, and Byung-in Lee for help with this protocol. The NIH grant GM46383 supports our work in this area.

References

1. Brummelkamp, T. R., Nijman, S. M., Dirac, A. M., and Bernards, R. (2003) Loss of the cylindromatosis tumour suppressor inhibits apoptosis by activating NF-kappaB. *Nature* **424,** 797–801.
2. Paddison, P. J., Silva, J. M., Conklin, D. S., et al. (2004) A resource for large-scale RNA-interference-based screens in mammals. *Nature* **428,** 427–431.
3. Berns, K., Hijmans, E. M., Mullenders, J., et al. (2004) A large-scale RNAi screen in human cells identifies new components of the p53 pathway. *Nature* **428,** 431–437.
4. Schwarz, D. S., Hutvagner, G., Du, T., Xu, Z., Aronin, N., and Zamore, P. D. (2003) Asymmetry in the assembly of the RNAi enzyme complex. *Cell* **115,** 199–208.
5. Elbashir, S. M., Harborth, J., Lendeckel, W., Yalcin, A., Weber, K., and Tuschl, T. (2001) Duplexes of 21-nucleotide RNAs mediate RNA interference in cultured mammalian cells. *Nature* **411,** 494–498.
6. Elbashir, S. M., Lendeckel, W., and Tuschl, T. (2001) RNA interference is mediated by 21- and 22-nucleotide RNAs. *Genes Dev.* **15,** 188–200.
7. Myers, J. W., Jones, J. T., Meyer, T., and Ferrell, J. E., Jr. (2003) Recombinant Dicer efficiently converts large dsRNAs into siRNAs suitable for gene silencing. *Nat. Biotechnol.* **21,** 324–328.
8. Zhang, H., Kolb, F. A., Brondani, V., Billy, E., and Filipowicz, W. (2002) Human Dicer preferentially cleaves dsRNAs at their termini without a requirement for ATP. *EMBO J.* **21,** 5875–5885.
9. Provost, P., Dishart, D., Doucet, J., Frendewey, D., Samuelsson, B., and Radmark, O. (2002) Ribonuclease activity and RNA binding of recombinant human Dicer. *EMBO J.* **21,** 5864–5874.
10. Kawasaki, H., Suyama, E., Iyo, M., and Taira, K. (2003) siRNAs generated by recombinant human Dicer induce specific and significant but target site-independent gene silencing in human cells. *Nucleic Acids Res.* **31,** 981–987.
11. Myers, J. W. and Ferrell, J. E., Jr. (2004) Dicer in RNAi: its roles *in vivo* and utility *in vitro*, in *RNA Interference Technology: From Basic Biology to Drug Development* (Appasani, K., ed.), Cambridge University Press, New York.
12. Fink, C. C., Bayer, K. U., Myers, J. W., Ferrell, J. E., Jr., Schulman, H., and Meyer, T. (2003) Selective regulation of neurite extension and synapse formation by the beta but not the alpha isoform of CaMKII. *Neuron* **39,** 283–297.
13. Jones, J. T., Myers, J. W., Ferrell, J. E., and Meyer, T. (2004) Probing the precision of the mitotic clock with a live-cell fluorescent biosensor. *Nat. Biotechnol.* **22,** 306–312.

14. Stark, G. R., Kerr, I. M., Williams, B. R., Silverman, R. H., and Schreiber, R. D. (1998) How cells respond to interferons. *Annu. Rev. Biochem.* **67,** 227–264.

15. Kostura, M. and Mathews, M. B. (1989) Purification and activation of the double-stranded RNA-dependent eIF-2 kinase DAI. *Mol. Cell Biol.* **9,** 1576–1586.

16. Manche, L., Green, S. R., Schmedt, C., and Mathews, M. B. (1992) Interactions between double-stranded RNA regulators and the protein kinase DAI. *Mol. Cell. Biol.* **12,** 5238–5248.

17. Minks, M. A., West, D. K., Benvin, S., and Baglioni, C. (1979) Structural requirements of double-stranded RNA for the activation of 2′,5′-oligo(A) polymerase and protein kinase of interferon-treated HeLa cells. *J. Biol. Chem.* **254,** 10,180–10,183.

18. Samuel, C. E. (2001) Antiviral actions of interferons. *Clin. Microbiol. Rev.* **14,** 778–809.

19. Yu, J. Y., DeRuiter, S. L., and Turner, D. L. (2002) RNA interference by expression of short-interfering RNAs and hairpin RNAs in mammalian cells. *Proc. Natl. Acad. Sci. USA* **99,** 6047–6052.

20. Brummelkamp, T. R., Bernards, R., and Agami, R. (2002) A system for stable expression of short interfering RNAs in mammalian cells. *Science* **296,** 550–553.

21. Castanotto, D., Li, H., and Rossi, J. J. (2002) Functional siRNA expression from transfected PCR products. *RNA* **8,** 1454–1460.

22. Paddison, P. J., Caudy, A. A., Bernstein, E., Hannon, G. J., and Conklin, D. S. (2002) Short hairpin RNAs (shRNAs) induce sequence-specific silencing in mammalian cells. *Genes Dev.* **16,** 948–958.

23. Paul, C. P., Good, P. D., Winer, I., and Engelke, D. R. (2002) Effective expression of small interfering RNA in human cells. *Nat. Biotechnol.* **20,** 505–508.

24. Lee, N. S., Dohjima, T., Bauer, G., et al. (2002) Expression of small interfering RNAs targeted against HIV-1 rev transcripts in human cells. *Nat. Biotechnol.* **20,** 500–505.

25. Miyagishi, M. and Taira, K. (2002) U6 promoter-driven siRNAs with four uridine 3′ overhangs efficiently suppress targeted gene expression in mammalian cells. *Nat. Biotechnol.* **20,** 497–500.

26. Reynolds, A., Leake, D., Boese, Q., Scaringe, S., Marshall, W. S., and Khvorova, A. (2004) Rational siRNA design for RNA interference. *Nat. Biotechnol.* **22,** 326–330.

27. Khvorova, A., Reynolds, A., and Jayasena, S. D. (2003) Functional siRNAs and miRNAs exhibit strand bias. *Cell* **115,** 209–216.

28. Sen, G., Wehrman, T. S., Myers, J. W., and Blau, H. M. (2004) Restriction enzyme-generated siRNA (REGS) vectors and libraries. *Nat. Genet.* **36,** 183–189.

29. Shirane, D., Sugao, K., Namiki, S., Tanabe, M., Iino, M., and Hirose, K. (2004) Enzymatic production of RNAi libraries from cDNAs. *Nat. Genet.* **36,** 190–196.

30. Luo, B., Heard, A. D., and Lodish, H. F. (2004) Small interfering RNA production by enzymatic engineering of DNA (SPEED). *Proc. Natl. Acad. Sci. USA* **101,** 5494–5499.

31. Zheng, L., Liu, J., Batalov, S., et al. (2004) An approach to genomewide screens of expressed small interfering RNAs in mammalian cells. *Proc. Natl. Acad. Sci. USA* **101,** 135–140.

32. Chi, J. T., Chang, H. Y., Wang, N. N., Chang, D. S., Dunphy, N., and Brown, P. O. (2003) Genomewide view of gene silencing by small interfering RNAs. *Proc. Natl. Acad. Sci. USA* **100,** 6343–6346.

33. Jackson, A. L., Bartz, S. R., Schelter, J., et al. (2003) Expression profiling reveals off-target gene regulation by RNAi. *Nat. Biotechnol.* **21,** 635–637.

34. Semizarov, D., Frost, L., Sarthy, A., Kroeger, P., Halbert, D. N., and Fesik, S. W. (2003) Specificity of short interfering RNA determined through gene expression signatures. *Proc. Natl. Acad. Sci. USA* **100,** 6347–6352.

35. Fire, A., Xu, S., Montgomery, M. K., Kostas, S. A., Driver, S. E., and Mello, C. C. (1998) Potent and specific genetic interference by double-stranded RNA in *Caenorhabditis elegans. Nature* **391,** 806–811.

36. Ford, N., ed. (1989) *Molecular Cloning: A Laboratory Manual.* Cold Spring Harbor Laboratory Press, Cold Spring Harbor, NY.

37. Gordon, J. A. (1991) Use of vanadate as protein-phosphotyrosine phosphatase inhibitor. *Methods Enzymol.* **201,** 477–482.

38. O'Reilly, D. R., ed. (1992) *Baculovirus Expression Vectors: A Laboratory Manual.* W. H. Freeman, New York, NY.

39. King, L. A., ed. (1992) *The Baculovirus Expression System: A Laboratory Guide.* Chapman & Hall, New York.

40. Reischt, U., ed. (1997) *Generation of Recombinant Baculovirus DNA in* E. coli *Using Baculovirus Shuttle Vector.* Humana, Totowa, NJ.

41. Polayes, D., Harris, R., Anderson, D., and Ciccarone, V. (1996) New baculovirus expression vectors for the purification of recombinant proteins from insect cells. *Focus* **18,** 10–13.

42. Luckow, V. A., Lee, S. C., Barry, G. F., and Olins, P. O. (1993) Efficient generation of infectious recombinant baculoviruses by site-specific transposon-mediated insertion of foreign genes into a baculovirus genome propagated in *Escherichia coli. J. Virol.* **67,** 4566–4579.

43. Carrington, J. C. and Dougherty, W. G. (1988) A viral cleavage site cassette: identification of amino acid sequences required for tobacco etch virus polyprotein processing. *Proc. Natl. Acad. Sci. USA* **85,** 3391–3395.

44. Carrington, J. C., Cary, S. M., and Dougherty, W. G. (1988) Mutational analysis of tobacco etch virus polyprotein processing: cis and trans proteolytic activities of polyproteins containing the 49-kilodalton proteinase. *J. Virol.* **62,** 2313–2320.

45. Westwood, J. A., Jones, I. M., and Bishop, D. H. (1993) Analyses of alternative poly(A) signals for use in baculovirus expression vectors. *Virology* **195,** 90–99.

46. Barry, G. F. (1988) A broad-host-range shuttle system for gene insertion into the chromosomes of gram-negative bacteria. *Gene* **71,** 75–84.

47. Handa, V., Saha, T., and Usdin, K. (2003) The fragile X syndrome repeats form RNA hairpins that do not activate the interferon-inducible protein kinase, PKR, but are cut by Dicer. *Nucleic Acids Res.* **31,** 6243–6248.

48. Doi, N., Zenno, S., Ueda, R., Ohki-Hamazaki, H., Ui-Tei, K., and Saigo, K. (2003) Short-interfering-RNA-mediated gene silencing in mammalian cells requires Dicer and eIF2C translation initiation factors. *Curr. Biol.* **13,** 41–46.

49. Matsuda, S., Ichigotani, Y., Okuda, T., Irimura, T., Nakatsugawa, S., and Hamaguchi, M. (2000) Molecular cloning and characterization of a novel human gene (HERNA) which encodes a putative RNA-helicase. *Biochim. Biophys. Acta* **1490,** 163–169.
50. Palmer, M. and Prediger E. (2003) Assessing RNA Quality. *TechNotes* **11,** 13–14.
51. Luckow, V. A. and Summers, M. D. (1988) Signals important for high-level expression of foreign genes in Autographa californica nuclear polyhedrosis virus expression vectors. *Virology* **167,** 56–71.
52. Wilfinger, W. W., Mackey, K., and Chomczynski, P. (1997) Effect of pH and ionic strength on the spectrophotometric assessment of nucleic acid purity. *Biotechniques* **22,** 474–476, 478–481.
53. Muratovska, A. and Eccles, M. R. (2004) Conjugate for efficient delivery of short interfering RNA (siRNA) into mammalian cells. *FEBS Lett.* **558,** 63–68.
54. Simeoni, F., Morris, M. C., Heitz, F., and Divita, G. (2003) Insight into the mechanism of the peptide-based gene delivery system MPG: implications for delivery of siRNA into mammalian cells. *Nucleic Acids Res.* **31,** 2717–2724.
55. Anonymous (2003) Whither RNAi? *Nat. Cell Biol.* **5,** 489–490.
56. Parrish, S., Fleenor, J., Xu, S., Mello, C., and Fire, A. (2000) Functional anatomy of a dsRNA trigger. Differential requirement for the two trigger strands in RNA interference. *Mol. Cell* **6,** 1077–1087.

9

Inhibition of Gene Expression In Vivo Using Multiplex siRNA

Sung-Suk Chae and Timothy Hla

1. Introduction

Angiogenesis, also known as new blood vessel formation, is a critical process during physiological and pathological conditions *(1,2)*. During normal development, the physiological wound healing process, and the female menstrual cycle, blood vessel growth is coordinated concomitantly with tissue growth. However, during pathological processes such as solid cancer development, diabetic retinopathy, rheumatoid arthritis, and psoriasis, abnormal angiogenesis contributes to disease progression. Thus, much effort is directed toward understanding of and the control of the angiogenesis process *(3)*. Angiogenesis involves the orderly migration, proliferation, and differentiation of vascular endothelial cells into nascent blood vessel sprouts. This is followed by vascular stabilization, which entails the coverage with mural cells (pericytes or vascular smooth muscle cells) *(4)*. Numerous factors regulate the various phases of angiogenesis. A major regulator of angiogenesis, vascular endothelial growth factor (VEGF), was upregulated in tumor angiogenesis, and inhibition of this factor with a specific monoclonal antibody was recently approved for clinical use in colorectal cancer *(3)*. However, angiogenesis is regulated by multiple redundant pathways; therefore, multiple approaches might be needed to effectively control angiogenesis.

Sphingosine 1-phosphate (S1P) is a lipid mediator produced from the metabolism of sphingomyelin, a membrane phospholipid. S1P binds to G-protein-coupled receptors (GPCR) on cell surface and transduce intracellular signals that regulate cell migration, proliferation, survival, and morphogenesis *(5)*. The prototypical S1P receptor, $S1p_1$ (also known as EDG-1), was isolated from

From: *Methods in Molecular Biology, vol. 309: RNA Silencing: Methods and Protocols*
Edited by: G. G. Carmichael © Humana Press Inc., Totowa, NJ

endothelial cells stimulated with the tumor-promoting phorbol ester *(6)*. S1P$_1$ is required for embryonic vascular stabilization because deletion of the gene for *S1p$_1$* resulted in embryonic lethality as a result of a defect in vascular stabilization *(7)*. However, the role of S1P$_1$ during angiogenesis in the adult, particularly, in tumor angiogenesis, has not been assessed. We have recently shown that S1P$_1$ is induced in tumor blood vessels in the mouse *(8)*. In addition, activation of S1P$_1$ with its ligand results in endothelial cell migration, an important first step in angiogenesis *(9)*. S1P$_1$ antagonists are not available at present; therefore, we have developed an RNA-interference-based procedure *(10)* to downregulate this receptor at the posttranscriptional level, with the aim of regulating angiogenesis.

RNA interference (RNAi) has been used successfully in mammalian cells by the application of small, interfering (si)RNA species into cultured cells *(11)*. Synthetic siRNAs have been used widely in downregulating many genes *(12–14)*. However, the synthetic siRNAs are costly for experiments in animals, especially for long-term experiments such as tumor growth models. In addition, not all siRNA species are effective at downregulating the genes of interest for reasons that are not fully understood *(10)*. Multiplex siRNA approach has been used, which might increase the chances of successful gene silencing. In this chapter, we discuss the use of *Escherichia coli* RNase III *(15,16)* to generate multiplex siRNA species effective at downregulating genes in vivo.

2. Materials

1. Oligofectamine (Invitrogen Inc.).
2. Opti-MEM tissue culture medium (Invitrogen Inc.).
3. Synthetic siRNA (e.g., Dharmacon Inc. or Ambion Inc.).
4. Endothelial cell growth medium (Medium 199, 10% fetal bovine serum [FBS], antibiotic antimycotic mixture, 150 μg/mL endothelial cell growth factor, 5 U/mL heparin).
5. siRNA annealing buffer: 30 mM HEPES–KOH (pH 7.4), 100 mM potassium acetate, 2 mM magnesium acetate.
6. Human umbilical vein endothelial cells, Lewis lung carcinoma cells (American Type Culture Collection).
7. *E. coli* RNase III (from Dr. Allan Nicholson, Wayne State University, Detroit, MI). The recombinant His$_6$-tagged RNaseIII enzyme was purified by Ni^{2+}-affinity chromatography as described in **ref. 15**.
8. RNase III digestion buffer: 30 mM Tris-HCl (pH 8.0), 160 mM NaCl, 0.1 mM DTT (dithiothreitol), 0.1 mM EDTA.
9. 1 M MgCl$_2$.
10. T7 RNA polymerase, RQ1 DNase (Promega Biotech).
11. Nude mice (female athymic, NCR *nu/nu*, 4–6 wk of age).
12. Anesthetics (e.g., Tribromoethanol, Avertin).
13. Matrigel (BD Bioscience Inc.).

14. Recombinant human basic FGF (BD Biosciences Inc.).
15. Rat anti-mouse CD31/ PECAM-1 antibody (Pharmingen Inc.).
16. 1,2-Dioleoyl-3-trimethylammonium-propane (DOTAP), cholesterol, mini-extruder (Avanti Polar Lipids Inc.).
17. Sense S1P$_1$ 3′ UTR primers:

    Forward:
    5′ TCCTAATACGACTCACTATAGGGCTGTTGATACTGAGGGAAGC 3′,
    Reverse:
    5′ TCCACAACCTCCTTCTGATGA 3′

18. Antisense S1P$_1$ 3′-UTR primers:

    Forward:
    5′ CTGTTGATACTGAGGGAAGC 3′,
    Reverse:
    5′ TCCTAATACGACTCACTATAGGGTCCACAACCTCCTTCTGATGA 3′.

3. Methods

The methods described in this section outline (1) the generation of the multiplex siRNA for S1P$_1$, (2) in vitro inhibition of gene expression in endothelial cells by multiplex siRNA, (3) inhibition of gene expression in vivo Matrigel plug assay using multiplex siRNA, and (4) inhibition of gene expression in growing tumors in mice by local injection of multiplex siRNA.

3.1. Generation of Multiplex siRNA for S1P

1. Use the mouse *S1p$_1$* 3′-UTR as a template for the preparation of the multiplex siRNA. Polymerase chain rection (PCR)-amplify a 500-bp region using the primer pairs for sense S1P$_1$-3′-UTR and antisense S1P$_1$-3′-UTR (*see* **Note 1**). Standard PCR methods are employed using the murine S1P$_1$ plasmid *(17)*.
2. Purify the PCR products using a QIAquick gel extraction kit (Qiagen Inc).
3. Use 1–5 µg of purified template DNA in a T7 RNA polymerase reaction to generate sense and antisense RNAs.
4. Treat reaction products with RQ1 DNase for 15 min at 37°C to remove the DNA template and purify products by phenol/chloroform extraction and ethanol precipitation.
5. Anneal these RNA fragments to produce double-stranded RNA (dsRNA) templates for the *E. coli* RNAse III reaction to produce multiplex siRNA. Solubilize and mix equal amounts of sense and antisense RNAs (125 µg/mL) in annealing buffer, heat at 95°C for 5 min, and then cool to room temperature to allow the formation of a dsRNA substrate.
6. Isolate hybridized RNA by ethanol precipitation (*see* **Note 2**).
7. Incubate samples with 500 n*M* purified RNase III in the digestion buffer without MgCl$_2$ (120 µg/mL) for 5 min at 37°C to allow the enzyme to bind substrate RNAs.
8. Add MgCl$_2$ (final, 10 m*M*) to induce the enzymatic cleavage and incubate for 0.25–4 h at 37°C. Complete digestion is obtained at 4 h of treatment (*see* **Note 3**).
9. Stop reaction by adding excess EDTA (final, 20 m*M*).

10. Remove enzymes and collect small duplex RNA species by phenol/chloroform extraction and ethanol precipitation.
11. Check the digestion status by running 0.5–1 µg of product on 12.5% nondenaturing polyacrylamide gel and stain with ethidium bromide.

3.2. Inhibition of Gene Expression in Endothelial Cells by Multiplex siRNA

1. Grow endothelial cells to subconfluent cultures (30–40% confluency) as described in **ref. 18**.
2. Complex siRNAs (200 n*M*) and Oligofectamine and add to cultures as described by the manufacturer.
3. Then, 24–48 h after treatment, trypsinize cells, neutralize the trypsin using soybean trypsin inhibitor, and then wash the cells.
4. Count cells in a hemocytometer and plate 50,000 cells in a Boyden chamber for cell migration assays *(18)*. Some cells can also be used for RNA or protein analysis by Northern or Western blot analysis, respectively *(19)*. In these assays, multiplex siRNA for S1P$_1$ potently suppresses S1P$_1$ mRNA expression and inhibits S1P-induced endothelial cell migration *(8)*.

3.3. Inhibition of Gene Expression In Vivo in the Matrigel Plug Using Multiplex siRNA

Matrigel angiogenesis assay was performed as previously described *(19)*.

1. Thaw frozen Matrigel at 4°C overnight.
2. Add siRNA (final concentration, 0–2 µ*M*) to 200–400 µL of Matrigel solution on ice and, if necessary, dilute with sterile plain Dulbecco's modified Eagle's medium (DMEM) or phosphate-buffered saline (PBS). The final concentration of matrix protein in the final mixture should be greater than 10 mg/mL to allow proper gel to form. Some Matrigel plugs contain basic FGF (4 µg/mL) and heparin (40 U/mL).
3. Thoroughly mix by pipetting up and down while avoiding the formation of bubbles.
4. Inject Matrigel mixtures subcutaneously into the dorsum of anesthetized FVB/N mice with a 27-gage needle (*see* **Note 4**). A small plug should be visible on the skin.
5. Allow the animals to recover and harvest the Matrigel plugs 7–10 d postimplantation. At the time of harvest, euthanize the animals and dissect the skin around the Matrigel plug.
6. Invert the skin flap, wash in sterile PBS, and fix in 10% formalin/PBS for 2 h at room temperature.
7. Wash the plugs in PBS for 30 min and then incubate in 70% ethanol for >24 h at 4°C and embed in paraffin.
8. Cut and process sections (5 µm) for routine histology with hematoxylin and eosin or PECAM immunostaining to reveal the extent of vascularization.
9. In some cases, Matrigel plugs can be freshly dissected and homogenized in chaos RNA isolation buffer to purify RNA (*see* **Note 4**), which is then analyzed by

Northern blot assays *(19)*. The results show that multiplex siRNA for S1P$_1$ potently inhibits S1P$_1$ gene expression and vascularity, whereas irrelevant siRNAs do not suppress these parameters *(8)*.

3.4. Inhibition of Gene Expression in Tumor Xenograft in Mice by Local Injection of Multiplex siRNA

3.4.1. Preparation of Cationic Liposome

1. Mix 11 : 9 molar ratio of DOTAP and cholesterol in chloroform until the solution is clear.
2. Evaporate solvents by using nitrogen stream at 42°C and placing the sample in vacuum for 1 h at room temperature.
3. Suspend dried lipids in 5% sucrose (10 mg/mL) at 42°C for 1 h with occasional agitation and sonicate in a bath sonicator for 20 min at 42°C.
4. Produce unilamellar liposome vesicles by passing the sonicated lipid vesicles at least 20 times through a mini-extruder *(8)*.
5. Store under argon or nitrogen gas at 4°C.

3.4.2. Inhibition of Gene Expression in Tumor Xenografts

1. Mix Lewis lung carcinoma cells (10^6) with 100 µL of Matrigel solution at 4°C and inject subcutaneously into the dorsal skin of nude mice.
2. Allowed these cells to establish into growing tumor xenografts. Tumor growth is visible around 3–5 d.
3. Make mixtures of siRNA/liposome complexes (60 µL) under sterile conditions by incubating 16 µg of siRNA and 25 µL of liposomes (10 mg/mL) in 5% glucose for 15 min at room temperature.
4. Inject siRNA/liposome complexes into four to six sites randomly into the growing tumor using a 28G insulin syringe. Repeat the injection protocol every 3 d.
5. Determine the tumor volume using a caliper to measure tumor diameter *(8)*. At various time periods, harvest tumors and analyze samples for gene expression using Northern blot, Western blot analysis, or immunohistochemical procedures.

Using this technology, multiplex siRNA for S1P$_1$ potently suppresses expression of the S1P$_1$ receptor, microvessel density, and tumor growth rate *(8)*. In contrast, injection of liposomes alone or irrelevant multiplex siRNA/ liposome complexes does not influence these parameters *(8)*.

4. Notes

1. The sense and antisense fragments contain fused T7 polymerase promoter sites so that sense and antisense RNAs can be readily generated with the T7 RNA polymerase using PCR-generated DNAs.
2. If necessary, obtained multiplex siRNA can be purified using QIAquick PCR purification kit (Qiagen Inc.) Products can be purified by phenol/chloroform extraction followed by ethanol precipitation. Nondenaturing gel electrophoresis of products indicated a size range of 12–20 bp.

3. Make sure injection is done subcutaneously. Do not inject into the muscle or intradermally.

4. Be cautious not to be contaminated with adjacent tissues when extracting Matrigel plugs for protein or RNA analysis.

Acknowledgments

The authors thank Dr. Allan Nicholson for the gift of the RNase III plasmid, and Dr. Henry Furneaux and Dr. Ji-Hye Paik for helpful discussions.

References

1. Risau, W. (1997) Mechanisms of angiogenesis. *Nature* **386**, 671–674.
2. Ferrara, N., Gerber, H. P., and LeCouter, J. (2003) The biology of VEGF and its receptors. *Nat. Med.* **9**, 669–676.
3. Ferrara, N., Hillan, K. J., Gerber, H. P., and Novotny, W. (2004) Discovery and development of bevacizumab, an anti-VEGF antibody for treating cancer. *Nat. Rev. Drug Discov.* **3**, 391–400.
4. Jain, R. K. (2003) Molecular regulation of vessel maturation. *Nat. Med.* **9**, 685–693.
5. Hla, T. (2001) Sphingosine 1-phosphate receptors. *Prostaglandins* **64**, 135–142.
6. Hla, T. and Maciag, T. (1990) An abundant transcript induced in differentiating human endothelial cells encodes a polypeptide with structural similarities to G-protein-coupled receptors. *J. Biol. Chem.* **265**, 9308–9313.
7. Allende, M. L., Yamashita, T., and Proia, R. L. (2003) G-Protein-coupled receptor S1P acts within endothelial cells to regulate vascular maturation. *Blood* **102**, 3665–3667.
8. Chae, S. S., Paik, J-H., Furneaux, H., and Hla, T. (2004) Requirement for sphingosine 1-phosphate receptor-1 in tumor angiogenesis demonstrated by in vivo RNA interference. *J. Clin. Invest.*, in press.
9. Lee, M. J., Thangada, S., Paik, J. H., et al. (2001) Akt-Mediated phosphorylation of the G protein-coupled receptor EDG-1 is required for endothelial cell chemotaxis. *Mol. Cell* **8**, 693–704.
10. Hannon, G. J. (2002) RNA interference. *Nature* **418**, 244–251.
11. Elbashir, S. M., Lendeckel, W., and Tuschl, T. (2001) RNA interference is mediated by 21- and 22-nucleotide RNAs. *Genes Dev.* **15**, 188–200.
12. Song, E., Lee, S. K., Wang, J., et al. (2003) RNA interference targeting Fas protects mice from fulminant hepatitis. *Nat. Med.* **9**, 347–351.
13. Rubinson, D. A., Dillon, C. P., Kwiatkowski, A. V., et al. (2003) A lentivirus-based system to functionally silence genes in primary mammalian cells, stem cells and transgenic mice by RNA interference. *Nat. Genet.* **33**, 401–406.
14. Dave, R. S. and Pomerantz, R. J. (2003) RNA interference: on the road to an alternate therapeutic strategy! *Rev. Med. Virol.* **13**, 373–385.
15. Amarasinghe, A. K., Calin-Jageman, I., Harmouch, A., Sun, W., and Nicholson, A. W. (2001) *Escherichia coli* ribonuclease III: affinity purification of hexahistidine-tagged enzyme and assays for substrate binding and cleavage. *Methods Enzymol.* **342**, 143–158.

16. Yang, D., Buchholz, F., Huang, Z., et al. (2002) Short RNA duplexes produced by hydrolysis with *Escherichia coli* RNase III mediate effective RNA interference in mammalian cells. *Proc. Natl. Acad. Sci. USA* **99,** 9942–9947.

17. Liu, C. H. and Hla, T. (1997) The mouse gene for the inducible G-protein-coupled receptor edg-1. *Genomics* **43,** 15–24.

18. Paik, J. H., Chae, S., Lee, M. J., Thangada, S., and Hla, T. (2001) Sphingosine 1-phosphate-induced endothelial cell migration requires the expression of EDG-1 and EDG-3 receptors and Rho-dependent activation of alpha vbeta3- and beta1-containing integrins. *J. Biol. Chem.* **276,** 11,830–11,837.

19. Ancellin, N., Colmont, C., Su, J., et al. (2002) Extracellular export of sphingosine kinase-1 enzyme. Sphingosine 1-phosphate generation and the induction of angiogenic vascular maturation. *J. Biol. Chem.* **277,** 6667–6675.

10

Gene Silencing by a DNA Vector-Based RNAi Technology

Guangchao Sui and Yang Shi

1. Introduction

Double-stranded RNA (dsRNA) can suppress gene expression by inducing mRNA degradation of the homologous gene, known as RNA interference (RNAi). First, an RNase III-like dsRNA-specific endonuclease, Dicer, cleaves long dsRNA into 21- to 23-nucleotide (nt) small, interfering RNAs (siRNAs). Second, each resulting siRNA is incorporated into an RNA-induced silencing complex (RISC), which consists of eIF2C1, eIF2C2, Gemin3 (an RNA helicase), Gemin4 and other proteins. The siRNA guides the RISC to the homologous mRNA. Finally, the endoribonuclease in the RISC cleaves the targeted mRNA at the vicinity of the binding site, followed by further degradation of the mRNA by an exoribonuclease. As a result, gene silencing is established.

The RNAi technology has been used to knock down gene expression in order to study its functions in multiple model organisms, including plants, *Caenorhabditis elegans*, and *Drosophila*, where large dsRNAs efficiently induce gene-specific silencing. However, in mammalian cells, introduction of dsRNAs longer than 30 nt will induce the interferon response that results in degradation of RNA transcripts and blockage of protein synthesis. This obstacle was overcome by introducing in vitro synthesized 21-nt siRNA into mammalian cells to suppress target gene expression *(1)*.

Recently, we and others *(2–9)* have also developed a DNA vector-based RNAi technology that produces functional double-stranded siRNAs to suppress gene expression in mammalian cells. This approach makes it possible to study gene function over a long period of time in cultured cells and in animals. Specifically, short-hairpin RNAs (shRNAs) are synthesized from a DNA template

From: *Methods in Molecular Biology, vol. 309: RNA Silencing: Methods and Protocols*
Edited by: G. G. Carmichael © Humana Press Inc., Totowa, NJ

under the control of U6 or H1 promoter that employs RNA polymerase III (Pol III). The resulting shRNAs are believed to be processed by Dicer to generate small RNA duplexes that resemble the active siRNA duplexes. Two alternative DNA vector-based RNAi approaches have also been reported. These include transcribing the sense and antisense strands by two tandem U6 promoters on the same plasmid *(8,9)* or on two separate plasmids *(7)*. However, these two approaches do not appear to work as efficiently as the hairpin strategy to repress gene expression *(7,10)*.

A distinct advantage of the DNA vector-based RNAi approach is the ability to generate stable cell lines in which the desired siRNAs can be either constitutively or inducibly expressed *(3,10–14)*. This enables analysis of gene function over an extended period of time. The inducible system also makes it possible to analyze genes that are essential for viability. In this chapter, we focus our discussion on the DNA vector-based shRNA strategy used in this lab.

2. Materials

1. pBluescript/U6 (pBS/U6) expression vector *(2)* (*see* **Fig. 1**) (i.e. pSilencer 1.1; Ambion).
2. Competent cells of *Escherichia coli* strain DH5α and Luria–Bertani (LB) broth medium with ampicillin.
3. Oligonucleotides (oligos).
4. Polymerase chain reaction (PCR) reagents, including the Platinum *pfx* DNA polymerase system or Platinum PCR SuperMix (Invitrogen, Corp., Carlsbad, CA).
5. Restriction enzymes (*Hind*III, *Eco*RI, *Kpn*I, *Eco*RV) and T4 DNA ligase.
6. Agarose gel.
7. Qiagen gel extraction, PCR purification and plasmid miniprep kits.
8. Isopropylthio-β-D-galactoside (IPTG) and 5-bromo-4-chloro-3-indolyl-β-D-galactoside (X-gal).
9. Transfection reagents: Lipofectamine 2000 (Invitrogen), Fugene 6 (Roche), or buffers used in calcium phosphate precipitation of DNA.
10. Equipments and reagents for sodium dodecyl sultate–polyacrylamide gel electrophoresis (SDS-PAGE) and Western blot.
11. DAPI (4′-6-diamidino-2-phenylindole) and immunofluorescence-conjugated secondary antibodies.
12. Immunofluorescence microscope.

3. Methods

The methods described in this section outline (1) the description of the pBS/U6, (2) the design of the hairpin DNA, (3) the preparation of the vector and inserts, (4) the ligation and transformation, (5) the identification of the recombinant siRNA plasmids, and (6) the introduction of siRNA constructs into mammalian cells.

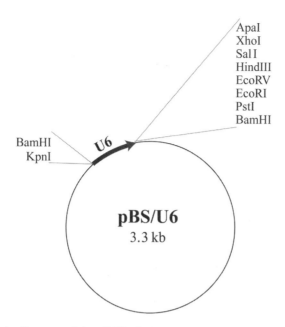

Fig. 1. Schematic diagram of the pBS/U6 plasmid used to generate siRNA plasmids.

3.1. The pBS/U6 Vector

The pBS/U6 (*see* **Fig. 1**) expression vector *(2)* was constructed by sub-cloning the mouse U6 promoter (–315 to –1) into the *Sac*I and *Apa*I sites of pBluescript (SK). The size of pBS/U6 is 3289 bp with the T_7 promoter located at the upstream of the U6 promoter and T3 at its downstream.

When recruited to the U6 promoter, RNA Polymerase III (Pol III) begins transcription at the first guanidine nucleotide (G) (designated as the +1 position). To obtain optimal transcription, we use this G as the first nucleotide of a target sequence (*see* **Fig. 2A**). When the H1 promoter is used, Pol III-directed transcription begins at an adenosine nucleotide (A) at the +1 position, but efficient transcription can also begin with other nucleotides at the +1 position *(15)*. The stop signal for Pol IIIs transcription is an uninterrupted string of four or five thymidines (Ts) *(16)* (*see* **Fig. 2A**).

3.2. Design of the Hairpin DNA

A pair of 21–23 nt of DNA (containing target sequence) with a palindrome symmetric structure linked by a short loop (6–9 nt) is inserted downstream of the U6 promoter. As shown in **Fig. 2A**, the sequence 1a (designated as the "target sequence") is chosen from the coding sequence of the target cDNA, whereas 2b is identical to 1a with the exception of some alterations indicated

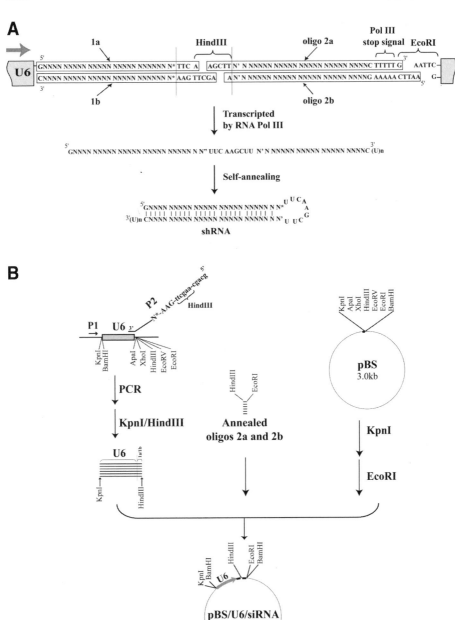

Fig. 2. Schematic diagram for the generation of a siRNA plasmid and the transcribed shRNA. (**A**) Appearance of the downstream region of U6 promoter in an siRNA plasmid and the transcribed shRNA. (**B**) Procedures to generate a pBS/U6/siRNA plasmid by three-fragment ligation.

below. Sequence 1b is complementary to the 1a, and 2a is complementary to 2b. At the 3′ end of 2a, there is a string of 5 Ts that serve as the stop signal for RNA Pol III. The loop in the middle has a TTC and a *Hin*dIII site used for subcloning (*see* **Fig. 2A**).

Target sequence selection is an important step in successfully knocking down of a target gene. However, the criteria given here are still largely empirical (*see* **Note 1**). Recent studies have begun to elucidate the basis for rational design of siRNAs *(17–21)*. Some of these principles are discussed in this subsection. In addition, computer programs have been reported to facilitate the selection of siRNA target sequences *(22)*. However, information is currently lacking with respect to the success rate of these prediction programs.

1. Find the regions of a cDNA to choose target sequences. A target sequence must be specific to the target gene and show no significant homology to any other genes. Using the BLAST search, regions of the target cDNA with no or low homology to other genes can be identified, from which candidate siRNA target sequences can be chosen.

2. A target sequence should start with a G. RNA Pol III always starts its transcription with a G from the U6 promoter. Therefore, one needs to find a region that begins with a G as a target sequence candidate.

3. Do not leave any string of four Ts in the designed hairpin. Four or five Ts is a stop signal for the transcription of Pol III and their presence in the designed hairpin will lead to premature transcriptional termination.

4. Avoid sequences containing *Kpn*I or *Hin*dIII site. *Kpn*I and *Hin*dIII are used to digest the PCR products later. Their presence in a target sequence will result in nonfunctional constructs.

5. Avoid sequences close to the ATG translational start codon. The region close to ATG on the mRNA might be associated with multiple proteins involved in translation that could interfere with RISC binding. A target sequence can also be selected from a 3′-UTR (untranslated region) *(23)*.

6. Avoid sequences with internal repeats or palindromes. The presence of these structures will reduce the production of functional hairpins *(20)*.

7. Use a sequence with a low G/C content, especially at its 3′ end. siRNAs with lower G/C content are believed to yield better silencing *(19)*. We recommend that the G/C content be kept between 35% and 55% (*see* **Note 2**).

MicroRNAs (miRNAs) accumulate in vivo as single-stranded species *(17)*, suggesting that there must be structural features within the endogenous siRNA and miRNA that favor the incorporation of one strand of the RNA duplex over the other into the RISC. Recent studies show that it is the stability of the basepairs at the 5′ ends of the two siRNA strands that determines which strand gets incorporated into the RISC and which strand gets degraded *(17)*. Reducing the strength of the base-pairing at the 5′ end of a strand promotes the incorporation of this strand into the RISC, presumably as a result of facilitated entry of an RNA helicase that helps unwind siRNA duplexes *(17,19)*. This asymmetry model helps to design better siRNAs that favor the incorporation of the antisense strand, which guides the RISC to the target mRNA *(24)*. Specifically, this can be achieved by placing the 5' end of the antisense siRNA strand (corresponding to the 3' end of the target sequence) in a mismatch (see the N*/N' pair in **Fig. 2A** and **Table 1**) *(17)*. The decreased incorporation of the sense strand is predicted to also have the benefit of reducing the "off-target" effects caused by the sense-strand RNA *(25,26)*.

8. Use a sequence with high specificity to the target gene. All target sequence candidates need to be analyzed using the NCBI/BLAST website to ensure that they do not significantly match any other gene sequence.

3.3. Preparation of the Vector and the Inserts

Human *yy1* will be used as an example throughout the rest of this chapter. A target sequence for *yy1* is selected that starts from the 617th base (with the A of the start codon ATG assigned as +1). The oligo set (P2-*yy1*, oligo 2a-*yy1*, and oligo 2b-*yy1*) used to generate pBS/U6/*yy1*-617 is shown in **Table 1**. As a control, a plasmid (pBS/U6/control; *see* **Table 1** for the oligo set) was also generated based on our previously reported enhanced green fluorescat protein (EGFP) target sequence *(2)* with three altered bases to ensure that it lacks homology with any known gene.

A "three-fragment" subcloning is conducted using linearized pBS (digested by *Kpn*I and *Eco*RI), a PCR product containing the entire U6 promoter and the target sequence (digested by *Kpn*I and *Hin*dIII), and the annealed oligos 2a/2b (with 5' and 3' ends ready to be ligated to the *Hin*dIII and *Eco*RI digested ends, respectively) (*see* **Fig. 2** and **Note 3**).

3.3.1. Digestion of pBS/U6 Vector

As shown in **Fig. 2B**, the pBS is first digested by *Kpn*I followed by purification using the Qiagen PCR purification kit. The purified DNA is then digested by *Eco*RI, followed by agarose gel purification using the Qiagen gel extraction kit. The purified pBS (*Kpn*I–*Eco*RI) is then ready for ligation (*see* **Fig. 3A**, lane 2 and **Note 3**).

3.3.2. Generating the First Pair of the Hairpin siRNA by PCR

To generate a DNA fragment containing the U6 promoter and the first pair of the hairpin, PCR is conducted with an upstream primer, P1, and a downstream primer, P2, using the pBS/U6 as a template (*see* **Fig. 2B** and **Table 1**). The 3' end of P2 can anneal to the last 28 nucleotides of the U6 promoter and contains the DNA sequence 1b (indicated in **Fig. 2A**) in the middle and a *Hin*dIII site at the 5' end. When using the *pfx* DNA polymerase system (Invitrogen), 50 µL of reaction contains 10 ng of pBS/U6 as a template, 150 pmol of each P1 and P2 primer, 1X *pfx* buffer, 1 mM MgSO$_4$, 0.15 mM of each dNTP, and 1.25 U of Platinum *pfx* DNA polymerase. The PCR was performed using the

Table 1
Primer Sequences

Name	Sequence	Use
P1	GCC AGG GTT TTC CCA GTC ACG ACG	Forward primer on upstream of U6 promoter.
P2[a]	GCA GC A **AGC TT** G AA N[b] NNN NNN NNN NNN NNN NN A AAC AAG GCT TTT CTC CAA GGG ATA TTT	Reverse primer whose 3′ end (underlined) can anneal the 3′ end of U6 promoter. HindIII site is in bold.
Oligo 2a[a]	**AGC TT** NNN NNN NNN NNN NNN NNN NNN T TTT G	To be annealed with oligo 2b. The T string is underlined and the sticky ends for HindIII and EcoRI are in bold.
Oligo 2b[a]	**AAT TC** A AAAA A NNN NNN NNN NNN NNN NNN NNN A	To be annealed with oligo 2a. The A string (complement to the T string above) is underlined and the sticky ends for EcoRI and HindIII are in bold.
P2-yy1	GCA GC A **AGC TT** G AA T[b] TCT GCA CCT GCT TCT GCT CCC A AAC AAG GCT TTT CTC CAA GGG ATA TTT	P2 oligo for yy1 siRNA construct.
Oligo 2a-yy1	**AGC TT** A TCT GCA CCT GCT TCT GCT CCC T TTT TG	Oligo 2a for yy1 siRNA construct.
Oligo 2b-yy1	**AAT TC** A AAAA A GGG AGC AGA AGC AGG TGC AGA T A	Oligo 2b for yy1 siRNA construct.
P2-control	GCA GC A **AGC TT** G AA G[b] TTG CCG TAC GTG CCA TGG CCC A AAC AAG GCT TTT CTC CAA GGG ATA TTT	P2 oligo for control siRNA construct.
Oligo 2a-control	**AGC TT** C TTG CCG TAC GTG CCA TGG CCC T TTT TG	Oligo 2a for control siRNA construct.
Oligo 2b-control	**AAT TC** A AAAA A GGG CCA TGG CAC GTA CGG CAA G A	Oligo 2b for control siRNA construct.

[a]P2, oligo 2a, and oligo 2b are templates for general designing.
[b]These bases are intentionally mutated to create a mismatched pair close to the loop of the hairpin (see **Fig. 2A**).

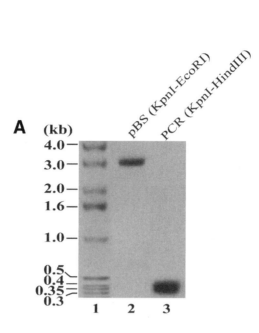

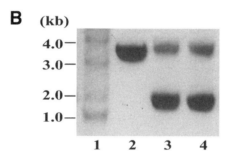

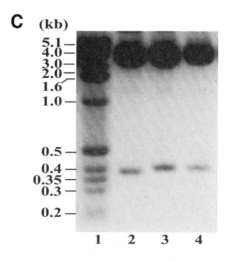

following parameters: 1 cycle of 94°C for 2 min; 30 cycles of 94°C for 30 s, 64°C for 40 s, 72°C for 45 s; 1 cycle of 72°C for 10 min. Platinum PCR SuperMix (Invitrogen) can also be used for PCR with the same conditions. The PCR products are purified and digested with *Hin*dIII and *Kpn*I, followed by agarose gel purification (*see* **Fig. 3A**, lane 3).

3.3.3. Annealing of Oligonucleotides

To anneal oligos 2a and 2b, mix 180 µL 10 m*M* Tris-HCl, pH 7.5, 10 µL oligo 2a, and 10 µL oligo 2b (each about 200 pmol/µL). Heat the tube of the oligo mix for 5 min in 600–700 mL of boiling water in a glass beaker. Keep the tube in the water and let the beaker slowly cool down to room temperature in 30 min to 1 h. The annealed oligos 2a/2b will form a *Hin*dIII and an *Eco*RI sticky end on the two termini (*see* **Fig. 2A**). The oligo 2a contains a string of five T that serves as the stop signal for RNA Pol III (*see* **Note 4**).

3.4. Ligation and Transformation

Twenty nanograms (0.01 pmol) of the pBS (*Kpn*I–*Eco*RI) (*see* **Subheading 3.3.1.** and **Note 5**), 23 ng (0.1 pmol) of the digested PCR fragment (*Kpn*I–*Hin*dIII) (**Subheading 3.3.2.**) and 2.1 ng (0.1 pmol) of the annealed oligos 2a/2b (**Subheading 3.3.3.**; *see* **Note 4**) are mixed in a 20-µL total volume containing 1X ligation buffer and 3 Weiss units of T4 DNA ligase (BioLabs). In the control ligation reaction, the pBS (*Kpn*I–*Eco*RI) and the PCR (*Kpn*I–*Hin*dIII) fragment are added and the annealed oligos 2a/2b are omitted. Ligation can be performed at 16°C or room temperature overnight. Five to ten microliters of the ligation product can be used to transform competent cells of *E. coli* DH5α that are then plated onto an ampicillin LB plate pretreated with 40 µL of X-gal (20 mg/mL) and 5 µL of IPTG (200 mg/mL) *(27)*. The bacteria containing the recombinant plasmid are white in color and the plasmids are analyzed further by restriction digestion (below).

3.5. Identification of pBS/U6/siRNA Plasmids

White colonies on the transformation plates are picked and used to inoculate in LB liquid medium containing 50 µg/mL of ampicillin. Plasmids are extracted using the Qiagen plasmid miniprep kit.

Fig. 3. (*Opposite page*) Analytical agarose gel for the subcloning and screening of siRNA plasmids. (**A**) DNA components used in ligation. Lane 1: 1-kb DNA ladder (Invitrogen); lane 2: *Kpn*I–*Eco*RI digested pBS; lane 3: *Kpn*I-*Hin*dIII digested PCR products. (**B,C**) Confirmation of the pBS/U6/*yy1* siRNA plasmids by *Eco*RV (**B**) and *Bam*HI (**C**) digestions.

The pBS/U6/siRNA and pBS/U6 can be distinguished by *Eco*RV and *Bam*HI digestions.

1. In pBS/U6/siRNA, two inserts (digested PCR fragment and annealed oligos 2a/2b) have replaced a part of the polylinker region and, as a result, is resistant to *Eco*RV digestion (*see* **Fig. 3B**, lanes 3 and 4), whereas the pBS/U6 vector is still cleavable by this enzyme (*see* **Fig. 3B**, lane 2).
2. Both pBS/U6/siRNA and pBS/U6 have two *Bam*HI sites (*see* **Figs. 1** and **2B**), but they do not generate the same fragment upon *Bam*HI digestion (397 bp or larger vs 376 bp). This small difference is distinguishable on 3% agarose gel (*see* **Fig. 3C**, lanes 3 and 4 vs lane 2). However, the hairpin construct is not easily verifiable by sequencing because the two inserts may form a tight stem loop that interferes with the sequencing reaction, although success has also been reported *(28)* (*see* **Note 6**).

3.6. Introducing siRNA into Mammalian Cells and Detecting the Knock Down of the Target Genes

3.6.1. Introduction of siRNA into Mammalian Cells

The siRNA plasmids can be introduced into cells using any currently available cell transfection approaches. Calcium phosphate precipitation is an economic method, but it does not provide high enough transfection efficiency in many cell lines tested. Commercial transfection reagents, such as Fugene 6 (Roche) and Lipofectamine (Invitrogen), give higher transfection efficiency in established cell lines, but they do not work as well for primary cells. Virus-based approaches, however, can significantly increase the efficiency of siRNA introduction into the cells *(29,30)*. Viral vectors that have been tested to date include retrovirus *(31,32)*, adenovirus *(33,34)*, and adeno-associated virus *(35)*. In most cases, gene silencing can be detected 2–3 d after transfection or infection, dependent on the abundance and the stability of the proteins encoded by the target genes.

3.6.2. Detection of Target Gene Expression

3.6.2.1. Immunofluorescence

Cells that are transfected by the siRNA plasmid are marked by a cotransfected plasmid expressing green fluorescence protein (GFP) or other autofluorescence proteins. The cells on the cover slip are stained with antibody recognizing the target protein, followed by blotting with fluorescence dye conjugated secondary antibody. If the siRNA plasmid is effective, the signal for the target gene will significantly decrease in the GFP positive cells. **Figure 4A** shows the knocking down of YY1 expression in HeLa cells.

3.6.2.2. Western Blot

If high transfection efficiency can be achieved, the silencing of the targeted endogenous gene can be visualized by Western blot using the antibody against

A

siRNA plasmids: **Control**   **YY1**

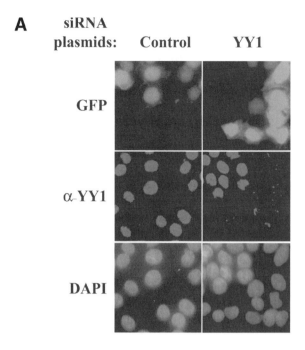

GFP

α-YY1

DAPI

B

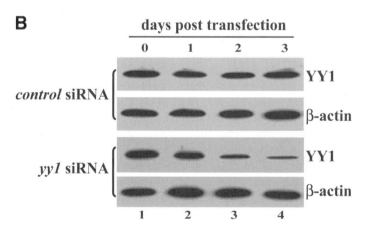

Fig. 4. Detection of YY1 knock down by RNAi in HeLa cells. (**A**) Immunofluorescence analysis of YY1 expression in HeLa cells transfected by pBS/U6/*control* and pBS/U6/*yy1*-617. pCMV/EGFP was cotransfected with each plasmid to mark the transfected cells. **Top panel**: Cells with GFP signals; **middle panel**: same field of cells detected by α-YY1 antibody; **bottom panel**: same field of cells stained by DAPI. (**B**) Western blot analysis of YY1 expression in HeLa cell transfected by pBS/U6/*control* and pBS/U6/*yy1*-617. Cells were harvested at different time points indicated on the top and analyzed by Western blot using YY1 (H-414, Santa Cruz) and β-actin antibodies (MAB1501; Chemicon International, Inc.).

the target protein. The Western blot in **Fig. 4B** shows the YY1 levels from the samples collected at different time points posttransfection.

If the transfection efficiency is low, Western blot might not be able to detect the expression difference of the endogenous target gene. However, the efficacy of siRNA construct can be determined by Western blot on the suppression of the expression of the cotransfected target gene tagged by an epitope. These experiments usually use a molar ratio of at least 5:1 to 20:1 (siRNA vs target gene expression plasmid).

4. Notes

1. It is advisable that two to three independent siRNA constructs with different targeted sequences for the same gene be generated in order to validate the phenotype. This can also be determined using rescue experiments by expressing the target cDNA with silent mutations that make it resistant to the siRNA.

2. If for some reason the above target sequence selection criteria cannot be satisfied upon examination of the coding region, the noncoding strand of the cDNA can also be used for selection of the "target sequence." However, lower G/C content at the 5′ end, instead of the 3′ end, should be considered to ensure that the antisense strand is favored for incorporation into the RISC.

3. We have improved the "two-step" subcloning strategy reported previously *(2)* and developed a one-step ligation protocol. The new strategy is faster and more efficient. We and other labs have tried to synthesize two long oligos by fusing the "1a–loop–2a" and "1b–loop–2b," respectively, and anneal them followed by subcloning them downstream of the U6 promoter. Although this procedure theoretically should work, in practice it can be problematic because each of the two long oligos might form stem-loop structures or self-anneal, which prevents the correct annealing.

4. The annealed oligos 2a/2b do not need to be phosphorylated before the ligation step because it might lead to multiple insertions.

5. The sequential digestion of pBS by *Kpn*I and *Eco*RI followed by agarose gel purification is recommended to ensure a complete digestion by the two restriction enzymes. This will significantly decrease the background in the screening.

6. Generally, the three-fragment ligation strategy should work well if the above-described procedure is followed. If difficulties are encountered, the two-step cloning procedure can be an alternative. First, subclone the *Kpn*I–*Hin*dIII-digested PCR product into *Kpn*I–*Hin*dIII-digested pBS followed by screening for the plasmid with the insert. Second, the annealed oligos 2a/2b can be subcloned into the *Hin*dIII–*Eco*RI-digested plasmid generated in the first step.

Acknowledgments

The authors are grateful to Dr. Sidney Altman for providing the U6 promoter plasmid pmU6. We thank our colleagues in the Shi lab for helpful discussion. We also thank Dr. Nathan R. Wall for critically reading the manuscript and

Dr. Peter Mulligan for working out the "three-fragment" ligation. G.C.S is supported by a Viral Oncology Training Grant from the NIH (T32CA09031). This work was supported by a grant from the NIH (GM53874) to YS.

References

1. Elbashir, S. M., Harborth, J., Lendeckel, W., Yalcin, A., Weber, K., and Tuschl, T. (2001) Duplexes of 21-nucleotide RNAs mediate RNA interference in cultured mammalian cells. *Nature* **411,** 494–498.

2. Sui, G., Soohoo, C., Affar el, B., et al. (2002) A DNA vector-based RNAi technology to suppress gene expression in mammalian cells. *Proc. Natl. Acad. Sci. USA* **99,** 5515–5520.

3. Brummelkamp, T. R., Bernards, R., and Agami, R. (2002) A system for stable expression of short interfering RNAs in mammalian cells. *Science* **296,** 550–553.

4. Paddison, P. J., Caudy, A. A., and Hannon, G. J. (2002) Stable suppression of gene expression by RNAi in mammalian cells. *Proc. Natl. Acad. Sci. USA* **99,** 1443–1448.

5. Paul, C. P., Good, P. D., Winer, I., and Engelke, D. R. (2002) Effective expression of small interfering RNA in human cells. *Nat. Biotechnol.* **20,** 505–508.

6. McManus, M. T., Petersen, C. P., Haines, B. B., Chen, J., and Sharp, P. A. (2002) Gene silencing using micro-RNA designed hairpins. *RNA* **8,** 842–850.

7. Yu, J. Y., DeRuiter, S. L., and Turner, D. L. (2002) RNA interference by expression of short-interfering RNAs and hairpin RNAs in mammalian cells. *Proc. Natl. Acad. Sci. USA* **99,** 6047–6052.

8. Miyagishi, M. and Taira, K. (2002) U6 promoter-driven siRNAs with four uridine 3′ overhangs efficiently suppress targeted gene expression in mammalian cells. *Nat. Biotechnol.* **20,** 497–500.

9. Lee, N. S., Dohjima, T., Bauer, G., et al. (2002) Expression of small interfering RNAs targeted against HIV-1 rev transcripts in human cells. *Nat. Biotechnol.* **20,** 500–505.

10. Gupta, S., Schoer, R. A., Egan, J. E., Hannon, G. J., and Mittal, V. (2004) Inducible, reversible, and stable RNA interference in mammalian cells. *Proc. Natl. Acad. Sci. USA* **101,** 1927–1932.

11. Kasim, V., Miyagishi, M., and Taira, K. (2004) Control of siRNA expression using the Cre-loxP recombination system. *Nucleic Acids Res.* **32,** e66.

12. Higuchi, M., Tsutsumi, R., Higashi, H., and Hatakeyama, M. (2004) Conditional gene silencing utilizing the lac repressor reveals a role of SHP-2 in cagA-positive *Helicobacter pylori* pathogenicity. *Cancer Sci.* **95,** 442–447.

13. Czauderna, F., Santel, A., Hinz, M., et al. (2003) Inducible shRNA expression for application in a prostate cancer mouse model. *Nucleic Acids Res.* **31,** e127.

14. Matsukura, S., Jones, P. A., and Takai, D. (2003) Establishment of conditional vectors for hairpin siRNA knockdowns. *Nucleic Acids Res.* **31,** e77.

15. Tuschl, T. (2002) Expanding small RNA interference. *Nat. Biotechnol.* **20,** 446–448.

16. Bogenhagen, D. F., Sakonju, S., and Brown, D. D. (1980) A control region in the center of the 5S RNA gene directs specific initiation of transcription: II. The 3′ border of the region. *Cell* **19,** 27–35.

17. Schwarz, D. S., Hutvagner, G., Du, T., Xu, Z., Aronin, N., and Zamore, P. D. (2003) Asymmetry in the assembly of the RNAi enzyme complex. *Cell* **115,** 199–208.
18. Khvorova, A., Reynolds, A., and Jayasena, S. D. (2003) Functional siRNAs and miRNAs exhibit strand bias. *Cell* **115,** 209–216.
19. Reynolds, A., Leake, D., Boese, Q., Scaringe, S., Marshall, W. S., and Khvorova, A. (2004) Rational siRNA design for RNA interference. *Nat. Biotechnol.* **22,** 326–330.
20. Mittal, V. (2004) Improving the efficiency of RNA interference in mammals. *Nat. Rev. Genet.* **5,** 355–365.
21. Ui-Tei, K., Naito, Y., Takahashi, F., et al. (2004) Guidelines for the selection of highly effective siRNA sequences for mammalian and chick RNA interference. *Nucleic Acids Res.* **32,** 936–948.
22. Wang, L. and Mu, F. Y. (2004) A Web-based design center for vector-based siRNA and siRNA cassette. *Bioinformatics* **4,** 4.
23. McManus, M. T., Haines, B. B., Dillon, C. P., et al. (2002) Small interfering RNA-mediated gene silencing in T lymphocytes. *J. Immunol.* **169,** 5754–5760.
24. Martinez, J., Patkaniowska, A., Urlaub, H., Luhrmann, R., and Tuschl, T. (2002) Single-stranded antisense siRNAs guide target RNA cleavage in RNAi. *Cell* **110,** 563–574.
25. Jackson, A. L., Bartz, S. R., Schelter, J., et al. (2003) Expression profiling reveals off-target gene regulation by RNAi. *Nat. Biotechnol.* **21,** 635–637.
26. Scacheri, P. C., Rozenblatt-Rosen, O., Caplen, N. J., et al. (2004) Short interfering RNAs can induce unexpected and divergent changes in the levels of untargeted proteins in mammalian cells. *Proc. Natl. Acad. Sci. USA* **101,** 1892–1897.
27. Sambrook, J., Fritsch, E. F., and Maniatis, T., eds. (1990) *Molecular Cloning, A Laboratory Manual.* Cold Spring Harbor Laboratory, Cold Spring Harbor, NY. p. 1.86.
28. Paddison, P. J., Silva, J. M., Conklin, D. S., et al. (2004) A resource for large-scale RNA-interference-based screens in mammals. *Nature* **428,** 427–431.
29. Tiscornia, G., Singer, O., Ikawa, M., and Verma, I. M. (2003) A general method for gene knockdown in mice by using lentiviral vectors expressing small interfering RNA. *Proc. Natl. Acad. Sci. USA* **100,** 1844–1848.
30. Rubinson, D. A., Dillon, C. P., Kwiatkowski, A. V., et al. (2003) A lentivirus-based system to functionally silence genes in primary mammalian cells, stem cells and transgenic mice by RNA interference. *Nat. Genet.* **33,** 401–406.
31. Matta, H., Hozayev, B., Tomar, R., Chugh, P., and Chaudhary, P. M. (2003) Use of lentiviral vectors for delivery of small interfering RNA. *Cancer Biol. Ther.* **2,** 206–210.
32. Devroe, E. and Silver, P. A. (2002) Retrovirus-delivered siRNA. BMC *Biotechnol.* **2,** 15.
33. Shen, C. and Reske, S. N. (2004) Adenovirus-delivered siRNA. *Methods Mol. Biol.* **252,** 523–532.
34. Zhao, L. J., Jian, H., and Zhu, H. (2003) Specific gene inhibition by adenovirus-mediated expression of small interfering RNA. *Gene* **316,** 137–141.
35. Tomar, R. S., Matta, H., and Chaudhary, P. M. (2003) Use of adeno-associated viral vector for delivery of small interfering RNA. *Oncogene* **22,** 5712–5715.

11

Plasmid-Based RNA Interference

Construction of Small-Hairpin RNA Expression Vectors

Scott Q. Harper and Beverly L. Davidson

1. Introduction

Recent work demonstrates that RNA interference (RNAi) helps coordinate the flow of information from transcription to protein expression, complicating tremendously our former understanding of how protein expression is regulated *(1,2)*. Inhibitory RNAs can be expressed naturally in cells as microRNAs (miRNAs) or introduced into cells as small, inhibitory RNAs (siRNAs). Both processes inhibit gene expression, but by different mechanisms. In general, miRNAs inhibit translation. One cellular process that benefits from "paused" translation is the nerve synapse *(3)*. siRNAs inhibit expression through targeted degradation of target mRNA. Both miRNAs and siRNAs can be used at the bench to silence mRNA expression. In this chapter, we provide a practical approach to accomplish siRNA-mediated gene silencing, through the generation and introduction of plasmids expressing short-hairpin RNAs (shRNAs) that are subsequently processed to siRNAs in vivo.

2. Materials

1. pAd5mU6 (polymerase chain reaction [PCR] template plasmid).
2. Oligonucleotide primers.
3. Deoxynucleotides (dNTPs).
4. *Escherichia coli* strains DH5α, TOP10 (Invitrogen, Carlsbad, CA).
5. pCR-BluntII TOPO cloning kits (Invitrogen, Carlsbad, CA).
6. Pfu DNA polymerase (Stratagene, Cedar Creek, TX).
7. Restriction enzymes (*Xho*I, *Dpn*I; New England Biolabs, Beverly, MA).
8. Agarose and ethidium bromide.
9. DNA sequencing capability.

From: *Methods in Molecular Biology, vol. 309: RNA Silencing: Methods and Protocols*
Edited by: G. G. Carmichael © Humana Press Inc., Totowa, NJ

10. Kanamycin.
11. Luria–Bertani (LB) broth.
12. Plasmid miniprep reagents (e.g., Qiagen Spin Miniprep kit or homemade reagents).
13. HEK293 cells.
14. Tissue culture media.
15. pVET$_L$CMV-βgal plasmid.
16. Lipofectamine-2000 (Invitrogen, Carlsbad, CA).
17. Fluoreporter *LacZ* quantitation kit (Molecular Probes, Eugene, OR).
18. Microplate fluorometer.
19. Lowry protein assay reagents (DC Protein Assay; Bio-Rad, Hercules, CA).
20. Spectrophotometer.
21. Sodium dodecyl sulfate–polyacrylamide gel electrophoresis (SDS-PAGE) equipment and semidry transfer apparatus (Bio-Rad, Hercules, CA).
22. Anti-Bgal antibody (Biodesign, Saco, ME).
23. Anti-actin antibody (Sigma, St Louis, MO).
24. Horseradish peroxidase (HRP)-coupled goat anti-rabbit secondary antibody (Jackson Immunologicals, West Grove, PA).
25. ECL Plus Western blotting detection reagents (Amersham Biosciences, Arlington Heights, IL).
26. Autoradiographic film (Kodak Biomax MR), film cassette, and developer.

3. Methods

The methods described below outline (1) a PCR-based approach for constructing short-hairpin RNA (shRNA) plasmids and (2) strategies for testing gene knockdown efficacy by shRNA plasmids.

3.1. shRNA Plasmid Construction: One-Step PCR Method

This subsection describes the PCR-based construction of an shRNA expression plasmid that can specifically reduce expression of cotransfected *E. coli* β-galactosidase in mammalian cells (*see* **Note 1**). This construction method employs standard molecular biological techniques, requires relatively little hands-on time, and can be completed quickly. The following subsections discuss the key components of this method, including (1) features of shRNA expression plasmids, (2) the general scheme of the PCR cloning method, (3) the PCR template plasmid and PCR cloning vector, (4) factors involved in oligonucleotide design and synthesis, (5) PCR reaction specifics and cloning of products, (6) screening plasmid DNA minipreps for positive clones, and (7) sequencing DNA hairpins.

3.1.1. Features of shRNA Expression Plasmids

In addition to commonly utilized antibiotic resistance genes and cis-acting elements required for DNA replication, all shRNA expression plasmids contain three key features: (1) a promoter, (2) the shRNA transcription template, and (3) a transcription termination signal.

1. *Promoter.* Various promoters directing both RNA polymerases II and III have been used to transcribe shRNA from plasmids in mammalian cells, including H1, U6, tRNA$_{val}$, and CMV *(4–7)*. Current data suggest that the production of functional shRNA requires proper positioning at or near the transcription start site, regardless of the promoter used *(7)*. The PCR-based method described here provides a means to insert the shRNA transcription template immediately downstream of the promoter transcription start site (*see* **Note 2**).

2. *shRNA Transcription Template.* This DNA sequence serves as the transcription template which gives rise to shRNA (*see* **Fig. 1**). The functional shRNA has a predicted stem-loop structure and leads to target mRNA degradation in a sequence-specific manner (*see* **Fig. 1B**). Therefore, production of an effective shRNA requires sequence specificity with a target mRNA of interest and base-pairing that allows for RNA hairpin formation (*see* **Fig. 1A,B**). As such, there are three key elements to consider when constructing the DNA sequences from which shRNAs arise: (1) Target site sequence: One-half of the stem will encode 19–22 nucleotides (nt) of sequence derived from the mRNA target (*see* **Fig. 1A**). Our laboratory generally uses a 21-nt sequence for shRNA construction. Because no free and reliable siRNA prediction algorithms have been published to date, we generally select a 21-nt target mRNA sequence randomly. (2) siRNA sequence: The siRNA sequence is the perfect reverse complement of the mRNA target site and constitutes the other half of the stem (*see* **Fig. 1B**). (3) Loop sequence: Various publications have described loops of varying sizes and sequence derivations, including loops of palindromic restriction enzyme sites *(4,7,8)*. Importantly, the loop sequence should be placed between the two halves of the stem sequence (*see* **Fig. 1B,C**).

3. *Transcription Termination Signal.* The transcription termination signal should be cloned at or near the 3′ end of the DNA sequence encoding the shRNA (*see* **Fig. 1C**). The PCR-based cloning method described here allows the termination sequence to be cloned immediately 3′ of the shRNA (*see* **Note 2**). The sequence of the termination signal will depend on the promoter used to drive shRNA expression. For promoters directing RNA polymerase III activity, the DNA sequence should be five to six thymidines (T). For promoters directing RNA polymerase II activity, the DNA sequence should be AATAAA.

3.1.2. PCR Method: General Scheme

The PCR method for generating shRNA expression plasmids does not significantly differ from a standard PCR reaction (*see* **Fig. 2A**). The reaction requires a DNA template containing the promoter of interest, a DNA oligonucleotide primer pair, a polymerase, standard buffers, and dNTPs. The most outstanding feature of this PCR reaction is the length of the reverse primer that encodes the shRNA transcription template (*see* **Fig. 2B**). Only approximately one-third of this primer

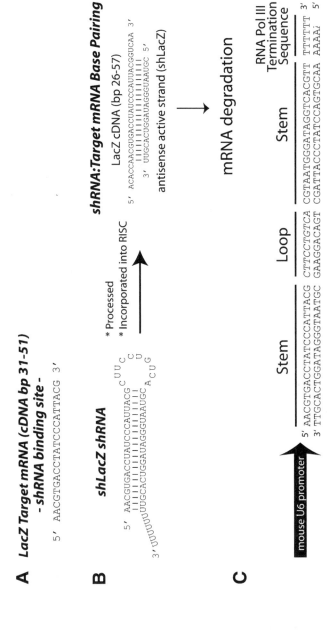

A *LacZ Target mRNA (cDNA bp 31–51)*
- shRNA binding site -

5′ AACGTGACCTATCCCATTACG 3′

B *shLacZ shRNA*

```
                              C U U C
5′ AACGUGACCUAUCCCAUUACG         C
   ||||||||||||||||||||          U
3′ UUUUUUUUGCACUGGAUAGGGUAAUGC  A C U G
```

* Processed
* Incorporated into RISC

shRNA:Target mRNA Base Pairing

LacZ cDNA (bp 26–57)

```
5′ ACACCAACGUGACCUAUCCCAUUACGGUCAA 3′
   |||||||||||||||||||||||
3′ UUGCACUGGAUAGGGUAAUGC 5′
```

antisense active strand (shLacZ)

→ mRNA degradation

C

mouse U6 promoter →

dsDNA plasmid sequence

Stem	Loop	Stem	RNA Pol III Termination Sequence
5′ AACGTGACCTATCCCATTACG	CTTCCTGTCA	CGTAATGGGATAGGTCACGTT	TTTTTT 3′
3′ TTGCACTGGATAGGGTAATGC	GAAGGACAGT	GCATTACCCTATCCAGTGCAA	AAAAi 5′

Fig. 1. (**A**) *LacZ* DNA sequence. The 21-nucleotide DNA sequence corresponds to basepairs 31–51 of the *E. coli LacZ* cDNA. (**B**) Schematic of shRNA-induced *LacZ* gene silencing. Depiction of an RNA hairpin with a stem containing cognate sequence to basepairs 31–51 of *LacZ*. The unpaired loop sequence is derived from mir-23 (*4*) and contains six terminal uridine ribonucleotides that represent an RNA polymerase III termination signal. The antisense strand of the sh*LacZ* hairpin anneals to the *LacZ* target site, catalyzing events resulting in mRNA degradation and suppression of *LacZ* gene expression. (**C**) Plasmid DNA sequence that serves as transcription template for sh*LacZ* 31–51. The 58-basepair DNA sequence encodes the sh*LacZ* RNA stem loop described in (B). This sequence is cloned immediately downstream of the U6 promoter (black arrow, labeled) transcription start site.

222

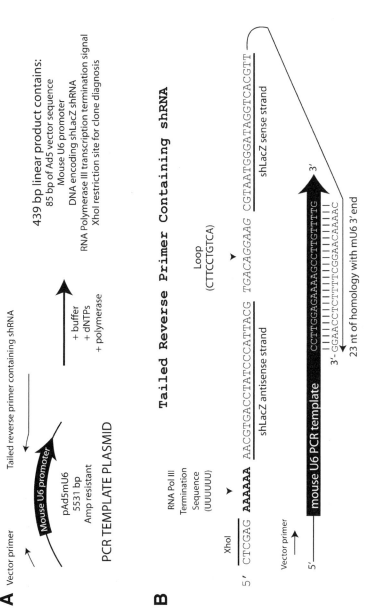

Fig. 2. (A) Overview of PCR method for creating shRNA expression plasmids. (B) Tailed reverse primer containing shRNA. Key features are noted in the text. Note that the primer sequence is the reverse complement of the transcription template DNA sequence depicted in **Fig. 1C**.

actually binds the 3′ end of the promoter, but the remaining tailed end of the primer encoding the shRNA becomes incorporated into the product (*see* **Fig. 2**). Subsequent amplification produces a distinct PCR product containing the entire promoter followed by the shRNA transcription template DNA. Following amplification, the product is cloned directly into a PCR cloning vector and transformed into bacteria, and clones are selected and analyzed using restriction digestion and DNA sequencing. The entire process from PCR reaction to sequence verification can take as little as 4 total days. However, the total hands-on time required to generate an shRNA plasmid can be less than 5 h over the course of those 4 d.

The following subsections describe the key components of this PCR reaction in greater detail. In the example that follows, we describe the construction of a plasmid containing the mouse U6 promoter (mU6; directs RNA polymerase III transcription) followed by stem-loop DNA encoding an shRNA that specifically reduces β-galactosidase expression (heretofore called sh*LacZ*; *see* **Fig. 1C**).

3.1.3. PCR Template Plasmid and PCR Cloning Vector

This method involves the PCR amplification of DNA sequences from a plasmid DNA template. The product is subsequently cloned into a PCR cloning vector. One way to simplify and expedite the cloning procedure is to ensure that the PCR cloning vector contains a different antibiotic resistance gene than the plasmid used as PCR template. Such a design allows the PCR product to be cloned directly from the PCR reaction without undertaking steps to remove contaminating DNA template plasmid. In our example, an ampicillin-resistant, adenovirus shuttle plasmid containing the mouse U6 promoter is used as PCR template (pAd5mU6; *see* **Fig. 2A**). Once amplified, the approx 439-bp product is cloned directly into the pCR-Blunt II TOPO vector (Invitrogen, Carlsbad, CA), which confers kanamycin resistance. Selection of transformed bacteria on kanamycin plates allows for selection of colonies containing only the shRNA expression cassette and eliminates the possibility of colony growth arising from the kanamycin-sensitive PCR template plasmid. If plasmids containing different antibiotic resistance genes cannot be used, additional steps can be taken to eliminate background-producing contaminants (*see* **Note 3**).

3.1.4. Factors Involved in Oligonucleotide Design and Synthesis

The forward PCR primer used in this reaction should encompass the 5′ end of the promoter of interest and ideally satisfy standard criteria for PCR primer design (e.g., 50–60°C melting temperature (T_m); 40–60% GC content; 17–25 nt in length). The forward primer used in this example (5′-TCGCGGGAAACTG AATAAGAGG-3′) is located in Ad5 sequence 85 bp upstream of the mU6 promoter. The reverse primer used to generate the shRNA is 87 nt in length (*see* **Fig. 2B**), and contains three distinct stretches of sequence that are described here in the $3'{\rightarrow}5'$ direction. The $3'$ end of the primer contains 23 nt of sequence that bind the top strand of the mU6 promoter. In the first few cycles of the PCR reaction, these 23 nt are the only sequences that actually participate in the annealing step. As such, we chose the minimum region of homology that provided a primer T_m of $>50°C$. The next 58 nt encode the shRNA transcription template (features described in **Subheading 3.1.1.**). The final 6 nt encode an *Xho*I restriction site that allows for selection of fully extended clones. Inclusion of a restriction site also eliminates the need to purify the primer prior to using it for PCR (discussed in greater detail below).

After the primer has been designed, there are a few issues to consider prior to placing an order with a vendor. Specifically, synthesis of an 87-nt DNA oligo will increase both the cost of the primer and the frequency of truncated products. The latter occurrence would cause the screening of a greater number of clones in order to identify one with the correct sequence. To reduce truncated products, many vendors recommend PAGE purification of large oligos (>50 nt), which can nearly double the cost of the primer. For example, synthesis of an 87-nt primer by a popular commercial vendor costs around $61 without, or $110 with, PAGE purification. We now discuss the application of two simple and practical strategies that lower the cost of a primer below $40. In addition, the second tactic helps to streamline the screening of recombinants.

1. Pay attention to synthesis scale. Most commercial vendors charge a certain rate per base that increases with synthesis scale. In other words, a company might charge a certain rate per base for oligos up to 35 bases in length, but significantly increase the price per base and scale of synthesis for oligos greater than 35 bases in length, and so on. Importantly, synthesis scales and size ranges vary depending on the vendor. Therefore, the same primer ordered from two different companies can vary significantly in price simply because the synthesis scales and size ranges set by the companies are different. To illustrate, Vendor A requires that an 87-nt primer be synthesized at no less than the company's third highest scale (250 n*M*), costing $61. The same primer can be synthesized at Vendor B's second highest synthesis scale (50 n*M*), costing $34. Although the primer obtained from Vendor B costs three times more per mole, the PCR reaction requires only a one-time use of picomolar amounts of primer, so the absolute cost is the most important determinant. If synthesizing only one shRNA plasmid, a $27 difference in price might seem small, but the cumulative cost could be significant if several plasmids are generated.
2. Add a restriction site to the 5′ end of the primer. As mentioned, long PCR primers generally contain a greater abundance of truncated products. PAGE purification allows for selection of full-length primers but also significantly increases the cost

and does not ensure that the entire primer will be extended during the PCR reaction. The simple addition of a restriction enzyme site at the 5′ end of the primer permits the screening of full-length clones without PAGE purification. This also allows fewer clones to be sequenced in order to identify a positive clone. In our example, an *Xho*I site (CTCGAG) is attached to the 5′ end of the primer. Following PCR, this site is located at the distal 3′ end of the double-stranded product. The product is then cloned into the pCR-Blunt II TOPO vector, which contains one *Xho*I site in the polylinker. The PCR product might be inserted such that the 3′ *Xho*I site is located distally or proximally to the polylinker *Xho*I site. *Xho*I digestion of the resulting clones followed by agarose gel electrophoresis permits identification of positive full-length clones. Digested recombinants will contain either one (approx 4 kb [kilobase]) or two visible bands (approx 3.5 and approx 0.5 kb). Clones showing the latter pattern possess an *Xho*I site that is distal to the polylinker site. Such clones are likely positive, but must be further verified by sequencing. *Xho*I digestion resulting in the former banding pattern does not allow discrimination between truncated primer products or full-length products that were cloned into the TOPO vector in an orientation placing the 3′ *Xho*I site proximal to the polylinker *Xho*I site. Because several clones with the correct two-band pattern are obtained, it is not necessary to pursue any clones showing the single-band digestion pattern. Therefore, this strategy provides an inexpensive and quick way to screen recombinants without broad-scale sequencing and without costly PAGE purification of long primers.

3.1.5. Protocol Specifics: PCR Construction of shRNA Plasmids

Up to this point, the important factors involved in getting started were discussed. The next subsections describe the protocol in detail.

3.1.5.1. PCR Amplification and TOPO Cloning Reactions

1. Setup a 50-μL PCR reaction as follows: 5 μL of 10X Pfu reaction buffer; 4 μL of 2.5 m*M* dNTP mix; 1 μL forward primer at 100 ng/μL; 1 μL reverse primer at 100 ng/μL; 0.5 μL Pfu polymerase at 2.5 units/μL; 1 μl pAd5mU6 template plasmid at 10 ng/μL; 37.5 μL water.
2. Amplify using the following thermocycling parameters:

Step	Time	Temperature	Cycles
Initial denaturation	3 min	94°C	1
Denaturation	30 s	94°C	
Annealing	30 s	50°C	30
Extension	30 s	72°C	
Final extension	7 min	72°C	1

3. During amplification, prepare a 1% agarose gel using standard procedures. Following thermocycling, load 4 μL of the PCR reaction and verify by agarose gel electrophoresis that a single 439-bp band was produced.

4. Following the manufacturer's protocol, add 4 µL of PCR product directly to a tube containing 1 µL pCR-Blunt II TOPO vector and 1 µL salt solution (1.2 *M* NaCl; 0.06 *M* MgCl$_2$; provided with TOPO kit). Incubate 30 min at room temperature.
5. Transform 2–3 µL of TOPO cloning reaction into chemically competent *E. coli* using standard procedures.
6. Spread 1/10 vol of the transformed bacteria onto kanamycin-selective plate. Centrifuge the remaining transformation at low speed (e.g., approx 2000 rpm/0.4 rcf for 1 min in an Eppendorf centrifuge model 5415) until bacteria form a pellet. Remove supernatant. Resuspend pellet in 50–100 µL of bacterial growth media (e.g., LB, SOC). Spread remaining cells on a second plate. Incubate plates overnight at 37°C. A typical cloning reaction yields hundreds of colonies.

3.1.5.2. RESTRICTION ANALYSIS OF POSITIVE CLONES

1. Pick 10 colonies into a culture tube containing 3 mL of LB medium and 50 µg/mL kanamycin. Grow overnight at 37°C.
2. Isolate plasmid DNA using standard procedures or a commercial miniprep kit. There are numerous methods to purify bacterial DNA.
3. Digest 5 µL of miniprep DNA with *Xho*I (New England Biolabs, Beverly, MA; *see* **Subheading 3.1.4.**): 2 µL of 10X NEB2 buffer, 1 µL of 2 mg/mL bovine serum albumin (BSA) (20X), 5 µL plasmid DNA, 0.3 µL *Xho*I (at 20 units/µL); 11.7 µL water. Incubate at 37°C for 2–4 h.

Clones producing two bands (3.5 kb and 479 bp) should contain the mU6 promoter and a full-length hairpin. Plasmids showing this restriction pattern are preliminarily correct and should be verified by sequencing.

3.1.5.3. SEQUENCING SHRNA CLONES

An automated sequencer that uses Sanger-based fluorescent chemistry will greatly expedite this procedure. The University of Iowa DNA Facility utilizes Perkin-Elmer–Applied Biosystems sequencers (Model 3700) and ABI Prism® Big Dye™ Terminator Cycle Sequencing reagents. If automated sequencing is not available, manual sequencing that utilizes similar di-deoxy terminator chemistry (e.g., ThermoSequenase cycle sequencing kit; Amersham Biosciences, Arlington Heights, IL) will suffice but might add to the time and labor involved.

Our initial attempts at sequencing shRNA plasmids using standard chemistry and cycling conditions (*see* **Note 4**) resulted in a high incidence of truncated or failed sequencing reads, probably the result of the inability of the polymerase to accurately read-through the shRNA hairpin. In collaboration with the University of Iowa DNA Facility, we modified the standard protocol to produce more reliable and consistent sequencing of shRNA hairpins.

1. Set up cycle sequencing reaction: 250 ng plasmid DNA template; 3 pmol of T7 or SP6 primer; 1 µL DMSO (dimethyl sulfoxide); 8 µL Perkin-Elmer dGTP Sequencing Dye Terminator; Water to 20 µL.

2. Cycle using the following conditions:

Step	Time	Temperature	Cycles
Initial denaturation	2 min	96°C	1
Denaturation	1 min*	94°C	
Annealing	5 s	50°C	30
Extension	4 min	60°C	

*See **Note 4**

3. Correct clones will contain the sequence presented in **Figs. 1C** and **3C**. Plasmids can now be tested for function (*see* **Subheading 3.2.**).

3.2. Functional Assessment of shRNA Expression Plasmids In Vitro

Not every shRNA sequence is capable of directing effective target gene silencing. Identification of a functional shRNA sequence is often the rate-limiting step when cloning shRNA plasmids. In our hands, no freely available algorithm has been proven to work any better than random selection for predicting the sequence of effective shRNAs. Furthermore, anecdotal evidence from our laboratory suggests the ease that a target transcript is silenced can vary greatly depending on the target mRNA or the context in which the target sequence is placed. As a result, the gene silencing capability of each shRNA must be tested empirically, and several shRNA plasmids might have to be screened prior to identifying one that effectively silences the gene of interest. The following subsections briefly discuss strategies to assess the gene silencing capability of an shRNA plasmid in vitro. As an example, two methods that demonstrate sh*LacZ*-induced gene silencing of *E. coli* β-galactosidase are presented.

3.2.1. Common Methods for Determining Gene Silencing by shRNA Plasmids

Several commonly used techniques can be employed to assay gene silencing by shRNA plasmids, including (1) Northern blot, (2) quantitative reverse transcription

---

Fig. 3. (*Opposite page*) (**A**) U6 sh*LacZ* PCR product (black box) cloned into the pCR-Blunt II TOPO vector (Invitrogen, Carlsbad, CA). Arrows indicate T7 and SP6 primer binding sites. Vector polylinker sequence spanning these two regions is shown. The position of the two *Xho*I restriction enzyme recognition sites is crucial for screening positive colonies (discussed in text). The U6 sh*LacZ* insert should be cloned such that the 3′ *Xho*I site is located at the SP6 side of the vector polylinker, thereby allowing for screening of full-length shRNA clones. (**B**) Sequence chromatogram of U6 sh*LacZ* plasmid DNA demonstrates the ability to sequence through shRNA expression plasmids using the methods described in the text. The U6 transcription start site is marked by a forward arrow. The raw sequence shown here perfectly matches the desired dsDNA plasmid sequence (*see* **Fig. 1C**).

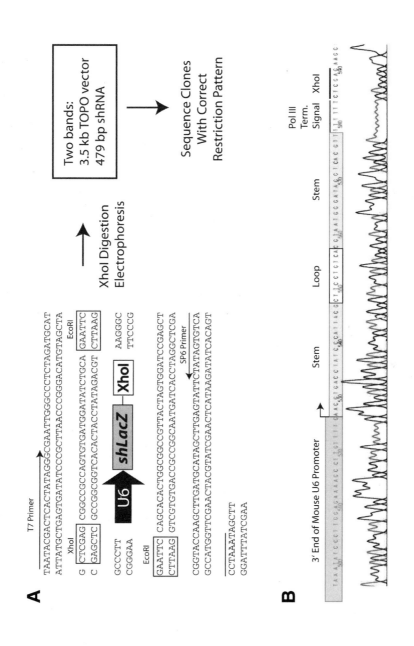

(RT)-PCR, (3) immunofluorescence, (4) fluorescent detection of target proteins with GFP fusions (or other fluorophores), and (5) functional protein assays (e.g., β-galactosidase assay). Several practical issues might impact the experimental design and choice of assay. Among these are the researcher's familiarity with techniques, availability of reagents, abundance of the target mRNA and protein product, and the function of the target gene. In the examples described in this section, we tested the ability of sh*LacZ* plasmids to reduce β-galactosidase expression in HEK293 cells using two different methods. Here, we discuss the logic involved in designing a quick screening assay using enzymatic assays and Western blotting.

The shRNA must be introduced into a cell line expressing the target gene of interest. Because our laboratory does not possess a mammalian tissue culture cell line that expresses β-galactosidase, we chose a transient cotransfection method to introduce two separate plasmids that express β-galactosidase or different shRNAs, respectively (*see* **Fig. 4A**). To this end, we used HEK293 cells, which are easily maintained and are readily transfected. Three different shRNAs were cotransfected with a plasmid expressing β-galactosidase from the CMV promoter. Gene silencing was initially determined by measuring the abundance of β-galactosidase protein 2 d posttransfection using the FluoReporter® LacZ/ Galactosidase Quantitation Kit (Molecular Probes, Eugene, OR). Representative data are presented in **Fig. 4B**.

A functional screen is ideal for assaying *LacZ* gene silencing, because the gene encodes a well-characterized enzyme whose activity can be easily measured using commercially available kits. However, enzymatic assays are not available for most gene products, and a more typical screening strategy might include the more broadly applicable technique of Western blotting. Therefore, we performed Western blots to demonstrate a more universal assay for plasmid-based shRNA gene silencing in vitro (*see* **Fig. 4C**). In addition, the second method serves to confirm results obtained in the initial enzyme assay screen. The details of the experiments are described next.

Fig. 4. (*Opposite page*) (**A**) Flowchart describing a general in vitro method for testing sh*LacZ* efficacy. HEK293 cells are cotransfected with plasmids encoding *LacZ* alone or *LacZ* with sh*LacZ* (1583–1603), sh*LacZ* (31–51), or shGFP expression cassettes. Samples are harvested 1–2 d posttransfection. Gene silencing can be assessed using RNA and protein-based assays. (**B**) β-Galactosidase enzymatic assay demonstrates that U6-transcribed sh*LacZ* (31–51) mediates *LacZ* gene silencing in HEK293 cells. Plasmids expressing shGFP or the nonfunctional sh*LacZ* (1583–1603) do not suppress *LacZ* gene expression. (**C**) Western blot showing reduced β-galactosidase protein in cell extracts expressing sh*LacZ* (31–51). Consistent with results described in (B), β-galactosidase protein levels in cells expressing sh*LacZ* (1583–1603) or shGFP were the same as those obtained from cells expressing no shRNA (target alone).

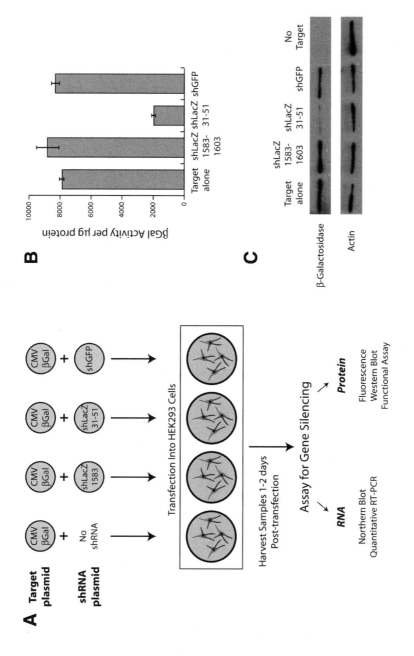

231

3.2.1.1. TRANSFECTION OF HEK293 CELLS WITH TARGET AND SHRNA PLASMIDS

1. The day before transfection, plate approx 200,000 HEK293 cells per well in 500 μL growth medium (*see* **Note 5**) on a 24-well tissue culture plate.
2. Transfect a 1 : 8 molar ratio of target and shRNA plasmids using Lipofectamine™ 2000 (Invitrogen, Carlsbad, CA) following the manufacturer's protocol.
3. Incubate the transfected cells at 37°C in a CO_2 incubator for 24–48 h.

3.2.1.2. PROTEIN EXTRACTION FROM TRANSFECTED CELLS

1. Remove growth media and wash the transfected cells with phosphate buffered saline (PBS), pH 7.2.
2. Remove PBS. Detach cells from the plate using either trypsin or physical disruption with a rubber policeman.
3. Resuspend disrupted cell contents in 500 μL of PBS. Do not boil the extract.
4. Transfer cells into a 1.5-mL Eppendorf tube, and perform three freeze–thaw cycles to lyse using a dry ice/ethanol bath or liquid nitrogen.
5. Quantify protein using the DC Protein Assay (Bio-Rad, Hercules, CA) following the manufacturer's instructions.

3.2.1.3. DETERMINE β-GALACTOSIDASE ACTIVITY

For these experiments, we used the FluoReporter® LacZ/Galactosidase Quantitation Kit (F-2905; Molecular Probes, Eugene, OR) and performed the assay following the manufacturer's instructions. β-Galactosidase activity was determined by measuring the emitted fluorescence of each protein sample using a microplate fluorometer. Fluorescence data are normalized to the total protein loaded per assay and β-galactosidase activity is presented as fluorescence per microgram of protein.

Figure 4B shows a typical result. Cells cotransfected with sh*LacZ* 31–51 contained 80% less β-galactosidase activity compared to cells transfected with a nonfunctional sh*LacZ* (sh*LacZ* 1583–1603), an shGFP control plasmid, or *LacZ* target alone.

3.2.1.4. WESTERN BLOT ASSAY FOR GENE SILENCING

1. Add 6X Laemmli loading dye to 10 μg of protein extracts obtained from transfected HEK293 cells.
2. Boil extracts for 5 min, cool on ice, and centrifuge briefly to force condensation back to the bottom of the tube.
3. Using standard Western blotting techniques, run the denatured protein extracts on a 10% SDS-PAGE gel.
4. Transfer proteins to nitrocellulose membrane using standard techniques.
5. Submerge membrane in blocking buffer (5% milk in PBS + 0.05% Tween; also known as PBS-T), for 2 h at room temperature or overnight at 4°C.

6. Remove blocking solution and incubate membrane with primary antibody solution for 2 h at room temperature with moderate agitation. Primary antibodies used were anti-β-galactosidase rabbit polyclonal antibody (Biodesign, Saco, ME) and anti-β-actin (Sigma, St Louis, MO) mouse monoclonal antibody. Both antibodies were diluted in PBS-T at 1 : 5000.

7. Perform three washes for 10 min each using PBS-T.

8. Remove the third wash and incubate membrane with secondary antibody solution for 2 h at room temperature with moderate agitation. Secondary antibodies used in this study were HRP-coupled goat anti-rabbit IgG (1 : 100,000 dilution in PBS-T) and HRP-coupled goat anti-mouse IgG (1 : 20,000 in PBS-T; both obtained from Jackson Immunologicals, West Grove, PA).

9. Perform three washes for 10 min each using PBS-T.

10. Incubate membrane with ECL-Plus Western Blotting Substrate (Amersham-Pharmacia, Arlington Heights, IL) following the manufacturer's protocols.

11. Immediately expose blot to standard autoradiographic film (e.g., Kodak Biomax MR film).

A typical result is presented in **Fig. 4C**. Cells cotransfected with sh*LacZ* 31–51 contained less β-galactosidase activity compared to cells transfected with a nonfunctional sh*LacZ* (sh*LacZ* 1583–1603), an shGFP control plasmid, or *LacZ* target alone, confirming our results obtained using a β-galactosidase assay.

3.3. Conclusions

The protocols provided should allow the researcher to develop plasmid-based silencing tools, using readily available reagents. Functional shRNA expressed from plasmids can be cloned easily into lentivirus, adenovirus, or adeno-associated virus expression systems, with retention of silencing characteristics.

4. Notes

1. An alternative method for cloning shRNA hairpins involves the direct ligation of two annealed oligonucleotides into a restriction-enzyme-linearized plasmid containing the promoter of interest. Instead of ordering one 87-nt oligonucleotide as described in **Subheadings 3.1.2.** and **3.1.3.** and **Fig. 2**, two slightly shorter (approx 70-nt) oligos are utilized. These oligos lack the 23-nt-binding site for the promoter of interest and, instead, contain restriction enzyme sites at both ends. The oligos are then annealed together and directly ligated into a vector of choice. The main advantage of this approach is that one set of oligonucleotides could be ligated into several different promoters, if the proper restriction enzyme sites are present. It has also been argued that another advantage of the direct-ligation method is that DNA sequencing of only the shRNA, and not the promoter, is required in clones generated from the direct-ligation approach. Conversely, because both the promoter and shRNA are amplified in the PCR-based method, it is necessary to sequence through both to ensure that no errors were introduced by

PCR. However, in our experience, a sequencing reaction originating at the 3′ end of the shRNA cassette and proceeding upstream toward the promoter typically produces enough sequence data to read through the entire shRNA and promoter. Thus, the same number of sequencing reactions is likely required regardless of the cloning method used. The major disadvantages of the direct-ligation method compared to the PCR-based method are (1) higher oligonucleotides costs and (2) the method is more labor-intensive. Regarding the first point, two 70-nt oligos are more expensive than one 87-nt oligo ($63 compared to $39 from one vendor; $70 compared to $61 from a second vendor). Regarding the second point, the direct-ligation method utilizes the traditional preparation of a vector plasmid (e.g., restriction digestion, phosphatase treatment, electrophoresis, gel purification, yield quantification, and ligation), which requires greater hands-on time commitment compared to the PCR-based approach.

2. Functional shRNAs cloned in our laboratory have contained up to 6 bp of sequence (i.e., one common restriction enzyme site) between the promoter's transcription start site and the 5′ end of the DNA sequence, giving rise to the shRNA *(7)*. Similarly, up to 6 bp of sequence can be present between the 3′ end of the shRNA transcription template and the transcription termination signal.

3. Cloning the PCR product using the procedures described herein will produce significant numbers of background colonies if the PCR template plasmid and the PCR cloning vector contain the same antibiotic selective genes. In the event that two different selective genes cannot be utilized, two methods can be used to eliminate background contaminants arising from the PCR template plasmid: (1) gel purify the PCR product using standard procedures or (2) digest the PCR product with *Dpn*I prior to cloning it into the pCR-Blunt II TOPO vector. *Dpn*I is a restriction enzyme that cleaves only when its restriction site is methylated. The PCR template plasmid is produced in bacteria and is, therefore, methylated and cleaved by *Dpn*I. Conversely, the unmethylated PCR product remains intact and can be cloned. To perform the *Dpn*I digestion, simply add 1 μL (20 units) of *Dpn*I (New England Biolabs, Beverly, MA) to the entire 50 μL PCR reaction; then, incubate 1 h at 37°C. Heat inactivate the digest at 80°C for 20 min and then clone 4 μL into the pCR-Blunt TOPO II vector as described in **Subheading 3.1.5.**

4. Standard reactions use ABI Prism® Big Dye™ Terminator Cycle Sequencing reagents with dITP following manufacturer's protocol. Cycling parameters are as follows: initial denaturation: 96°C for 2 min, 1 cycle; then 94°C for 10 s, 50°C for 5 s, 60°C for 4 min, 30 cycles. The longer denaturation step (1 min) sufficiently melted the hairpin to allow extension of nucteotides through the shRNA sequence.

5. HEK293 growth medium: 10% fetal bovine serum (FBS), 1% penicillin/streptomycin (P/S) in Dulbecco's modified eagle's medium (DMEM).

Acknowledgments

For outstanding work contributing to this chapter, the authors thank the University of Iowa staff Patrick D. Staber (Davidson Laboratory) and Kristy Johnson (DNA Sequencing Facility).

References

1. Hannon, G. J. (2002) RNA interference. *Nature* **418,** 244–251.
2. Davidson, B. L. and Paulson, H. L. (2004) Molecular medicine for the brain: silencing disease genes with RNA interference. *Lancet Neurol.* **3,** 145–149.
3. Kim, J., Krichevsky, A., Grad, Y., et al. (2004) Identification of many microRNAs that copurify with polyribosomes in mammalian neurons. *Proc. Natl. Acad. Sci. USA* **101,** 360–365.
4. Kawasaki, H. and Taira, K. (2003) Short hairpin type of dsRNAs that are controlled by tRNA(Val) promoter significantly induce RNAi-mediated gene silencing in the cytoplasm of human cells. *Nucleic Acids Res.* **31,** 700–707.
5. Elbashir, S. M., Harborth, J., Lendeckel, W., Yalcin, A., Weber, K., and Tuschl, T. (2001) Duplexes of 21-nucleotide RNAs mediate RNA interference in cultured mammalian cells. *Nature* **411,** 494–498.
6. Paul, C. P., Good, P. D., Winer, I., and Engelke, D. R. (2002) Effective expression of small interfering RNA in human cells. *Nat. Biotechnol.* **20,** 505–508.
7. Xia, H., Mao, Q., Paulson, H. L., and Davidson, B. L. (2002) siRNA-mediated gene silencing *in vitro* and *in vivo. Nat. Biotechnol.* **20,** 1.
8. Paddison, P. J., Caudy, A. A., Bernstein, E., and Hannon, G. J. (2002) Short hairpin RNAs (shRNAs) induce sequence-specific silencing inmammalian cells. *Genes Dev.* **16,** 948–958.

12

siRNA Delivery In Vivo

Mouldy Sioud

1. Introduction
1.1. The Problem of Delivery

Inhibition of gene expression at the mRNA levels can be accomplished by several methods, including ribozymes, DNAzymes, and small interfering RNAs (siRNAs) *(1–3)*. This is now driven predominantly by siRNAs, as they are not technically demanding as traditional antisense and ribozyme technologies *(4,5)*. The success of siRNAs as therapeutics, however, is largely dependent on the development of a delivery vehicle that can efficiently deliver them in vivo.

Whether being used as experimental tools and/or pharmaceutical drugs, siRNAs need to be able to cross cell membranes and to be incorporated into the silencing complex. Negatively charged oligonucleotides will not pass through a lipid layer such as cell membranes. Similar to antisense oligonucleotides and ribozymes, the delivery of synthetic siRNAs can be improved via the use of various delivery systems, which include synthetic carriers, composed primary of lipids. At the moment, cationic-lipid-based carriers have emerged as the most popular nonviral method to deliver nucleic acids in therapeutic applications *(6)*. However, there are some questions to consider when combining a siRNA with a delivery system for in vivo application. Can the carrier help to stabilize the molecule and will the siRNAs stay associated with the carrier for an appropriate period of time and be released at an appropriate rate? In the case of synthetic siRNAs, it will be preferable to use carriers that can also function as drug reservoirs, because of the transient activity of siRNAs. Larger carriers such as liposomes (approx 50–200 nm in diameter) localize the drugs mainly to the blood compartment. However, angiogenic blood vessels in most tissues have gaps as

From: *Methods in Molecular Biology, vol. 309: RNA Silencing: Methods and Protocols*
Edited by: G. G. Carmichael © Humana Press Inc., Totowa, NJ

large as 600–800 nm between adjacent endothelial cells; therefore, liposome/ siRNA complexes can extravagate via these gaps into the tumor. In addition, most solid tumors possess an enhanced vascular permeability and impaired lymphatic drainage, which leads to the accumulation of most liposomes within the tumor tissues *(7)*. Regarding vascular permeability, many similarities between inflammation and cancer were found, so encapsulating siRNAs in liposomes should enable targeted delivery to tumour tissues as well as to inflammatory sites.

Small interfering RNA can be administered via different routes such as direct injection, intravenous, intraperitoneal, intra-arterial, and intracranial delivery. Recently, we have found that an efficient siRNA delivery in vivo can be achieved via intravenous or intraperitoneal administration of siRNAs formulated in cationic liposomes. From the perspective of human cancer therapy, cationic liposomes have been shown to be safe and effective for in vivo gene delivery and are currently being used in many approved clinical trials. Despite some encouraging results, however, the liposomes still have not the characteristics to be perfect carriers because of toxicity, short circulation time, and limited intracellular delivery for target cells. These potential problems are also relevant for the delivery of plasmid-expressing short-hairpin RNAs (shRNAs).

As mentioned earlier, one of the drawbacks of synthetic siRNAs is the transient nature of the inhibition in mammalian cells. The activity is mainly dependent on the rate of cell growth and the turnover of the targeted proteins. Usually, it lasts 3–6 d. To analyze gene function overtime, several plasmid and viral vectors were developed to express shRNAs that are driven by a RNA polymerase (pol) III promoter (including U6, H1, and tRNA promoters) *(8–11)*. In one strategy, the sense and antisense strands are expressed as two independent transcripts that hybridize within the cells to form functional siRNA duplexes. In the second strategy, the sense and antisense strands are expressed as single transcripts separated by a short loop sequence.

Both H1 and U6 promoters efficiently express shRNAs, which undergo Dicer processing into siRNAs, which are then incorporated into the silencing complex. Gene expression was more stably inhibited than with the transient knockdown obtained with synthetic siRNAs. Although the H1 and U6 promoters could be controlled with respect to their activity to express siRNAs by modification the promoter to a tetracycline-responsive versions *(12,13)*, they are active in all mammalian tissues and are not easily adapted for cell- or tissue-specific knockdown. However, the developments of pol-II-based vectors that can produce hairpin siRNAs in vivo have provided an alternative method for the tissue-specific gene suppression *(14)*. To further facilitate the transfection of mammalian cells (in particular, primary cells), several groups have also developed viral vectors *(15,16)*. Lentiviral-based vectors have also been developed in this regard. Lentivectors make use of the ability of human immunodeficiency virus (HIV)-1 to infect nondividing cells such as hematopoietic progenitor cells. Although viral vectors are commonly used as carriers for gene transfer because of their high in vitro infection efficiency, safety issues and the toxicity of these vectors will likely hamper their use in humans *(17)*. Thus, improved viral-based vectors to overcome some of the unwanted side effects and improve upon the efficiency of gene transfer into specific target tissues are required before the siRNA technology can be tried clinically.

1.2. Improving the Efficacy of siRNAs via Chemical Modifications

The ability to modify RNA oligonucleotides so that they are more stable in vivo is likely necessary before adopting siRNA in humans. Regarding RNA stability, early studies indicated that the 2′-hydroxyl groups of pyrimidines are the primary sites of ribonucleases. Thus, sugar and base alterations might increase nuclease stability significantly. In this connection, the substitution of pyrimidine with 2′-fluoro, 2′-amino, 2′-*O*-methyl, 2′-*O*-allyl, and/or 2′-*C*-allyl analogs has provides stable ribozymes *(1)*. In addition, several chemical modifications can also be applied to the phosphodiester linkage. Three of the more used modifications are the phosphorothioates (PS), the N3′-P5′ phosphoramidates, and the peptide nucleic acids (PNAs). In principle, most of the chemical modification that have been developed for ribozymes and antisense oligonucleotides can be incorporated into the siRNA molecules, provided that the RNAi pathway was not affected. Therefore, chemical modifications of siRNA should not interfere with the incorporation of the siRNAs into RNA-induced silencing complex (RISC), unwinding of the siRNA duplex by helicase activities, and/or the rate of target cleavage and dissociation. Although siRNA molecules with either one or both strands consisting of 2′-*O*-methyl residues were not able to induce RNAi in mammalian cells, siRNA molecules with partial 2′-*O*-methyl modification showed an increased stability when compared to unmodified siRNAs *(18)*. Modifications like the 2′-fluoro (F) and PS linkages both increased the half-life of siRNAs upon exposure to cytoplasmic extracts. In vivo studies with 2′-FU and 2′-FC siRNAs showed that increasing the half-life of siRNA did prolong the effects of RNAi *(19)*. Although PS linkage is an efficient means to stabilize DNA, a recent study found that they do not significantly stabilize the siRNAs in serum as compared to their unmodified versions *(20)*. Given that PS substitutions are known to improve the pharmacokinetics of traditional antisense oligonucleotides by increasing binding to serum proteins, there still is a role of using PS in optimizing the pharmacokinetics of siRNAs. The association of RNA or DNA with serum proteins is expected to reduce clearance rates and improve the in vivo half-life in circulation. However, once across the membrane, the intracellular distribution of siRNAs will depend on the chemistry of the molecules that should be properly chosen.

1.3. Cationic Liposome-Mediated Intravenous Delivery of Nucleic Acids

Synthetic siRNAs are generally delivered to cells via liposome-based transfection reagents *(4)*. These reagents offer the possibility of developing pharmaceutical siRNAs for local or systemic applications. In addition, when catalytic RNAs such as siRNAs are made in vitro, a variety of 2′ and backbone chemical modifications can be introduced into the molecules to increase their half-life in serum. In the past, many groups have shown that lipid mediated intravenous (iv) delivery of gene can produce high-level, systemic expression of biologically and therapeutically relevant genes *(21)*. In addition, DNA has been successfully complexed with cationic, anionic, and neutral liposomes, as well as various mixtures thereof.

Being unstable in biological fluids, the in vivo stability and uptake of DNA and RNA oligonucleotides have been improved by the use of cationic and anionic liposomes after iv administration. In this respect, iv pretreatment of mice with cationic lipids and antisense against intercellular adhesion molecule-1 (ICAM) significantly decreased its expression in the lung following lipopolysaccharide (LPS) challenge *(22)*. Systemic administration of cationic liposomes and a plasmid encoding for a ribozyme against the transcription factor necrosis factor (NF)-κB suppressed NF-κB expression in metastatic melanoma cells and significantly reduced metastatic spread *(23)*. Recently, we have found that iv injection of liposome-formulated siRNAs against GFP-inhibited GFP gene expression in various organs such as the liver and spleen *(24,25)*. In the liver and spleen, most of the uptake was in endothelial, Kupffer cells and macrophages. Taken together, the data demonstrated the utility of in vivo gene targeting via iv delivery of nucleic acids such as antisense oligonucleotides, ribozymes, and siRNAs. However, lipid-based systems still have important drawbacks, including the lack of specific targeting and variation arising during fabrication.

To evaluate the ability of liposomal carriers such as DOTAP, we have assessed the iv delivery of a FITC-labeled siRNA. Tissues from a group of mice receiving siRNA showed a significant uptake by some organs such as the spleen (*see* **Fig. 1**). A substantial majority of the siRNA molecules were found to be localized around the vessels 6 h after iv injection via the tail vein. Endothelial cells are more likely to be the most targeted by the siRNA complexes.

1.4. Cationic-Liposome-Mediated Intraperitoneal Nucleic Acid Delivery

The appropriate delivery route to assess the efficacy of siRNAs can facilitate the development of therapeutic siRNAs. Malignant ascites is a major cause of

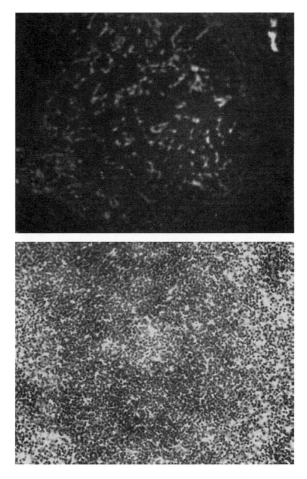

Fig. 1. Distribution of a FITC-labeled siRNA in mouse spleen cells 6 h following iv injection of FITC-labeled siRNA liposome complexes. Six hours after vein tail injection, mice were killed and organs were taken and then frozen in embedding medium for frozen tissue specimens with liquid nitrogen. Ten-micrometer cryosections were cut, fixed in ethanol, washed twice with phosphate-buffered saline, and then analyzed by an epifluorescence microscope. FITC-labeled siRNAs were chemically synthesized and purified by Eurogentec. The siRNAs were annealed in transfection buffer (20 mM HEPES, 150 mM NaCl, pH 7.4). (From **ref. 25**).

morbidity in patients with intra-abdominal dissemination of various neoplasms such as ovarian, colorectal, and breast cancer (*26*). Treatment of patients with this troubling clinical condition would involve systemic or intra-abdominal chemotherapy, generally not successful because of the drug resistance of such tumors. Notably, the peritoneal cavity represents the largest cavity of the human body and its anatomic structures, along with residual leukocyte populations,

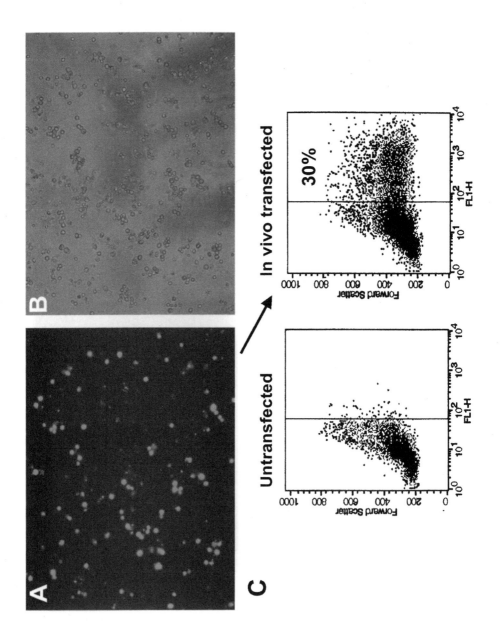

play an important role in the defense against invading micro-organisms, in particular, those breaching the gut integrity. Thus, the development of efficient intraperitoneal (ip) delivery agents would represent a new potential to explore the efficacy of new therapeutics in vivo. In this respect, we have found that cationic liposomes can enhance the uptake of anti-tumor necrosis factor-α (TNF-α) ribozymes and siRNAs by peritoneal macrophages and were able to downregulate the expression of TNF-α in vivo *(24,27)*. **Figure 2** documents the uptake of a FITC-labeled siRNA by peritoneal cells 20 h after ip delivery of 100 µg siRNAs. A substantial number of cells have taken up the siRNA–liposome complexes. However, the uptake could be underestimated, because FITC might get quenched at the low pH in the endosomes.

1.5. Adherent Peritoneal Cells Are More Susceptible to In Vivo Transfection Than Their Nonadherent Counterparts

The in vivo uptake of siRNAs can differ dramatically with cell types as well as with the status of cell differentiation. Different cell types might require significantly different transfection conditions to yield optimal uptake of siRNAs. For biological activity, liposomes must deliver their contents into the cytoplasm, which can be cell dependent. In this respect, we have analyzed the in vivo transfection capacity of adherent and nonadherent peritoneal cells. As shown in **Fig. 3**, a large proportion of adherent peritoneal cells were in vivo transfected with a FITC-labeled siRNA. In contrast, most of the nonadherent peritoneal cells were not transfected. Thus, there might be some mechanism in adherent macrophages for effectively taking up the liposome-formulated siRNA molecules. These macrophages could be the major source of TNF-α production because ip delivery of siRNAs against mouse TNF-α reduced the expression of TNF-α *(24)*, Mice pretreated with anti-TNF-α siRNA prior to LPS challenge showed less severe clinical symptoms than those treated with an inactive siRNA. These data would suggest that liposome-mediated delivery of anti-TNF-α siRNAs might be valuable in a number of conditions such as rheumatoid arthritis, in which TNF-α is a pathogenic mediator *(28)*. In contrast to mouse cells, however, certain synthetic siRNAs activated the production of TNF-α and interleukin (IL)-6 in human freshly isolated monocytes via the activation of NF-κB signaling pathway *(25)*. In accordance with our observations,

---

Fig. 2. (*Opposite side*) Analysis of mouse peritoneal cells by an epifluorescence microscope after ip delivery of 100 µg FITC-labeled siRNA. Twenty hours after delivery, mice were killed and peritoneal cells were prepared washed with RPMI medium, and then analyzed by an epifluorescence microscope **(A)**. The light image of the same field is shown **(B)**. Dot plots of flow cytometric analysis of the same cells are shown in **C**. Nearly 30% of the cells show a significant siRNA uptake. (From **ref**. *25*.)

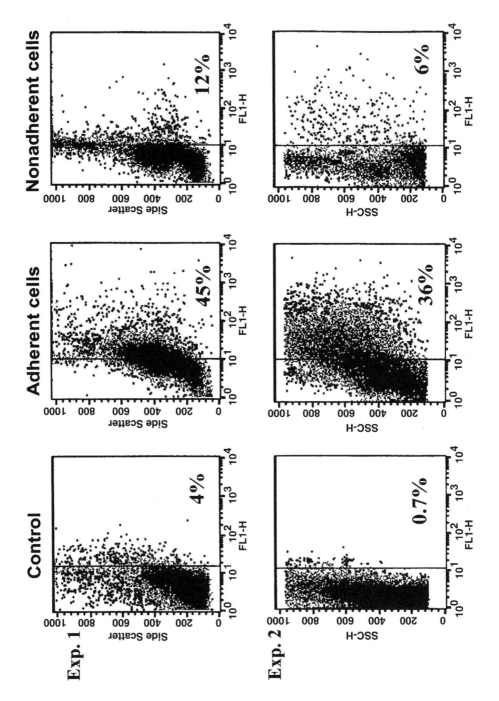

Control   Adherent cells   Nonadherent cells

Exp. 1

Exp. 2

244

recent studies showed that, indeed, siRNAs do activate the interferon pathway *(29,30)*. Thus, appropriate controls must be used in siRNA-mediated knockdowns. Notably, naked siRNAs can be delivered in vivo via iv injection *(31,32)*.

2. Materials

2.1. Equipment

1. Coulter cell counter.
2. Fluorescence microscope with appropriate filters.
3. Flow cytometer and related reagents.
4. Tissue culture incubator.
5. Eppendorf centrifuge.
6. Falcon tube centrifuge.

2.2. Animals

BALB/C mice, 6–8 wk of age and weighing 20–25 g.

2.3. Reagents

1. Chemically made siRNAs.
2. Cationic liposomes for transfection (e.g., DOTAP).
3. Transfection buffer 5X: 100 mM HEPES, 750 mM NaCl, pH 7.4.
4. RPMI culture medium supplemented with 10% fetal bovine serum (FBS) and antibiotics.
5. Pasteur pipets.
6. Sterile syringes 1, 5, and 10 mL.
7. 25G Needle.
8. Glass tubes (5 and 10 mL).
9. Polystyrene and Eppendorf tubes.
10. Tissue culture dishes (25 and 75 cm^3).

3. Methods

3.1. Peritoneal Lavage (See Note 1)

1. Kill the mouse by cervical dislocation and place the animal into 95% ethanol.
2. Restrain the mouse in the supine position and make an incision in the abdominal wall and gently lift the abdominal wall.

Fig. 3. (*Opposite side*) Analysis of adherent and nonadherent peritoneal cells by flow cytometry. Eighteen hours after iv delivery of a FITC-labeled siRNA, mice were killed and peritoneal cells were prepared, washed, resuspended in RPMI medium supplemented with 10% fetal calf serum (FCS), and then plated in 75-cm^3 tissue culture dishes. After 2 h, the cultures were gently washed with RPMI to remove nonadherent cells. Adherent cells were harvested by gentle scrapping. Both cell populations were analyzed by flow cytometry. (From **ref**. *25*.)

3. With a syringe and 25G needle, inject 5–10 mL of saline or PBS buffer to the abdominal cavity and massage the abdomen for 1–2 min.
4. Gently aspirate the fluid by a syringe with the 25G needle and place the fluid into a 5 to 10-mL glass tube.
5. Spin down and wash the cells with RPMI medium.

3.2. Preparation of Adherent Peritoneal Cells for Flow Cytometry Analysis

1. Plate peritoneal cells at 10×10^6/20 mL complete medium in a 75-cm^3 culture dish and allow adhering for 90 min at 37°C.
2. Gently aspirate nonadherent cells, add 20 mL complete medium, and continue incubation at 37°C for 2–4 h.
3. Harvest the adherent cells by gentle scrapping.

3.3. In Vivo Delivery of siRNAs

We have been exploring the use of liposomes for delivering synthetic ribozymes and siRNAs to animals and eventually to patients. As mentioned above, a variety of cationic liposomes have been used and shown to deliver DNA and RNA oligonucleotides into cells. However, whatever the type cationic liposome used, the formulated complexes should have a net positive charge. Therefore, it is desirable to test a range of cationic-liposome concentrations and siRNA ratio.

3.4. Intraperitoneal Delivery

1. In a sterile microcentrifuge tube, mix 100 µg of siRNA (up to 20 µL) and 80 µL of 1X transfection buffer.
2. In a separate polystyrene tube, mix 100–200 µL (1 µg/µL) of DOTAP and 300–600 µL of transfection buffer. The ratio of siRNAs to DOTAP is 1:1 and 2:1, respectively.
3. Transfer the siRNA mixture to the polystyrene tube containing the DOTAP (*see* **Note 2**).
4. Mix gently by pipetting several times and incubate at room temperature for 30 min. The final volume should be around 500 or 800 µL.
5. Complete to 1 mL with 1X transfection buffer, mix gentle, and then inject intraperitoneally into mice.
6. After the desired time of postinjection (12–48 h), investigate the in vivo uptake or biological activities (*see* **Notes 3–7**).

3.5. In Vivo Uptake of siRNA by Peritoneal Cells

Using the protocol described above, a FITC-labeled siRNA (100 µg) was delivered to peritoneal cells. The siRNA to DOTAP ratio was 1:1. **Figure 1** shows the uptake of the siRNA by peritoneal cells 20 h after ip injection. Flow cytometry analysis indicates that nearly 30% of peritoneal cells were transfected. Most of the transfected cells are adherent cells (*see* **Fig. 3**).

3.6. Intravenous Delivery

1. In a sterile microcentrifuge tube, mix 50–100 μg siRNA (up to 20 μL) and 40 μL of transfection buffer.
2. In a separate polystyrene tube, mix 50–100 μL (1 μg/μL) of DOTAP and 90 μL transfection buffer. The charge ratio of siRNAs and DOTAP is 1:1 (*see* **Note 8**).
3. Transfer the siRNA mixture to the polystyrene tube containing the DOTAP.
4. Mix gently by pipetting several times and incubate at room temperature for 30 min. The final volume should be around 200 μL.
5. Inject the mixture via the tail vein.

4. Notes

1. In a normal situation, peritoneum contain some residual cells. However, they are not sufficient for in vitro studies. Liposomes such as DOTAP can recruit cells into the peritoneal cavity. During in vivo experiments, the number of recruited cells could vary between animals. Thus, in vivo siRNA effects should be adjusted to the number of recruited cells.
2. Unlike many transfection reagents, DOTAP does not need to be washed from the cells after transfection. Prolonged exposure does not induce cytotoxicity in most cells tested. However, transfection conditions should be evaluated for each cell type.
3. Before analyzing siRNA activity in peritoneal cells or other organs, transfection efficiency should be analyzed. For this purpose, it will be desirable to use 3′-FITC- or Cy5-labeled siRNAs. A small amount of labeled DNA oligonucleotide can also be used as an indicator for transfection efficiency.
4. The maximum period available for observation of the effects of suppression of a protein after a single treatment depends on the stability of the protein after elimination of mRNA. Therefore, it is important to determine the half-life of the target proteins after a single injection.
5. Changes in gene expression can be directly monitored from extract of peritoneal cells by Western and/or Northern blot analysis. If desired, cytokine contents can be measured in the peritoneal lavage fluids using commercially available enzyme-linked immunosorbent assay (ELISA).
6. The difference between in vivo and in vitro experiments: In vivo cells are mostly nondividing, so there is no siRNA dilution with cell division. Thus, the in vivo effect of a single injection of siRNA could be longer than that seen in tissue culture.
7. The condensation that occurs during complex formation is progressive, and within minutes, it might result in the precipitation of large aggregates that are not suitable for iv delivery. Small particles are desirable. In this respect, cationic polyspermine could be a good carrier system for iv delivery.
8. Recently, we have found that siRNAs can activate genes involved in innate immunity such as TNF-α. These unwanted effects seem be concentration and siRNA dependent. Therefore, for each siRNA preparation, it would be desirable to identify the minimal active concentration that would not induce such unwanted effects.

Acknowledgments

The Norwegian Cancer Society supported this work. We thank Dr. Anne Dybwad for critical reading of the manuscript.

References

1. Sioud, M. (2001) Nucleic acid enzymes as a novel generation of anti-gene agents. *Curr. Mol. Med.* **1,** 575–588.
2. Hannon, G. J. (2002) RNA interference. *Nature* **418,** 244–251.
3. Sioud, M. (2004) Therapeutic siRNAs. *Trends Pharmacol. Sci.* **25,** 22–28.
4. Elbashir, S. M., Harborth, J., Lendeckel, W., Yalcin, A., Weber, K., and Tuschl, T. (2001) Duplexes of 21-nucleotide RNAs mediate RNA interference in cultured mammalian cells. *Nature* **411,** 494–498.
5. Caplen, N. J., Parrish, S., Imani, F., Fire, A., and Morgen, R. A. (2001) Specific inhibition of gene expression by small double-stranded RNAs in invertebrate and vertebrate systems. *Proc. Natl. Acad. Sci. USA* **98,** 9742–9747.
6. Templeton, S. N. (2002) Liposomal delivery of nucleic acids in vivo. *DNA Cell Biol.* **21,** 859–867.
7. Maeda, H., Fang, J., Inutsuka, T., and Kitamoto, Y. (2003) Vascular permeability enhancement in solid tumor: various factors, mechanisms involved and its implications. *Int. Immunopathol.* **3,** 319–328.
8. Brummelkamp, T. R., Bernards, R., and Agami, R. (2002) A system for stable expression of short interfering RNAs in mammalian cells. *Science* **296,** 550–553.
9. Miyagishi, M. and Taira, K. (2002) U6 promoter-driven siRNA with four uridines 3′ overhangs efficiently suppress targeted gene expression in mammalian cells. *Nat. Biotechnol.* **19,** 497–500.
10. Lee, N. S., Dohjima, T., Bauer, G., et al. (2002) Expression of small interfering RNAs targeted against HIV-I rev transcripts in human cells. *Nat. Biotechnol.* **20,** 500–505.
11. Paul, C. P., Good, P. D., Winer, I., and Engelke, D. R. (2002) Effective expression of small interfering RNA in human cells. *Nat. Biotechnol.* **19,** 505–508.
12. Gossen, M. and Bujard, H. (1992) Tight control of gene expression in mammalian cells by tetracycline-response promoters. *Proc. Natl. Acad. Sci. USA* **89,** 5547–5551.
13. Czauderna, F., Santel, A., Hinz, M., et al. (2003) Inducible shRNA expression for application in a prostate cancer mouse model. *Nucleic Acids Res.* **31,** e127.
14. Xia, H., Mao, Q., Paulson, H. L., and Davidson, B. L. (2002) siRNA-mediated gene silencing *in vitro* and *in vivo*. *Nat. Biotechnol.* **20,** 1006–1010.
15. Shen, C., Buck, A., Liu, X., Winkler, M., and Reske, S. N. (2003) Gene silencing by adenovirus-delivered siRNA. *FEBS Lett.* **539,** 111–114.
16. Rubinson, D. A., Dillon, C. P., Kwiatkowski, A. V., et al. (2003) A lentivirus-based system to functionally silence genes in primary mammalian cells, stem cells and transgenic mice by RNA interference. *Nat. Genet.* **33,** 401–406.
17. Thomas, C. E., Ehrhardt, A., and Kay, M. A. (2003) Progress and problems with the use of viral vectors for gene therapy. *Nat. Rev. Genet.* **4,** 346–358.

18. Czauderna, F., Fechtner, M., Dames, S., et al. (2003) Structural variations and stabilizing modifications of synthetic siRNAs in mammalian cells. *Nucleic Acids Res.* **31,** 2705–2716.

19. Chiu, Y.-L. and Rana, T. M. (2003) siRNA function in RNAi: a chemical modification analysis. *RNA* **9,** 1034–1048.

20. Braasch, D. A., Jensen, S., Liu, Y., et al. (2003) RNA interference in mammalian cells by chemically modified RNA. *Biochemistry* **42,** 7967–7975.

21. Liu, Y., Liggitt, D., Zhong, W., Tu, G., Gaensler, K., and Debs, R. (1995) Cationic liposome-mediated intravenous gene delivery. *J. Biol. Chem.* **270,** 24,864–24,870.

22. Ma, Z., Zhang, J., Alder, S., et al. (2002) Lipid-mediated delivery of oligonucleotide to pulmonary endothelium. *Am. J. Respir. Cell Mol. Biol.* **27,** 151–159.

23. Kashani-Sabet, M., Liu, Y., Fong, S., et al. (2002) Identification of gene function and functional pathways by systemic plasmid-based ribozyme targeting in adult mice. *Proc. Natl. Acad. Sci. USA* **99,** 3878–3883.

24. Sørensen, D. R., Leirdal, M., and Sioud, M. (2003) Gene silencing by systemic delivery of synthetic siRNAs in adult mice. *J. Mol. Biol.* **327,** 761–766.

25. Sioud, M. and Sørensen, D. R. (2003) Cationic liposome-mediated delivery of siRNAs in adult mice. *Biochem. Biophys. Res. Commun.* **26,** 1220–1225.

26. Parsons, P., Lang, M., and Steele, R. (1996) Malignant ascites: a 2-year review from a technical hospital. *J. Surg. Oncol.* **22,** 237–239.

27. Sioud, M. (1996) Ribozyme modulation of lipopolysaccharide-induced tumor necrosis factor- production by peritoneal cells *in vitro* and *in vivo. Eur. J. Immunol.* **26,** 1026–1031.

28. Beutler, B. A. (1999) The role of tumour necrosis factor in health and diseases. *J. Rheumatol.* **57,** 16–21.

29. Sledz, C. A., Holko, M., de Veer, M. J., Silverman, R. H., and Williams, B. R. G. (2003) Activation of the interferon system by short-interfering RNAs. *Nat. Cell Biol.* **5,** 834–839.

30. Persengiev, S. P., Zhu, X., and Green, M. R. (2004) Nonspecific, concentration-dependent stimulation and repression of mammalian gene expression by small interfering RNAs (siRNAs). *RNA* **10,** 12–18.

31. Lewis, D. L., Hagstom, G., Haley, B., and Zamore, P. D. (2002) Efficient delivery of siRNA for inhibition of gene expression in postnatal mice. *Nat. Genet.* **32,** 107–108.

32. McCaffrey, A. P., Meuse, L., Pham, T. T., Conklin, D. S., Hannon, G. J., and Kay, M. A. (2002) RNA interference in adult mice. *Nature* **418,** 38–39.

13

Peptide-Based Strategy for siRNA Delivery into Mammalian Cells

Federica Simeoni, May C. Morris, Frederic Heitz, and Gilles Divita

1. Introduction

The potential to control and alter gene expression constitutes an essential strategy in both fundamental and pharmaceutical research. The recent discovery of the RNA interference pathway in a wide variety of eukaryotic organisms has provided a novel means of characterizing gene function in mammalian cells and new perspectives in both molecular biology and future therapeutic developments *(1–3)*. Short, interfering RNAs (siRNAs) constitute a powerful tool to silence gene expression posttranscriptionally *(1–3)*. However, the major limitation of siRNA application, as for most antisense or nucleic acid-based strategies, remains their poor cellular uptake associated with low permeability of the cell membrane to nucleic acids *(4,5)*. Several viral *(6–9)* and nonviral *(6,10)* strategies have been proposed to improve the delivery of either siRNAs expressing vectors or synthetic siRNAs, both in cultured cells and in vivo *(6)*. So far, although siRNA transfection can be achieved with classical laboratory-cultured cell lines using lipid-based formulations, siRNA delivery remains a major challenge for many cell lines and there is still no reasonably efficient method for in vivo application *(6)*. The most efficient method for in vivo applications is the nonviral "hydrodynamic" tail-vein injection of mice with high doses of siRNA *(11–13)*. Cell-penetrating peptides are powerful carriers for cellular uptake of a variety of macromolecules, including proteins, peptides, and oligonucleotides *(14–17)*. Several peptide-based strategies have been developed to improve the delivery of oligonucleotides both in vitro and in vivo using either covalent or complex approaches *(18–20)*. We have described a new peptide-based gene delivery system, MPG, which forms stable noncovalent complexes with several nucleic acids (plasmid DNA, oligonucleotides) and

From: *Methods in Molecular Biology, vol. 309: RNA Silencing: Methods and Protocols*
Edited by: G. G. Carmichael © Humana Press Inc., Totowa, NJ

promotes their delivery into a large panel of cell lines without the need for prior chemical covalent coupling *(21–23)*. We recently demonstrated that a variant containing a single mutation of this peptide carrier constitutes an excellent tool for the delivery of siRNA into different cell lines *(10)*. MPG is a peptide of 27 residues (GALFLGFLGAAGSTMGAWSQPKKKRKV) consisting of 3 domains with specific functions: a hydrophobic motif (GALFLGFLGAAGSTMGA)-derived from the fusion sequence of the human immunodeficiency virus (HIV) protein gp41 required for efficient targeting to the cell membrane and internalization, a hydrophilic lysine-rich domain (PKKKRKV) derived from the Nuclear Localization Sequence (NLS) of simian virus 40 (SV-40) large T antigen, involved in the main interactions with nucleic acids and required to improve intracellular trafficking of the cargo, as well as a spacer domain (WSQP), which improves the flexibility and the integrity of both the hydrophobic and the hydrophilic domains *(23)*. Given that the mechanism through which MPG delivers siRNA does not involve the endosomal pathway, the degradation of macromolecules delivered is significantly limited, and rapid dissociation of the MPG/cargo particle is favored as soon as it crosses the cell membrane. This peptide-based oligonucleotide delivery strategy presents several advantages, including rapid delivery of siRNA into cells with very high efficiency, stability in physiological buffers, lack of toxicity, and lack of sensitivity to serum. MPG technology constitutes a powerful tool for basic research, and several studies have demonstrated that this technology is extremely powerful for targeting specific genes in vitro as well as in vivo *(10,23–25)*.

This chapter will describe several protocols for the use of the noncovalent MPG technology for the delivery of siRNA into mammalian cells.

2. Materials

2.1. Cell Culture

1. Phosphate-buffered saline (PBS) (Invitrogen Life Technologies, cat. no. 14190-169).
2. Dulbecco's modified Eagle's medium (DMEM) (Invitrogen Life Technologies, cat. no. 41965-062).
3. Glutamine, streptomycin/penicillin (Invitrogen Life Technologies, cat. no. 15140-130).
4. Fetal bovine serum (FBS) (PERBIO, lot 3264EHJ, cat. no. CH30160-03).

2.2. Oligonucleotide and Transfection Reagents

1. siRNAs targeting the GAPDH gene (Silencer™ kit) (Proligo and/or Ambion).
2. Oligofectamine™ reagent (cat. no. 12252-011; Invitrogen Life Technologies, Carsbad, CA, USA).

2.3. Peptide Carrier MPG

1. MPG is a peptide of 27 residues: GALFLGFLGAAGSTMGAWSQPKSKRKV (molecular weight: 2908 Daltons). MPG can be synthesized in house or obtained

from commercial sources. Protocols for the synthesis and purification of MPG and derivatives are described in **refs. 23** and **26**. The cysteamide group was shown to be essential for the transfection mechanism and stabilization of the MPG/siRNA particle *(10)*. MPG and derivatives are stable for at least 1 yr when stored at −20°C in a lyophilized form.

2. The sequence variant of MPG used for siRNA delivery contains a single mutation in the NLS sequence (K^{23}S), GALFLGFLGAAGSTMGAWSQPKSKRKV (molecular weight: 2867 Daltons) (*see* **Note 6**). The peptide is acetylated at its N-terminus and carried a cysteamide group at its C-terminus, both of which are essential for the stability of the peptide and the transfection mechanism *(10)*.

3. Methods

The methods described below outline (1) the formation of MPG/siRNA complexes and handling (2) protocol for the delivery of siRNA into adherent cell lines, (3) protocol for the delivery of siRNA into suspension cell lines, and (4) a control lipid-based method for siRNA transfection. The different procedures were modified according to **refs. 10** and **23**.

3.1. Preparation of MPG/siRNA Complexes

The procedure for formation of MPG/siRNA complexes constitutes an important factor in the success of MPG technology and should be followed carefully (*see* **Notes 1** and **2**).

3.1.1. Preparation of Solution of MPG

1. Take the vial containing the peptide powder out of the freezer and equilibrate for 30 min at room temperature without opening the vial. This step is essential to limit hydration of the peptide powder. Resuspend the peptide at a concentration not higher than 1 mg/mL (concentration: 0.35 mM) in water.
2. Mix gently by tapping the tube or by vortexing at low speed for 20 s.
3. Repeated freeze–thaw cycles can induce peptide aggregation, therefore, it is recommended to aliquot the MPG stock solution into tubes containing the amount you expect to use in a typical experiment prior to freezing. The MPG stock solution is stable for about 2 mo when stored at −20°C.

3.1.2. Formation of MPG/siRNA Complexes

1. Prepare a stock solution of annealed duplex at 20 μM in water. Dilute the amount of siRNA duplex in 100 μL of water or PBS for each reaction (*see* **Table 1**). The siRNA can be used at concentrations varying from 5 to 100 nM, depending on the biological response expected. From our experience, a concentration ranging from 20 to 50 nM of the siRNA is sufficient for a silencing response higher that 80%.
2. The MPG solution must be diluted 1:10 in sterile H$_2$O (concentration: 35 μM). At this stage, sonication of the peptide solution is recommended (to limit aggregation) for 5 min in a water bath sonicator. Alternatively, a probe sonicator can also be used: Place the tube in water and sonicate for 1 min at an amplitude of

Table 1
Recommended Conditions for MPG/siRNA Complex Formation

siRNA (n*M*)	MPG/siRNA (molar ratio)					
	10/1		20/1		40/1	
	μ*M*	μL of stock	μ*M*	μL of stock	μ*M*	μL of stock
5	0.05	1[a]	0.1	1.7[a]	0.2	3.5[a]
10	0.1	1.7[a]	0.2	3.5[a]	0.4	7.0[a]
50	0.5	1.0	1.0	1.7	2.0	3.5
100	1.0	1.7	2.0	3.5	4.0	7.0

Note: The concentration of the stock solutions of siRNA and MPG are of 20 μ*M* and 350 μ*M*, respectively. The conditions reported in the table are for transfection of a 35-mm culture vessel (six-well), including a volume of 100 μL of MPG and a total transfection volume of 600 μL.

[a] For low concentration of siRNA (5 and 10 n*M*), the stock solution of MPG is diluted 10-fold to 35 μ*M*.

    30%. In a tube, dilute the appropriate volume of MPG into 100 μL of sterile water for each reaction. For optimal transfection, the molar ratio between MPG and siRNA is generally 10 : 1 to 40 : 1 (*see* **Table 1**).

3. Add 100 μL of diluted siRNA to 100 μL diluted MPG. Mix gently by tapping the tube. It is necessary to make the MPG/siRNA complex in a concentrated solution, which will be added to the cells and then be diluted to the final transfection volume (600 μL for a 35-mm culture plate).

4. For multiple assays, a mix for six transfections can be used. Dilute the corresponding MPG volume into 600 μL of water and the siRNA into 600 μL of PBS. Mix the two solutions and then proceed as described in **Subheading 3.2., steps 1–6**. Do not exceed the volume required for six transfections, as this might cause aggregation.

5. Incubate at 37°C for 20 min to allow the MPG/siRNA complex to form; then, proceed immediately to the transfection experiments. At this stage, MPG/siRNA complexes should not be stored.

3.2. MPG-Mediated siRNA Delivery into Adherent Cell Lines

    The protocol is described for HeLa and human fibroblast (HS-68) cell lines cultured in 35-mm culture plates, using a siRNA targeting the GAPDH gene. SiRNA concentration dependence of the silencing response was analyzed by Western blot (*see* **Fig. 1A**) and the kinetics of siRNA-induced degradation of GAPDH mRNA were followed by Northern blotting (*see* **Fig. 1B**). The amount of siRNA, MPG, transfection volume, and the number of cells should be adjusted accordingly to the size of the culture plate used (*see* **Notes 3** and **4**).

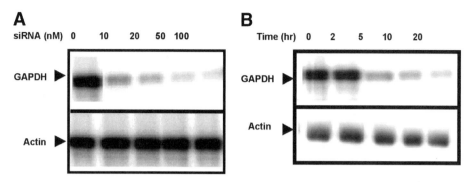

Fig. 1. MPG-mediated delivery of siRNA targeting GAPDH gene in HeLa cells. (**A**) MPG-mediated delivery of siRNA targeting GAPDH gene: Western blot analysis. Different concentrations of siRNA (25, 50, and 100 n*M*) were transfected with MPG at a molar ratio of 10:1 and the levels of GAPDH protein were analyzed by Western blotting 30 h posttransfection. Actin was used as a control to normalize protein loading. (**B**) MPG-mediated delivery of siRNA targeting GAPDH gene: Northern blot analysis. The kinetics of siRNA (50 n*M*)-induced degradation of GAPDH mRNA following transfection with MPG (molar ratio,10:1) were analyzed by Northern blotting. Actin was used as a control to normalize mRNA levels in each sample. (Reproduced from **ref. *10*** with permission from Oxford University Press.)

1. In a six-well or a 35-mm tissue culture plate, seed 0.3×10^6 cells per well in 2 mL of complete growth medium. Incubate the cells at 37°C in a humidified atmosphere containing 5% CO_2 until the cells are 50–70% confluent. It is recommended to pass the cells the day before treatment for a better response following transfection (*see* **Notes 2** and **4**).
2. Aspirate the medium from the cells to be transfected and wash the cells twice with PBS.
3. Overlay the cells with the 200 μL MPG/siRNA complex. Add 400 μL of serum-free medium to the overlay to achieve the final transduction volume of 600 μL for a 35-mm plate.
4. Incubate at 37°C in a humidified atmosphere containing 5% CO_2 for 30 min.
5. Add 1 mL of complete growth medium to the cells and adjusted the level of serum to 10% or 20%, depending on the culture condition required. Do not remove the MPG/siRNA complex. Then, incubate at 37°C in a humidified atmosphere containing 5% CO_2 for 24–48 h, depending on the cellular response expected (*see* **Note 5**). The siRNA are fully released in the cells 1 and 2 h later, respectively.
6. Process the cells for observation or detection assays. GAPDH protein and mRNA levels were analyzed by Western and Northern blots, respectively. GAPDH protein was probe using an anti-GAPDH (Sigma). For Northern blotting, total RNAs were isolated from cells using TriReagent™ (Sigma, Saint Louis, MO) according to the manufacturer's recommendations. RNAs were then purified by phenol extraction followed by ethanol precipitation. RNA samples (10 μg) were separated by

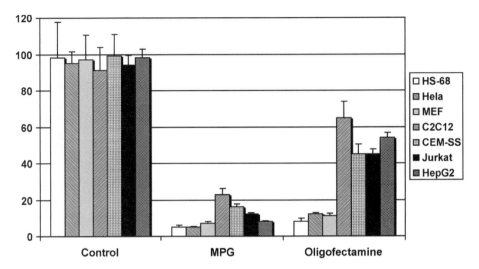

Fig. 2. MPG-mediated delivery of siRNA targeting GAPDH in adherent and suspension cell lines. MPG/siRNA targeting GAPDH gene complexes were formed at a molar ratio of 10 : 1 and then transfected on several adherent and suspension cell lines grown to 60% confluence. The levels of GAPDH protein were analyzed by Western blotting 30 h posttransfection. Control transfection experiments were performed using Oligofectamine™.

electrophoresis in formaldehyde agarose gels (1.2%), transferred to a Nylon membrane (Hybond N⁺; Amersham Pharmacia Biotech) and hybridized with ^{32}P-labeled GAPDH and actin probes (the latter used to normalize RNA loading). Signals were detected by a PhosphorImager (Molecular Dynamics) and quantified using ImageQuant software. For morphology assays, the cells can be fixed or observed directly, using live imaging technology.

3.3. MPG-Mediated siRNA Delivery into Suspension Cells

The protocol of MPG-mediated siRNA delivery was optimized on Jurkat T-cells using siRNA targeting GAPDH (*see* **Note 6**). However, efficient transfection might require optimization of MPG concentration, cell numbers, and exposure time of cells to the MPG/siRNA complex. A comparative study of different cell lines using a siRNA targeting GAPDH is reported in **Fig. 2**.

1. The same number of cells recommended for seeding adherent cells is recommended for suspension cells (confluency between 50% and 70%). Cells are cultured in standard medium in 35-mm dishes or six-well plates (*see* **Notes 2** and **4**).
2. The MPG/siRNA complexes are formed as described for adherent cells (**steps 1–4** of **Subheading 3.1.**).
3. Collect the suspension cells by centrifugation at 400*g* for 5 min. Remove the supernatant and wash the cells twice with PBS.

4. Centrifuge at $400g$ for 5 min to pellet the cells. Remove the supernatant.
5. Solubilize the cell pellet in the MPG/siRNA complex solution (200 μL). Add serum-free medium to achieve a final transduction volume of 600 μL.
6. Incubate at 37°C in a humidified atmosphere containing 5% CO_2 for 30 min to 1 h depending on the cell lines.
7. Add complete growth medium to the cells and adjust serum levels according to culture requirements. Do not remove the MPG/siRNA complex. Continue to incubate at 37°C in a humidified atmosphere containing 5% CO_2 for 24–48 h depending on the expected cellular response. As described for adherent cells, siRNA are fully released in to cells 1–2 h later, respectively.
8. Process the cells for observation or detection assays (**step 6** of **Subheading 3.2.**).

3.4. Lipid-Based Formulation for siRNA Delivery

Several lipid or cationic polymer-based formulations for oligonucleotide and siRNA delivery are commercially available (for review *see* **ref. 6**). Control transfection experiments were performed for all the cell lines using the most commonly used Oligofectamine™ (Invitrogen). Transfection protocols for six-well plates were performed according to the manufacturer's guidelines using a concentration of siRNA of 50 nM. Assay for silencing was measured 48 h after transfection and conditions used lead to typical transfection efficiencies of about 80% for HeLa cell lines.

4. Notes

1. It is essential to perform a complex formation between MPG and the siRNA in the absence of serum to limit degradation of siRNA and interactions with serum proteins. However, the transfection process itself is not affected by the presence of serum, which is a considerable advantage for most biological applications *(10)*.
2. Although this protocol was tested on several cells lines, conditions for efficient siRNA delivery should be optimized for every new cell line, including reagent concentration, cell number, and exposure time of cells to the MPG/siRNA complexes. A well-characterized siRNA should always be used as a positive control of transfection.
3. A large variety of antisense DNA or siRNA have been successfully introduced into different cell lines using MPG-based strategy and have been shown to induce significant silencing activity *(10,23–28)*. Recently, MPG formulation was adapted to condition required for in vivo applications and successfully used for intra-tumoral injection of siRNA *(24)*.
4. Low efficiency might be associated with several parameters: (1) Cell confluency: For adherent cells the optimal confluence is of about 50–60%; higher confluence (90%) dramatically reduces the transduction efficiency. (2) Formation of MPG/siRNA complexes: Conditions for the formation of MPG/siRNA complexes are critical and should be respected. Special attention should be paid to the

recommended volumes, incubation times for the formation of the complexes, and time of exposure of these complexes to cells.

5. The advantages of MPG technology are directly associated with the mechanism through which this carrier promotes delivery of siRNA into cells. The independence of MPG-mediated siRNA transfection on the endosomal pathway significantly limits degradation and preserves the biological activity of internalized cargoes for prolonged time periods. Silencing effects were observed up to 7 d after transfection.

6. An important property of MPG-based siRNA delivery is that it allows for the relatively rapid introduction of oligonucleotides into cells. The lack of requirement of covalent coupling for formation of MPG/oligonucleotide particles favors rapid release of siRNA into the cytoplasm as soon as the cell membrane has been crossed *(29)*. Entry of MPG/siRNA complexes into the cell occurs in a short time (10 min), and release of macromolecules takes place within the first 2 h. The rapid release of siRNA into the cytoplasm favors rapid degradation of target mRNA. However, nuclear targeting can be achieved thanks to an MPG variant containing an integral NLS. This latter version of MPG was shown deliver siRNA efficiently but exhibits slower degradation kinetics of mRNA. For long terms effects, the release of siRNA can be controlled by using this version of MPG leading to a more stable formulation *(10)*.

Acknowledgments

This work was supported in part by the Centre National de la Recherche Scientifique (CNRS) and by grants from the Agence Nationale de Recherche sur le SIDA (ANRS), the European Community (QLK2-CT-2001-01451), the Association pour la Recherche sur le Cancer to M.C.M. (ARC-4326) and to G.D. (ARC-5271). F.S. was supported by a grant from La Ligue de Recherche Contre le Cancer.

References

1. Hannon, G. J. (2002) RNA interference. *Nature* **418,** 244–251.
2. McManus, M. T. and Sharp, P. A. (2002) Gene silencing in mammals by small interfering RNAs. *Nat. Rev. Genet.* **3,** 737–747.
3. Elbashir, S. M., Harborth, J., Lendeckel, W., Yalcin, A., Weber, K., and Tuschl, T. (2001) Duplexes of 21-nucleotide RNAs mediate RNA interference in cultured mammalian cells. *Nature* **411,** 494–498.
4. Luo, D. and Saltzman, M. W. (2000) Synthetic DNA delivery system. *Nat. Biotechnol.* **18,** 33–37.
5. Niidome, T. and Huang, L. (2002) Gene therapy progress and prospects: non viral vectors. *Gene Ther.* **10,** 991–998.
6. Rozema, D. B. and Lewis, D. L. (2003) siRNA delivery technologies for mammalian systems. *Target* **2,** 253–260.
7. Hommel, D. J., Sears, R. M., Georgescu, D., Simmons, D., and Dileone, R. J. (2003) Local gene knockdown in the brain using viral-mediated RNA interference. *Nat. Med.* **9,** 1539–1543.

8. Brummelkamp, T. R., Bernards, R., and Agami, R. (2002) Stable suppression of tumorigenicity by virus-mediated RNA interference. *Cancer Cell* **2,** 243–247.

9. Xia, H., Mao, Q., Paulson, H. L., and Davidson, B. L. (2002) siRNA-mediated gene silencing in vitro and in vivo. *Nat. Biotechnol.* **20,** 1006–1010.

10. Simeoni, F., Morris, M. C., Heitz, F., and Divita, G. (2003) Insight into the mechanism of the peptide-based gene delivery system MPG: implication for delivery of siRNA into mammalian cells. *Nucleic Acids Res.* **31,** 2717–2727.

11. McCaffrey A. P., Meuse, L., Phan, T. T., Conklin, D. S., Hannon, G. J., and Kan, M. A. (2002) RNA interference in adult mice. *Nature* **418,** 38–39.

12. Song, E., Lee, S., Wang, J., et al. (2003) RNA interference targeting Fas protects mice from fulminant hepatitis. *Nat. Med.* **9,** 347–351.

13. Lewis, D. L., Hagstrom, J. E., Loomos, A. G., Wolff, J. A., and Herweijer, H. (2002) Efficient delivery of siRNA for inhibition of gene expression in postnatal mice. *Nat. Genet.* **32,** 107–108.

14. Gariepy, J. and Kawamura, K. (2000) Vectorial delivery of macromolecules into cells using peptide-based vehicles. *Trends Biotechnol.* **19,** 21–26.

15. Morris, M. C., Depollier, J., Mery, J., Heitz, F., and Divita, G. (2001) A peptide carrier for the delivery of biologically active proteins into mammalian cells. *Nat. Biotechnol.* **19,** 1173–1176.

16. Wadia, J. S. and Dowdy, S. F. (2002) Protein transduction technology. *Curr. Opin. Biotechnol.* **13,** 52–56.

17. Langel, U. (2002) *Cell Penetrating Peptides: Processes and Application.* (Langel, U., ed.), Pharmacology & Toxicology Series, CRC, Boca Raton, FL.

18. Morris, M. C., Chaloin, L., Heitz, F., and Divita, G. (2000) Translocating peptides and proteins and their use for gene delivery. *Curr. Opin. Biotechnol.* **11,** 461–466.

19. Järver, P. and Langel, U. (2004) The use of cell-penetrating peptides as a tool for gene regulation. *Drug. Discov. Today* **9,** 395–402.

20. Gait. M. J. (2003) Peptide-mediated cellular delivery of antisense oligonucleotides and their analogues. *Cell. Mol. Life Sci.* **60,** 1–10.

21. Morris, M. C., Vidal, P., Chaloin, L., Heitz, F., and Divita G. (1997) A new peptide vector for efficient delivery of oligonucleotides into mammalian cells. *Nucleic Acids Res.* **25,** 2730–2736.

22. Vidal, P., Morris, M. C., Chaloin, L., Heitz, F., and Divita G. (1997) New strategy for RNA vectorization in mammalian cells. Use of a peptide vector. *C.R. Acad. Sci. III* **320,** 279–287.

23. Morris, M. C., Chaloin, L., Mery, J., Heitz, F., and Divita G. (1999) A novel potent strategy for gene delivery using a single peptide vector as a carrier. *Nucleic Acids Res.* **27,** 3510–3517.

24. Morris M. C., Heitz, F., and Divita G. Personal communication.

25. Marthinet, E., Divita, G., Bernaud, J., Rigal, D., and Baggetto, L. G. (2000). Modulation of the typical multidrug resistance phenotype by targeting the MED-1 region of human MDR1 promoter. *Gene Ther.* **14,** 1224–1233.

26. Mery, J., Granier, C., Juin, M., and Brugidou, J. (1993) Disulfide linkage to polyacrylic resin for automated Fmoc peptide synthesis. Immunochemical applications of peptide resins and mercaptoamide peptides. *Int. J. Pept. Protein Res.* **42,** 447–452.

27. Morris, M. C., Chaloin, L., Choob, M., Archdeacon, J., Heitz, F., and Divita, G. (2004) Combination of a new generation of PNAs with a peptide-based carrier enables efficient targeting of cell cycle progression. *Gene Ther.* **11,** 757–764.

28. Morris, K. V., Chan, S. W., Jacobsen, S. E., and Looney, D. J. (2004) Small interfering RNA-induced transcriptional gene silencing in human cells. *Science* **305,** 1289–92.

29. Deshayes, S., Plenat, T., Aldrian-Herrada, G., Divita, G., Le Grimellec, C., and Heitz, F. (2004) Primary amphipathic cell-penetrating peptides: structural requirements and interactions with model membranes. *Biochemistry* **43,** 7698–7706.

14

Lentiviral Vector Delivery of siRNA and shRNA Encoding Genes into Cultured and Primary Hematopoietic Cells

Mingjie Li and John J. Rossi

1. Introduction

Small interfering RNAs (siRNAs) mediate RNA interference (RNAi), providing a powerful new tool for gene silencing in a sequence-specific manner. RNAi is a useful tool for studying gene function and regulation, but it also has potential for therapeutic applications. In many cases, a long-term effect of RNAi is required, such as in human immunodeficiency virus (HIV)-1-infected individuals. Lentiviral vectors are able to transduce nondividing cells, with sustained long-term expression of genes contained in the vector backbone.

The majority of target cell types for gene therapy are nondividing or slowly dividing. These include brain, liver, muscle and hematopoietic stem cells. Each of these cell types have been efficiently transduced by lentiviral vectors *(1–3)*. We and others *(4–6)* have reported that lentiviral vectors can efficiently deliver si/shRNA (short-hairpin RNA) expression cassettes into a variety of cells with sustained expression and potent function of the encoded siRNAs.

The current third-generation replication-defective and self-inactivating (SIN) lentiviral vectors have minimized the potential risk of generating replication-competent helper virus. We used pHIV-7-GFP, a typical third-generation replication-defective SIN vector as the transfer vector for delivering shRNAs. As illustrated in **Fig. 1**, this vector contains a hybrid 5′ LTR (long terminal repeat), in which the U3 region is replaced with the cytomegalovirus (CMV) promoter and enhancer sequence. Use of this promoter to make the vector RNA makes the transcription of the vector sequence independent of HIV-1 Tat function, which is normally required for HIV gene expression *(7)*. The packaging signal (ψ) is essential for encapsidation and the Rev-responsive element

From: *Methods in Molecular Biology, vol. 309: RNA Silencing: Methods and Protocols*
Edited by: G. G. Carmichael © Humana Press Inc., Totowa, NJ

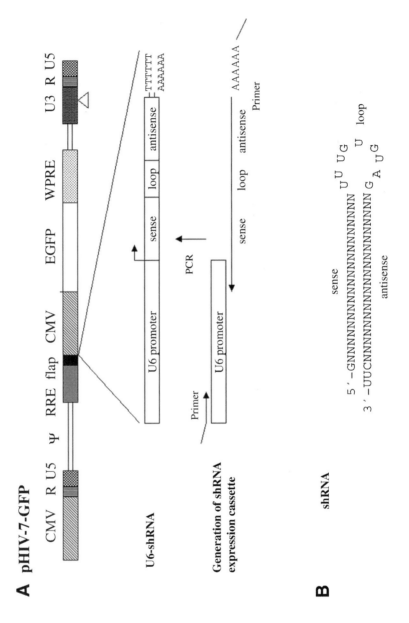

(RRE) is required for producing high-titer vectors. The flap sequence or poly-purine tract (cPPT) and the central termination sequence (CTS) are important for nuclear import of the vector DNA, a feature required for transducing non-dividing cells *(8)*. The enhanced green fluorescent protein (EGFP) reporter gene is driven by an internal CMV promoter. This is useful for vector titration, measurements of transduction efficiency, and selection of transduced cells. In the 3′ LTR, the cis-regulatory sequences were completely removed from the U3 region. This deletion is copied to 5′ LTR after reverse transcription, resulting in transcriptional inactivation of both LTRs. To produce the packaged vectors, the vector DNA **(Fig. 1)** along with three other plasmids are cotransfected into 293T cells. The Gag/Pol proteins are supplied by the pCHGP-2 plasmid; pCMV-Rev encodes Rev, which binds to the RRE for efficient RNA export from the nucleus, and pCMV-G encodes the vesicular stomatitis virus glycoprotein (VSV-G) that replaces HIV-1 Env. VSV–G expands the tropism of the vectors and allows concentration via ultracentrifugation. The siRNA sequence(s) along with a pol III promoter are inserted directly upstream of the CMV–EGFP sequence.

The human U6 small nuclear RNA promoter and human H1 promoter are among the common pol III promoters used for expressing siRNAs. These promoters have relatively small size and the transcription is conveniently terminated within a stretch of four or more uridines. In our experience, at least three pol III expression cassettes can be delivered by a single lentiviral vector backbone. The sense and antisense sequences of siRNAs can be expressed from separate promoters or from a single promoter directing a shRNA structure. We previously demonstrated that anti-rev siRNAs expressed from separate

---

Fig. 1. (*Opposite page*) Construction of lentiviral vector expressing shRNA. (**A**) *Top*: The transfer vector, pHIV-7-GFP contains a hybrid 5′ LTR in which the U3 region is replaced with the CMV promoter, the packaging signal (ψ), the RRE sequence, the flap sequence, the EGFP gene driven by CMV promoter, the woodchuck hepatitis virus posttranscriptional regulatory element (WPRE) and the 3′ LTR in which the cis regulatory sequences are completely removed from the U3 region. The genes of interest along with a pol III promoter can be inserted directly upstream of the CMV promoter of EGFP in pHIV-7-GFP vector. *Bottom*: The U6-shRNA expression cassette is constructed by polymerase chain reaction with primers shown. The PCR product contains restriction sites at both ends for cloning into the lentiviral vector. The shRNA expression cassette, including the U6 pol III promoter, the sense and antisense sequence of the shRNA separated by a 9-base loop, and a terminator composed of six thymidines is inserted directly upstream of the CMV promoter of EGFP in pHIV-7-GFP vector. Arrows indicate the orientation of transcription for a given gene. (**B**) The putative structure of the shRNA expressed in target cells.

promoters showed marked target downregulation as well as anti-HIV-1 activity *(9)*. We subsequently found that the shRNAs exhibited even more potent RNAi activity than the separately expressed siRNAs designed to inhibit the same target sequences *(4)*. In addition, shRNA expression cassettes are easier to construct than the siRNA counterparts. Therefore, we have focused on shRNA constructs in the lentiviral vector delivery system incorporating the human U6 pol III promoter to direct shRNA expression. As an example of lentiviral vector-mediated transduction of an shRNA gene, we provide an example of RNAi-mediated protection against HIV-1 challenge in stably transduced hematopoietic cells.

2. Materials

2.1. Construction of Lentiviral Vectors Carrying shRNA Sequences

1. Plasmid DNA: lentiviral vector backbone pHIV-7-GFP, U6 promoter containing plasmid pTZ-U6+1.
2. Polymerase chain reaction (PCR) primers.
3. *Taq* DNA polymerase, restriction enzymes, T4 DNA ligase.
4. QIAEX II gel extraction kit (QIAGEN, Valencia, CA).
5. QIAGEN Plasmid Maxi Kit (QIAGEN).

2.2. Production of Lentiviral Vectors

2.2.1. Packaging

1. 293T Cell (human embryonic kidney cells containing SV40 large T antigen).
2. Dulbecco's modified Eagle's medium (DMEM) with high glucose (4500 mg/L) supplemented with 10% fetal bovine serum (FBS), 100 units/mL penicillin, and 100 µg/mL streptomycin.
3. Plasmid DNA: the transfer vector containing shRNA sequences, pCHGP-2, pCMV-rev, and pCMV-G.
4. TE 79/10: 1 mM Tris-HCl, 0.1 mM EDTA, pH 7.9.
5. 2 M CaCl$_2$.
6. 2X HBS: 0.05 M HEPES, 0.28 M NaCl, 1.5 mM Na$_2$HPO$_4$, pH 7.12.
7. 0.6 M Butyric acid.

2.2.2. Concentration

1. 30-mL syringe.
2. 0.2-µL Syringe filter.
3. Ultracentrifuge and SW28 rotor (Beckman).
4. 1 × 3 ½-in. polyallomer centrifuge tubes (Beckman).

2.2.3. Titration

1. Lentiviral vector stock.
2. HT1080 (a human fibrosarcoma cell line).

3. DMEM supplemented with 10% FBS, 100 units/mL penicillin, and 100 μg/mL streptomycin.
4. 4 mg/mL Polybrene stock.
5. 3.7% Formaldehyde in phosphate-buffered saline (PBS).

2.3. Transduction of Lentiviral Vectors to Target Cells

1. Packaged vectors.
2. Target cells.
3. Culture media.
4. Polybrene.
5. Cytokines.
6. RetroNectin (Takara Mirus Bio Inc., Madison, WI).

2.4. Detection of the Expression of shRNAs

1. STAT-60 reagent (Tel-Test, Inc., Friendswood, TX).
2. Polyacrylamide gel containing 7 *M* urea.
3. RNA loading buffer: 95% deionized formamide, 0.025% bromophenol blue, 0.025% xylene cyanol, 0.5 m*M* EDTA, 0.025% sodium dodecyl sulfate (SDS).
4. 0.5 *M* TBE buffer.
5. Hybond-N nylon membrane (Amersham, Arlington Heights, IL).
6. Electrotransfer apparatus.
7. 20X SSPE stock: 3 *M* NaCl, 0.2 *M* NaH$_2$PO$_4$, 0.02 *M* EDTA, pH 7.4.
8. Prehybridization buffer: 6X SSPE, 5X Denhardt's, 0.5% SDS, and carrier DNA.
9. γ-^{32}P-Labeled oligonucleotide probe.

3. Methods

This section covers (1) construction of the lentiviral vectors containing the shRNA genes, (2) production of the vectors, (3) transduction of the vectors in target cells, (4) detection of the expression of the shRNA, and (5) verification of shRNA functions.

3.1. Construction of Lentiviral Vectors Carrying shRNA Sequences

3.1.1. The Lentiviral Vector Backbone and the Packaging System

The vector backbone we used as a transfer vector of shRNAs is pHIV-7-GFP *(10)*, which is a typical third-generation, replication-defective, and SIN vector. As illustrated in **Fig. 1**, this vector contains a hybrid 5′ LTR, in which the U3 region is replaced with the CMV promoter and enhancer sequence. This strong promoter makes the transcription of the vector sequence Tat independent. The packaging signal (ψ) is essential for encapsidating the vector sequence. The RRE is required for producing high-titer vectors. The flap sequence composed of cPPT and the CTS is important for nuclear import, a feature required for transducing nondividing cells *(8)*. The enhanced EGFP reporter gene driven by an internal CMV promoter facilitates titration, detection of transduction efficiency, and selection of transduced cells. The woodchuck hepatitis virus post-transcriptional regulation element (WPRE) *(11)* following the EGFP sequence improves the expression of the reporter gene by promoting RNA nuclear export and/or polyadenylation. In the 3′ LTR, the cis-regulatory sequences were complete removed from the U3 region. This deletion will be copied to 5′ LTR after reverse transcription and resulting in transcriptional inactivation of both LTRs. To produce transducible vector, a cotransfection with the transfer vector and three other plasmids into 293T cells is carried out. The pCHGP-2 encodes the Gag/Pol protein products. The pCMV-Rev provides Rev protein, which binds to RRE for efficient RNA export from the nucleus. The pCMV-G provides VSV-G that replaces HIV-1 env for increasing tropism and allowing ultracentrifugation for concentration of the vector supernatant.

3.1.2. Construction of shRNA Expression Cassette

The following steps describe construction of a shRNA cassette with human U6 promoter (*see* **Note 1**). The human U6 small nuclear RNA promoter is among the common pol III promoters used for expressing shRNAs. This promoter is relatively small in size and the transcription is conveniently terminated within a stretch of four or more uridines. In our experience, at least three Pol III expression cassettes can be delivered by a single lentiviral vector backbone *(4)*.

1. PCR amplification of plasmid containing human U6 promoter sequence following standard procedures. Use a plasmid containing the U6 promoter sequence (e.g., pTZ U6+1) as the template. The upstream primer is complementary to the 5′ end of the U6 promoter. This can serve as a universal primer for all shRNA gene constructs. The downstream primer should be complementary to 3′ end of the U6 promoter and include the complement to the sense, loop, antisense, and transcriptional terminator (**Fig. 1**) (*see* **Note 2**) *(12)*. Include a restriction site at the 5′ end of both primers for ligating the PCR products to the vector (*see* **Note 3**).
2. Gel-purify the PCR product with a QIAEX II gel extraction kit following the manufacturer's instructions.

3.1.3. Cloning the shRNA Gene into the Transfer vector

1. Digest the PCR product and the transfer vector with appropriate restriction enzymes and ligate the insert into the multiple cloning site located between the flap and CMV–EGFP in the lentiviral vector backbone.
2. Verify the cloning product by restriction analysis and DNA sequencing.
3. Amplify and purify the correct cloning product and the three packaging-required plasmids with the Plasmid Maxi Kit (QIAGEN).

3.2. Production of Lentiviral Vectors

3.2.1. Packaging of the Vectors

1. The packaging cell line, 293T, is maintained in DMEM, supplemented with 10% FBS, 100 units/mL penicillin, and 100 µg/mL streptomycin in 37°C incubator with 10% CO_2. Plate the 293T cells in 100-mm tissue culture dishes 24 h before transduction. We usually prepare at least five dishes per vector to obtain reasonable amount vector for repeatable experiments. The cell density is 30–40% confluence when seeding and will be about 80% confluence when transfection.
2. Change the culture medium with 10 mL of fresh medium 5 h before transfection.
3. Prepare 1 mL of calcium phosphate–DNA suspension for each 100-mm plate of cells as follows:
   a. Set up two sterile tubes for transfection of one plate. Label the tubes 1 and 2.
   b. Add 0.5 mL of 2X HBS to tube 1.
   c. Add TE 79/10 to tube 2. The volume of TE 79/10 = 440 µL − the volume of DNA.
   d. Add 15 µg of the transfer vector containing the shRNA expression cassette, 15 µg of pCHGP-2, 5 µg of pCMV-Rev, and 5 µg of pCMV-G to tube 2 and mix.
   e. Add 10 µL of 2 *M* $CaCl_2$ solution to tube 2; gently mix.
   f. Add 50 µL of 2 *M* $CaCl_2$ solution to tube 2; gently mix.
   g. Transfer of contents from tube 2 to tube 1, dropwise with gently mix.
   h. Allow the suspension to sit at room temperature for at least 30 min.
4. Mix the precipitation well by pipetting or vortexing.
5. Add 1 mL of suspension to a 100-mm plate containing cells. The suspension must be added slowly, dropwise, while gently swirling the medium in the plate. Return the plates to the incubator and leave the precipitation for 5–6 h.
6. Replace the medium with 6 mL of complete medium. Add 60 µL of 0.6 M butyric acid. Return to culture.
7. After 24 h culture, collect supernatant and freeze it at −80°C. Add 6 mL complete medium to each plate. Add 60 µL of 0.6 *M* butyric acid. Return to culture.
8. After 12 h culture, collect supernatant. Freeze it at −80°C or go to next step.

3.2.2. Concentration of the Vectors

1. Centrifuge the supernatant freshly collected or thawed from freezer 900*g* for 10 min to remove any cell debris in the supernatant. Filter the supernatant with a 0.2-µ*M* syringe filter (*see* **Note 4**). Transfer the supernatant to polyallomer tubes (*see* **Note 5**). Concentrate the supernatant by ultracentrifugation at 24,500 rpm and 4°C for 1.5 h with Beckman SW28 swing rotor.
2. Remove the supernatant and resuspend the pellet into an appropriate amount of culture medium (e.g., 150 µL for 30 mL of original supernatant if a 200-fold concentration is desired).
3. The concentrated vector should be divided into 10- to 50-µL aliquots and stored at −80°C until use. Avoid freeze–thaw cycle.

3.2.3. Titration of the Vectors

1. Seed 1×10^5 per well of HT1080 in six-well plate in DMEM medium supplemented with 10% FBS and culture overnight.
2. Add serial diluted vector stock and 4 μL/mL of polybrene to the cultured cells. Continue culture for 48 h.
3. Trypsinize the cells. Following centrifuge, remove the supernatant and resuspend the pellet with 300 μL of 3.7% formaldehyde in PBS.
4. Determine the percentage of EGFP positive cells by flow-activated cell sorting (FACS) analysis (*see* **Note 6**). The titer will be represented as transduction unit per milliliter concentrated vector (TU/mL):

$$\text{Titer} = \frac{\text{Cell number} \times \text{Percentage of EGFP}^+ \text{ cells} \times \text{Dilution}}{\text{Vector volume (mL)} \times 100}$$

In this formula, the cell number stands for the cell count when the vector was added.

3.3. Transduction of Lentiviral Vectors to Target Cells

The following are transduction protocols for some commonly used target cells in shRNA research and therapeutic applications.

3.3.1. Transducing Monolayer Cultured Cells

For monolayer-cultured cells, seed cells to culture plates 24 h before transduction. At the time of transduction, the cell density should be 30–40% confluent. Add a vector at an appropriate multiplicity of infection (MOI) and polybrene at a final concentration of 4 μg/mL and return the cells into incubator. After overnight culture, replace the culture medium. For many monolayer-cultured cell lines, an MOI of 5 can achieve over 90% transduction efficiency. For a new cell line to be used, a series of different MOIs should be tested to find a minimum effective MOI.

3.3.2. Transducing Suspension-Cultured Cell Lines

To transduce suspension-cultured cell lines, seed 2×10^5 per cells well into a 24-well plate. Add appropriate amount of vector and 4 μg/mL polybrene. After overnight incubation, replace the medium. For the K562 cell line (a leukemia cell line from CML patient), the transduction efficiency is close to 100% at a MOI of 10. For some suspension-cultured cells such as CEM (a human T-cell line), centrifugation can remarkably enhance transduction efficiency. Place 2×10^5 cells in 1 mL culture medium in 15-mL centrifuge tube. Add vector and 4 μg/mL polybrene. Centrifuge at 900*g* and 20°C for 30 min. Resuspend the cell pellet with pipet and transfer the cells to the culture plate. After overnight culture, replace the medium. Transduction efficiency can be determined by FACS analysis 48 h after transduction.

3.3.3. Transducing Hematopoietic Stem Cells

For transduction of hematopoietic stem cells, the CD34$^+$ cells were enriched from umbilical cord blood or bone marrow by anti-CD34 antibody-coupled magnetic beads (Miltenyi Biotech, Aubum, CA). Forty-eight hours prior to transduction, the CD34$^+$ cells were cultured in Iscove's modified Dulbecco's medium (IMDM) supplemented with 20% BIT9500 (Stem Cell Technology, Vancouver, BC, Canada), 40 µg/mL human low-density lipoproteins, 10^{-4} M of 2-mercaptoethanol, 100 ng/mL SCF, 100 ng/mL flt3-ligand, 10 ng/mL TPO (PeproTech, Rocky Hill, NJ), 20 ng/mL interleukin (IL)-3, and 20 ng/mL IL-6 (R & D Systems, Minneapolis, MN). Coat 24-well nontissue-culture-treated plate with RetroNectin (Takara Mirus Bio Inc., Madison, WI) following the manufacturer's instruction. The lentiviral vector stock was adjusted to 40 MOI in 200 µL culture medium and loaded to the wells of the coated plate. After incubation at 32°C for 4 h, the vector supernatant was removed and the well was washed with PBS. The prestimulated CD34$^+$ cells were added to the well at 5×10^4/mL in the growth medium. After overnight culture, centrifuge the cells, change the medium, and put the cells back in the culture.

3.4. Detecting the Expression of shRNA Delivered by Lentiviral Vector

Northern blotting is among the most reliable methods to verify whether the cloned shRNA sequence can be expressed and the level of expression in the target cells.

1. Extract total RNA from transduced cells with STAT-60 reagent (Tel-Test, Inc., Friendswood, TX) according to the manufacturer's protocol.
2. Prepare polyacrylamide gel containing 7 M urea. A 15% polyacrylamide gel is commonly used for detecting shRNA. An 8% polyacrylamide gel can resolve the shRNA band equally well and is more efficient to transfer. In addition, an 8% polyacrylamide gel allows shRNAs and other larger RNAs up to several hundred nucleotides length to be detected in a single blotting.
3. Fifteen micrograms of total RNA solution (we were able to detect shRNA using 3.5 µg of total RNA extracted from sorted EGFP$^+$ population of CD34$^+$ cells) is mixed with an equal volume of RNA loading buffer. Heat the samples at 95°C for 4 min and then put on ice.
4. Load samples onto the gel and carry out electrophoresis in 0.5X TBE buffer until the bromophenol blue dye migrates to about two-thirds of the gel plate.
5. Transfer the RNA to Hybond-N nylon membrane by electroblotting in 0.5X TBE buffer.
6. Fix the RNA to the membrane with an ultraviolet (UV) crosslinker.
7. Prehybridize in buffer containing 6X SSPE, 5X Denhardt's, 0.5% SDS, and carrier DNA for 2 h.

8. Add γ-^{32}P-labeled oligonucleotide probe complementary to the antisense sequence of the shRNA and hybridize overnight at 37°C.
9. The membrane is washed with 6X SSPE and 0.1% SDS, at 37°C for 10 min, and then washed with 2X SSPE and 0.1% SDS twice at 37°C for 10 min each and exposed to a X-ray film.

3.5. Determining the Efficacy of RNA Cleavage by shRNAs in Lentiviral Vector Transduced Cells

For detecting the cleavage of endogenous mRNAs by shRNAs, reverse transcription (RT)-PCR, real-time PCR and Northern blotting are among the choices. To determine the protein-level knockdown, Western blotting and immunofluorescence are most common approaches. For functional assay, in the case of anti-HIV shRNAs, assays for HIV-1 p24 antigen and reverse transcriptase are used. For shRNA targeting fusion mRNA in malignant cells, cell proliferation assay and apoptosis assay would be good choices.

3.6. Experimental Example

In the following example, human embryonic cord-blood-derived CD34$^+$ hematopoietic progenitor cells were transduced with HIV-7 harboring an shRNA targeting HIV-1 rev or the HIV-7 vector backbone as a control. Eleven days after transduction, the cells were sorted by FACS and EGFP$^+$ cells were collected. After recovery from sorting, 5×10^5 EGFP$^+$ cells were exposed to a monocytotropic JR-FL strain of HIV-1 at an MOI of 0.01 overnight. The infected cells were washed four times with HBSS and the cultures were in media for CD34$^+$ cells as described earlier. The culture supernatant was collected on a weekly bases. The p24 antigen analyses were performed using Coulter HIV-1 p24 Antigen Assay (Beckman Coulter, Brea, CA) according to the manufacturer's instructions. From the data of **Fig. 2**, it can be seen that the cells expressing only EGFP were highly susceptible to HIV-1 replication, whereas the shRNA transduced cells were completely protected from HIV-1 replication, as indicated by the baseline levels of p24 antigen production. The protection lasted for over 1 mo before the experiment was terminated. This type of protection is commonly observed in anti-HIV shRNA expressing primary cells. There are many potential target sites in HIV for RNAi, and we highly recommend multiplexing shRNAs targeting two or more different HIV sequences to minimize viral escape mutants.

In summary, lentiviral vectors can be effective vehicles for the delivery of shRNA genes into a variety of cultured and primary cells. Our example demonstrates that human hematopoieitc progenitor cells are good targets for lentiviral-vector-mediated transduction, thus providing the foundation for future hematopoietic stem cell gene therapy against viral diseases and perhaps some leukemias.

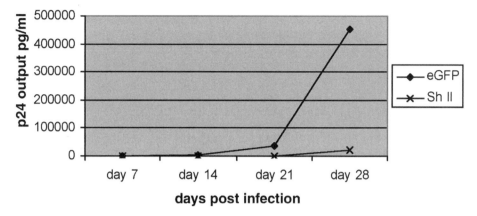

Fig. 2. HIV-1 challenge assay of anti-HIV rev shRNA transduced cells. Embryonic cord blood derived CD34⁺ hematopoietic progenitor cells were transduced with either the parental HIV-7 vector or HIV-7U6-shRNArev. Transduced cells were FACS sorted for EGFP expression and following a short period of recovery challenged with HIV-1 JRFL at a multiplicity of infection of 0.01. The supernatants were withdrawn on the indicated days and assayed for HIV-1 secreted p24 antigen, as described in the text.

4. Notes

1. In addition to the human U6 promoter, other pol III promoters, such as the human H1 promoter *(13)* and tRNA promoter *(14)*, can also be used for expressing siRNAs or shRNAs.
2. The first position in the transcript of U6 promoter should be guanine (G) to ensure efficient transcription. If the first base in the target sequence is not G, an extra G can be added to the sense strand transcript without affecting the function of the shRNA.
3. Alternatively, the restriction sites in the primers can be omitted by ligating the PCR product to pCR2.1 using TA cloning kit (Invitrogen, Carlsbad, CA).
4. It could also be filtered with a 0.2-μm cellulose acetate bottle-top filter for preparing large quantity and reducing the loss of the vector.
5. The polyallomer tubes are autoclavable. Using autoclaved centrifuge tubes and a 0.2-μL filter can minimize contamination during handling the vector supernatant. This is especially important if the transduced cells are used for long-term culture or engrafting animals.
6. If a transfer vector used does not contain a reporter gene, the vector titer could be determined as the number of integrated vector DNA/mL of vector by real time PCR with primers for psi element of the vector *(15)*.

Acknowledgments

We thank Dr. Jiing-Kaun Yee for providing pHIV-7-GFP and the packaging plasmids. This work was supported by NIH grants AI29329 and AI42552 and HL074704 to J.J.R.

References

1. Naldini, L., Blomer, U., Gage, F. H., Trono, D., and Verma, I. M. (1996) Efficient transfer, integration, and sustained long-term expression of the transgene in adult rat brains injected with a lentiviral vector. *Proc. Natl. Acad. Sci. USA* **93,** 11,382–11,388.
2. Kafri, T., Blomer, U., Peterson, D. A., Gage, F. H., and Verma, I. M. (1997) Sustained expression of genes delivered directly into liver and muscle by lentiviral vectors. *Nat. Genet.* **17,** 314–317.
3. Uchida, N., Sutton, R. E., Friera, A. M., et al. (1998) HIV, but not murine leukemia virus, vectors mediate high efficiency gene transfer into freshly isolated G0/G1 human hematopoietic stem cells. *Proc. Natl. Acad. Sci. USA* **95,** 11,939–11,944.
4. Li, M.-J., Bauer, G., Michienzi, A., et al. (2003) Inhibition of HIV-1 infection by lentiviral vectors expressing Pol III-promoted anti-HIV RNAs. *Mol. Ther.* **8,** 196–206.
5. Qin, X. F., An, D. S., Chen, I. S., and Baltimore, D. (2003) Inhibiting HIV-1 infection in human T cells by lentiviral-mediated delivery of small interfering RNA against CCR5. *Proc. Natl. Acad. Sci. USA* **100,** 183–188.
6. Scherr, M., Battmer, K., Ganser, A., and Eder, M. (2003) Modulation of gene expression by lentiviral-mediated delivery of small interfering RNA. *Cell Cycle* **2,** 251–257.
7. Arya, S. K., Guo, C., Josephs, S. F., and Wong-Staal, F. (1985) Trans-activator gene of human T-lymphotropic virus type III (HTLV-III). *Science* **229,** 69–73.
8. Sirven, A., Pflumio, F., Zennou, V., et al. (2000) The human immunodeficiency virus type-1 central DNA flap is a crucial determinant for lentiviral vector nuclear import and gene transduction of human hematopoietic stem cells. *Blood* **96,** 4103–4110.
9. Lee, N. S., Dohjima, T., Bauer, G., et al. (2002) Expression of small interfering RNAs targeted against HIV-1 rev transcripts in human cells. *Nat. Biotechnol.* **20,** 500–505.
10. Yam, P. Y., Li, S., Wu, J., Hu, J., Zaia, J. A., and Yee, J. K. (2002) Design of HIV vectors for efficient gene delivery into human hematopoietic cells. *Mol. Ther.* **5,** 479–484.
11. Zufferey, R., Donello, J. E., Trono, D., and Hope, T. J. (1999) Woodchuck hepatitis virus posttranscriptional regulatory element enhances expression of transgenes delivered by retroviral vectors. *J. Virol.* **73,** 2886–2892.
12. Castanotto, D., Li, H., and Rossi, J. J. (2002) Functional siRNA expression from transfected PCR products. *RNA* **8,** 1454–1460.
13. Brummelkamp, T. R., Bernards, R., and Agami, R. (2002) A system for stable expression of short interfering RNAs in mammalian cells. *Science* **296,** 550–553.
14. Kawasaki, H. and Taira, K. (2003) Short hairpin type of dsRNAs that are controlled by tRNA(Val) promoter significantly induce RNAi-mediated gene silencing in the cytoplasm of human cells. *Nucleic Acids Res.* **31,** 700–707.
15. Sastry, L., Johnson, T., Hobson, M. J., Smucker, B., and Cornetta, K. (2002) Titering lentiviral vectors: comparison of DNA, RNA and marker expression methods. *Gene Ther.* **9,** 1155–1162.

15

Exon-Specific RNA Interference

A Tool to Determine the Functional Relevance of Proteins Encoded by Alternatively Spliced mRNAs

Alicia M. Celotto, Joo-Won Lee, and Brenton R. Graveley

1. Introduction

The majority of metazoan genes encode pre-mRNAs that are subject to alternative splicing. For example, it has recently been estimated that as many as 74% of human genes encode alternatively spliced mRNAs *(1)*. An alternatively spliced gene can generate anywhere from 2 different isoforms to as many as 38,016 isoforms in the case of the *Drosophila Dscam* gene *(2)*. Thus, alternative splicing serves to greatly expand the diversity of the proteins encoded by a genome *(3)*.

In many cases, it has been shown that different protein isoforms synthesized from a single gene have distinct functions *(3)*. In our efforts to determine the functional significance of alternative splicing, we developed a variant of RNA interference (RNAi) we call exon-specific RNAi *(4)*. This technique involves the selective design of a double-stranded RNA (dsRNA) trigger that is complementary to a specific alternative exon (*see* **Fig. 1**). We have shown that in cultured *Drosophila* cells, these exon-specific dsRNA triggers specifically induce the degradation of mRNA isoforms containing the targeted exon, but they do not affect the stability of mRNA isoforms synthesized from the same gene that lack the targeted exon. Thus, it should be possible to use this technique to determine the function of proteins encoded by alternatively spliced mRNAs.

It is important to note that exon-specific RNAi is unlikely to work in organisms that contain RNA-dependent RNA polymerases (RdRPs), such as worms and plants *(5)*. This is because these enzymes are responsible for a phenomenon referred to as transitive RNAi *(6)*. RdRPs can use the siRNA triggers to catalyze the synthesis of dsRNA complementary to the mRNA upstream of the targeted

From: *Methods in Molecular Biology, vol. 309: RNA Silencing: Methods and Protocols*
Edited by: G. G. Carmichael © Humana Press Inc., Totowa, NJ

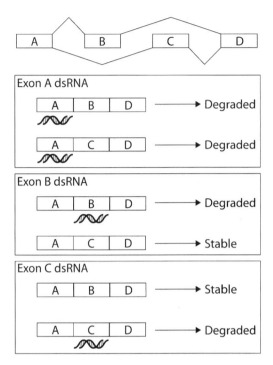

Fig. 1. Overview of exon-specific RNAi. The hypothetical gene shown contains four exons, two of which—B and C—are alternatively spliced in a mutually exclusive manner. dsRNAs corresponding to a common exon, such as exon A, would induce the degradation of all mRNA isoforms synthesized from this gene. In contrast, dsRNAs corresponding to the alternative exons would specifically induce the degradation of mRNAs containing the targeted exon without affecting the stability of mRNAs lacking the targeted exon.

sequence. This, in turn, induces the degradation of mRNAs complementary to the secondary siRNAs. Thus, if the dsRNA trigger is homologous to an alternative exon, the RdRPs will synthesize secondary siRNAs complementary to the upstream constitutive exon and, consequently, all mRNA isoforms from the targeted gene will be degraded *(7)*. However, exon-specific RNAi can be used in organisms such as *Drosophila*, mouse, and humans, which lack RdRPs *(5)*. In this chapter, we discuss the methods and approaches to perform exon-specific RNAi with a specific emphasis on using this technique in *Drosophila* cells.

2. Materials

2.1. Equipment

The equipment required for these experiments are common to most labs routinely performing molecular biology experiments.

1. Electrophoresis supplies: Power supplies and standard agarose gel electrophoresis units will be needed.
2. Thermal cycler: This is required for polymerase chain reaction (PCR), but can also be used for other reaction incubations if needed.
3. Cell culture incubator: An incubator capable of maintaining a constant temperature of 25–27°C is required. It is not necessary to use an incubator with a CO_2 supply line.
4. Cell culture hood.

2.2. Supplies

1. Cell culture plasticware: 100-mL plastic Erlenmeyer flasks for maintaining cell stocks, six-well dishes for performing RNAi.
2. Schneider (S2) cells: Dmel-2 cells (Invitrogen, cat. no. 10831-014) or S2 cells (Invitrogen, cat. no. R690-07). Cells are also available from the ATCC (CRL-1963).
3. Cell culture medium: Drosophila-SFM (Invitrogen, cat. no. 10797-017), supplemented with 1X penicillin–streptomycin (Invitrogen, cat. no. 15140-122, 100X stock), and 2 mM glutamine (Invitrogen, cat. no. 25030-081, 100X stock).
4. Oligonucleotides: T7 primer, TAATACGACTCACTATAGGG; SP6 primer, ATT-TAGGTGACACTATAG, gene-specific primers to clone complementary DNA (cDNA) fragments.
5. Cloning vector: pCRII-TOPO cloning kit (Invitrogen, cat. no. K4600-01).
6. Transcription reagents: Ampliscribe High Yield Transcription Kits (Epicentre, SP6 kit, cat. no. AS3106, T7 kit, cat. no. AS3107).
7. Electrophoresis chemicals: agarose, ethidium bromide, and so forth.
8. Reverse transcription–polymerase chain reaction (RT-PCR) reagents: standard reagents (i.e., dNTPs, reverse transcriptase, *Taq* DNA polymerase, reaction buffers, etc.).
9. RNA isolation reagents: Trizol reagent (Invitrogen).

2.3. Solutions

1. 10X Annealing buffer: 1 M NaCl, 200 mM Tris-HCl, pH 8.0, 10 mM ethylenediamine tetraacetic acid.
2. Phenol : chloroform : isoamyl alcohol (25 : 24 : 1, v : v : v).
3. Ethanol.

3. Methods

The methods described in this section discuss (1) designing the exon-specific dsRNA trigger, (2) cloning the transcription template for the exon-specific dsRNA trigger, (3) synthesis of the exon-specific dsRNA trigger, (4) application of the exon-specific dsRNA trigger, and (5) testing the efficiency and specificity of the exon-specific dsRNA trigger.

3.1. Design of the Exon-Specific dsRNA Trigger

When performing RNAi in *Drosophila* cells, it is preferable to use long dsRNAs, as you can add these directly to the culture medium. The dsRNAs enter the cells and are processed into siRNAs by Dicer. We, therefore, typically use exon-specific dsRNA triggers that encompass the entire alternative exon (*see* **Note 1**). As we will describe later, it is important to ensure that the exon-specific dsRNA does not share enough homology with other regions of the same gene, let alone other genes, such that a subset of the siRNAs liberated from the dsRNA by Dicer induce the degradation of untargeted mRNA isoforms. This length is typically about 18 nt (nucleotides).

3.2. dsRNA Trigger Transcription Template

Once a target sequence has been selected, a transcription template must be synthesized. We typically do this by designing oligonucleotide primers to be used for RT-PCR that encompass the targeted sequence. We then perform RT-PCR on total cellular RNA and clone the PCR product into a vector containing bacteriophage RNA polymerase promoters flanking the cloning site. The transcription template is then generated by PCR using oligonucleotide primers complementary to the RNA polymerase promoters.

3.2.1. Reverse Transcription of Total Cellular RNA

We typically isolate total cellular RNA to be used as a template for RT-PCR from the source that exon-specific RNAi will be performed on (i.e., cells, animals, etc.). The reverse transcription reaction is carried out as follows:

1. Assemble a 20-μL reaction containing 1 μL of RNase inhibitor (Invitrogen), 1 μL of a 265 ng/μL stock of Random Hexamers, 5 μg of total *Drosophila* cellular RNA, 4 μL of First Strand Buffer (provided with Superscript II), 2 μL of 0.1 M dithiothreitol (DTT), 1 μL of 10 mM dNTPs, and 1 μL of Superscript II (Invitrogen).
2. Incubate the reaction at 42°C for 1 h.
3. At this point, the reaction can be stored at –20°C indefinitely.

3.2.2. Polymerase Chain Reaction to Amplify the DNA Encoding the dsRNA Trigger

Next, the cDNA synthesized by reverse transcription is used as a template for PCR. The PCR is carried out as follows:

1. Assemble a 50-μL reaction containing 3 μL cDNA from **Subheading 3.2.1.**, 1 μL of 10 mM dNTPs, 4 μL of a 2-μM stock of the primer complementary to the 5′ end of the dsRNA target sequence, 4 μL of a 2-μM stock of the primer complementary to the 3′ end of the dsRNA target sequence, 5 μL of 10X PCR buffer without MgCl$_2$, 1.5 μL of 50 mM MgCl$_2$, 0.25 μL of *Taq* DNA polymerase (1 unit/μL), and 31.25 μL H$_2$O.

2. The reactions are cycled 35 times at 94°C for 45 s, 55°C for 45 s, 72°C for 1 min (extension time is determined as 1 min/kb), followed by a 3-min incubation at 72°C. The annealing temperature and extension time should be optimized for each DNA fragment being amplified.
3. Run an aliquot of each reaction on an agarose gel to verify that a PCR product of the correct size was produced.

3.2.3. Cloning of the PCR Product

Next, the PCR product obtained in **Subheading 3.2.2.** should be cloned into the pCRII-TOPO (Invitrogen) vector.

1. Assemble a 1.9-µL reaction containing 1.3 µL of the PCR product, 0.3 µL of salt solution (included in the TOPO cloning kit), and 0.3 µL of pCRII-TOPO vector.
2. Incubate the reaction at room temperature for 5–30 min.
3. Use the entire reaction to transform *Escherichia coli* using standard procedures and plate onto Luria–Bertani (LB)-Amp plates containing X-gal.
4. The next day, pick several colonies to grow up and screen for the presence of an insert by restriction mapping.
5. Sequence one representative clone to ensure that it contains the cDNA of interest. The cDNA will next be used to generate transcription templates.

3.2.4. Generating the Transcription Template

Once clones containing the target sequence have been obtained, you will need to generate the actual transcription template. To do this, PCR is performed using T7 and SP6 primers, which anneal the plasmid on either side of the cloning site. The resulting PCR product will contain the dsRNA trigger sequence flanked by RNA polymerase promoters (*see* **Note 2**).

1. In 0.2-mL tubes, assemble a 100-µL reaction containing 5 µL of DNA template (1 ng/µL), 2 µL of 10 mM dNTPs, 8 µL of 2 µM SP6 primer, 8 µL of 2 µM T7 primer, 10 µL of 10X PCR buffer without MgCl$_2$, 3 µL of 50 mM MgCl$_2$, 0.5 µL of *Taq* DNA polymerase (1 U/µL), and 63.5 µL H$_2$O.
2. Cycle the reactions 35 times at 94°C for 45 s, 50°C for 45 s, and 72°C for 2 min, followed by a 3-min incubation at 72°C.
3. Run a portion of the reaction on an agarose gel to verify that a product was generated.
4. Clean up the PCR products by extraction with an equal volume of phenol : chloroform : isoamyl alcohol (25 : 24 : 1) and precipitation with 2.5 vol of ethanol.
5. Resuspend the pellet in 10 µL of H$_2$O and quantify by spectrophotometry.

3.3. dsRNA Synthesis

The next step is to synthesize RNA from the transcription template. This is done by transcribing the two RNA strands and then annealing them together.

3.3.1. RNA Synthesis

Separate transcription reactions are required to synthesize the top and bottom strands of RNA because the transcription template is a PCR product containing a T7 promoter on one end and an SP6 promoter on the other. A variety of methods can be used for large-scale synthesis, but we typically use the Ampliscribe High Yield transcription kits from Epicentre.

1. Set up a 20-μL reaction T7 transcription reaction containing 1 μg of the transcription template, 2 μL of 10X AmpliScribe T7 reaction buffer, 1.5 μL of 100 mM ATP, 1.5 μL of 100 mM CTP, 1.5 μL of 100 mM GTP, 1.5 μL of 100 mM UTP, 2 μL of 100 mM DTT, and 2 μL of the Ampliscribe T7 Enzyme Solution.
2. Set up a separate 200-μL SP6 transcription reaction containing 1 μg of the transcription template, 2 μL of 10X AmpliScribe SP6 reaction buffer, 1 μL of 100 mM ATP, 1 μL of 100 mM CTP, 1 μL of 100 mM GTP, 1 μL of 100 mM UTP, 2 μL of 100 mM DTT, and 2 μL of the AmpliScribe SP6 Enzyme Solution.
3. Incubate both reactions at 37°C for 2 h.
4. Add 5 μL of DNase I to each reaction and continue the incubation for 15 additional minutes at 37°C to degrade the transcription template.
5. Extract the reactions with and equal volume of phenol : chloroform : isoamyl alcohol (25 : 24 : 1) and precipitate the RNA by adding 1/10 vol of 3 M sodium acetate, pH 5.2, and 2.5 vol of ethanol.
6. Resuspend the pellet in 50 μL of H$_2$O.

3.3.2. RNA Annealing

After the two RNA strands are synthesized, they are annealed together to produce dsRNA. The annealing reaction is carried out as follows.

1. Assemble a 50-μL reaction containing 20 μg of each RNA strand and 5 μL of 10X annealing buffer.
2. Heat the samples to 78°C for 10 min, followed by incubation at 37°C for 30 min (*see* **Note 3**).

3.4. dsRNA Application

Once you have made the dsRNA, you can use it to perform RNAi on the cultured cells. We typically use the *Drosophila* Dmel-2 cell line (Invitrogen), which is a variety of S2 cells that have been optimized for growth in serum-free medium. Serum interferes with dsRNA uptake. Thus, using Dmel-2 cells significantly streamlines the RNAi process (*see* **Note 4**).

1. Dilute S2 cells to a concentration of 1×10^6 cells/mL in *Drosophila*-SFM media supplemented with penicillin/streptomycin and glutamine.
2. Add 2 mL of the diluted cells to each well of a six-well dish.
3. Add the dsRNA directly to the culture medium in each well.
4. Incubate the cells at 27°C for 3 d (*see* **Note 5**).

3.5. Determining the Efficiency and Specificity of the dsRNA Trigger

The exact method that is used to determine the efficiency and specificity of exon-specific RNAi will depend on the reagents available to you. For example, if antibodies exist that are specific to each isoform produced from the target gene, Western blots of the cell lysate could be performed. However, it is unlikely that such reagents will be available. Thus, the most common method will be to analyze the abundance of the different mRNA isoforms from the target gene. We typically check this by extracting total cellular RNA from the cells, perform RT-PCR to amplify the different mRNA isoforms, and examine the abundance of each by electrophoretic methods.

3.5.1. RNA Isolation

We typically isolate RNA using Trizol reagent (Invitrogen).

1. Completely remove the culture medium from the six-well dishes.
2. Add 1 mL of Trizol to each well.
3. Gently pipet the solution up and down a few times and then transfer to a 1.5-mL microcentrifuge tube.
4. Add 100 µL of chloroform to each tube and briefly vortex.
5. Incubate the tubes on ice for 5 min.
6. Centrifuge the samples at 12,000g for 15 min at 4°C.
7. Remove the aqueous layer to a new tube and add 500 µL of isopropanol to precipitate the RNA.
8. Centrifuge the samples at 12,000g for 15 min at 4°C.
9. Remove the supernatant, briefly dry the pellet on the bench, and then dissolve the pellet in 50 µL of H_2O.
10. Quantify the RNA by spectrophotometry and analyze by RT-PCR using protocols optimized for your target gene. If the RNA is not to be used immediately, it should be stored at –80°C.

Figure 2 shows an example of the efficiency and specificity of exon-specific RNAi. In this case, we have attempted to degrade different isoforms of the *Drosophila Dscam* mRNA. The portion of the *Dscam* gene we targeted, the exon 4 cluster, contains 12 alternative exons that are included in the mRNA in a mutually exclusive manner (*see* **Fig. 2A**). We treated Dmel2 cells with each of 12 dsRNAs containing the entire sequence of each exon 4 variant. In nearly every case, the abundance of the mRNA containing the targeted exon was reduced while the abundance of the other 11 isoforms remained unchanged. One exception is seen with the exon 4.1 and 4.8 dsRNA triggers, which each induced the degradation of transcripts containing both exons (*see* **Fig. 2B,** lanes 1 and 8). For example, the exon 4.1 dsRNA degrades both exon 4.1 and 4.8 containing transcripts as does the exon 4.8 dsRNA trigger (the exon 4.8 band is difficult to observe because of its normal low abundance). A similar phenomenon

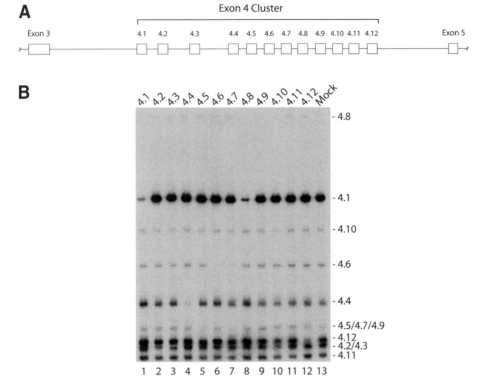

Fig. 2. Example of exon-specific RNAi in action. (**A**) Diagram of the exon 4 region of the *Drosophila Dscam* gene. This gene contains 12 variants of exon 4, which are alternatively spliced in a mutually exclusive manner. (**B**) An SSCP gel demonstrating the efficiency and specificity of exon-specific RNAi. *Drosophila* cells were treated with dsRNA to each of the 12 exon 4 variants for 3 d. Total cellular RNA was isolated and the abundance of each *Dscam* isoform analyzed by RT-PCR and SSCP gel electrophoresis.

is observed for the exon 4.6 and 4.7 dsRNAs (lanes 6 and 7). Finally, the exon 4.11 dsRNA trigger does not efficiently induce degradation of the exon 4.11 containing mRNAs (lane 11). However, all of the other dsRNA triggers efficiently and specifically induce degradation of the targeted mRNA isoforms.

The basis of and solution to the crossreactivity of some of the dsRNA triggers is best seen with the exon 4.1 and 4.8 dsRNAs. **Figure 3A** depicts a nucleotide sequence alignment of these two exons. These two exons share a region of 23 consecutive identical nucleotides. Thus, a subset of the siRNAs liberated from each of these dsRNAs would be complementary to both exon 4.1 and 4.8 containing mRNAs. To circumvent this problem, we synthesized smaller dsRNA triggers containing only the first 79 nucleotides of each exon. As shown in **Fig. 3B,** each of these shorter dsRNAs specifically induced degradation of the

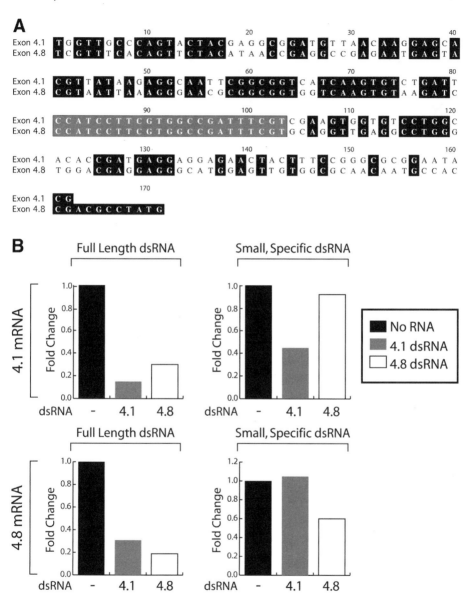

Fig. 3. Circumventing crossreacting dsRNA triggers. (**A**) An alignment of the nucleotide sequences of exons 4.1 and 4.8. The region likely responsible for the cross-reaction is shaded in gray. (**B**) Smaller dsRNA triggers result in isoform-specific mRNA degradation. Small dsRNAs corresponding to the first 79 nt of exon 4.1 and 4.8 were synthesized and used to treat *Drosophila* cells. The abundance of each *Dscam* isoform was analyzed as described in Fig. 2 and the results for exon 4.1 and 4.8 are shown graphically.

targeted mRNA without affecting the stability of the nontargeted mRNA. These results stress the importance of ensuring that the sequence of the dsRNA trigger is specific to the targeted mRNA isoforms. In addition, these results show that even for highly similar exons, it is possible to design specific dsRNA triggers.

4. Notes

1. We have successfully used dsRNAs as large as 1500 bp and as small as 79 bp.
2. Alternatively, primers specific to the target sequence can be designed that each contain a T7 promoter such that the PCR product can be bidirectionally transcribed with T7 RNA polymerase.
3. You can check for the efficiency of RNA annealing by running an aliquot of both single strands and the dsRNA on an agarose gel. The dsRNA should migrate more slowly than each single strand.
4. Detailed protocols performing RNAi on cell lines that require serum for growth are available in **ref. *14*** or the Dixon lab website (http://cmm.ucsd.edu/Lab_Pages/Dixon/).
5. There are at least three variables that determine the efficiency of RNAi: the length of time the cells are exposed to the dsRNA, the amount of dsRNA added to the cells, and the frequency of dsRNA addition. Each of these values must be determined empirically, but we have found that a single treatment of 20 µg of dsRNA for 3 d is sufficient for most targets. However, in some cases, we have needed to treat cells every day for 6 d to efficiently deplete the target protein.

Acknowledgments

The authors thank members of the Graveley lab for discussions. This work was supported by NIH grants to B.R.G.

References

1. Johnson, J. M., Castle, J., Garrett-Engele, P., et al. (2003) Genome-wide survey of human alternative pre-mRNA splicing with exon junction microarrays. *Science* **302,** 2141–2144.
2. Schmucker, D., Clemens, J. C., Shu, H., et al. (2000) *Drosophila Dscam* is an axon guidance receptor exhibiting extraordinary molecular diversity. *Cell* **101,** 671–684.
3. Graveley, B. R. (2001) Alternative splicing: increasing diversity in the proteomic world. *Trends Genet* **17,** 100–107.
4. Celotto, A. M. and Graveley, B. R. (2002) Exon-specific RNAi: a tool for dissecting the functional relevance of alternative splicing. *RNA* **8,** 718–724.
5. Hutvagner, G. and Zamore, P. D. (2002) RNAi: nature abhors a double-strand. *Curr. Opin. Genet. Dev.* **12,** 225–232.
6. Sijen, T., Fleenor, J., Simmer, F., et al. (2001) On the role of RNA amplification in dsRNA-triggered gene silencing. *Cell* **107,** 465–476.
7. Nishikura, K. (2001) A short primer on RNAi: RNA-directed RNA polymerase acts as a key catalyst. *Cell* **107,** 415–418.

16

Immunoprecipitation of MicroRNPs and Directional Cloning of MicroRNAs

Elisavet Maniataki, Maria Dels Angels De Planell Saguer, and Zissimos Mourelatos

1. Introduction

MicroRNAs (miRNAs) comprise a class of approx 22-nucleotide (nt) regulatory RNAs, found in plants and animals *(1–8)*. miRNAs contain 5′ phosphates and 3′ hydroxyls and are processed by the Dicer nuclease from one of the stems of longer precursors (pre-miRNAs) that form stem-loop structures *(9–11)*. In animals, pre-miRNAs are themselves processed by the nuclease Drosha from longer transcripts, termed primary miRNAs (pri-miRNAs) *(12)*. miRNAs bind to Argonaute proteins and typically associate with additional proteins to form microribonucleoproteins (miRNPs), the effector complexes that mediate translational repression or endonucleolytic cleavage of their cognate mRNAs *(6,10, 13,14)*. Another class of approx 22-nt RNAs, termed short, interfering RNAs (siRNAs), is inextricably linked to miRNAs *(15,16)*. siRNAs are processed by Dicer from double-stranded RNAs (dsRNAs), and siRNAs are also bound to Argonaute proteins and might assemble with additional proteins to form complexes termed RNA-induced silencing complexes (RISCs) *(10)*. miRNPs and RISCs are functionally equivalent and their core components are Argonaute proteins and miRNAs or siRNAs *(17)*. siRNAs might also silence chromatin *(9,18)*. The development of strategies to clone small RNAs has dramatically accelerated the discovery of miRNAs and our understanding of the function of siRNAs and miRNAs. We describe a method for immunoprecipitation of miRNPs and isolation and cloning of associated small RNAs *(6)*. This approach combines techniques developed for the immunoprecipitation of RNPs *(19,20)*, with techniques developed for the directional cloning of small RNAs *(15)*. In

From: *Methods in Molecular Biology, vol. 309: RNA Silencing: Methods and Protocols*
Edited by: G. G. Carmichael © Humana Press Inc., Totowa, NJ

principal, this methodology could be used for immunopurification and directional cloning of RNAs from any RNP that contains small RNAs.

2. Materials

1. HeLa cells.
2. Phosphate-buffered saline (PBS) (Fisher).
3. Lysis buffer: 20 mM Tris-HCl, pH 7.4, 200 mM sodium chloride, 2.5 mM magnesium chloride, 0.05% NP-40, 1 tablet of EDTA-free protease inhibitors per 50 mL of lysis buffer.
4. RNasin (Promega).
5. Recombinant protein-G agarose beads (Invitrogen).
6. Sonicator (Sonics Vibra-Cell or equivalent).
7. RNAse-free, DNAse I (Roche).
8. Dimethyl sulfoxide (DMSO) (Sigma).
9. Millipore water.
10. T4 RNA Ligase (New England Biolabs [NEB]).
11. pBR322 DNA–Msp I digest (DNA markers; NEB).
12. DNA Polymerase I, large (Klenow) fragment (NEB).
13. T4 polynucleotide kinase (T4 PNK; NEB).
14. Calf intestinal alkaline phosphatase (CIP; NEB).
15. Protease inhibitor tablets, EDTA-free (Roche).
16. Nonimmune mouse serum or mouse IgG.
17. Anti-miRNP antibodies.
18. Phenol, buffer saturated (pH 7.5–7.8; Invitrogen).
19. Phenol/chloroform/isoamyl alcohol (25 : 24 : 1); pH 7.9 (Fisher).
20. RNA loading buffer (2X): 95% formamide, 18 mM EDTA, xylene cyanol, bromophenol blue (Ambion).
21. Glycogen (Ambion).
22. Ethanol, 100%.
23. 3 M sodium acetate, pH 5.2.
24. 5 M ammonium acetate.
25. Superscript II reverse trascriptase (RT; Invitrogen).
26. Adapter oligos (Dharmacon; for sequences see below).
27. Primers for reverse transcription–polymerase chain reaction (RT-PCR) (Invitrogen; for sequences see below).
28. dNTPs (Roche).
29. Pfu Turbo, thermostable DNA polymerase (Stratagene).
30. SE 400 Sturdier Gel electrophoresis apparatus with 18×24 cm glass plates (Amersham).
31. Urea (Ambion).
32. Acrylamide/bis 19 : 1 40% (w/v) solution (Ambion).
33. 10X TBE (Ambion).
34. 10% (w/v) Ammonium persulfate (APS; dissolved in H$_2$O).
35. TEMED (Bio-Rad).

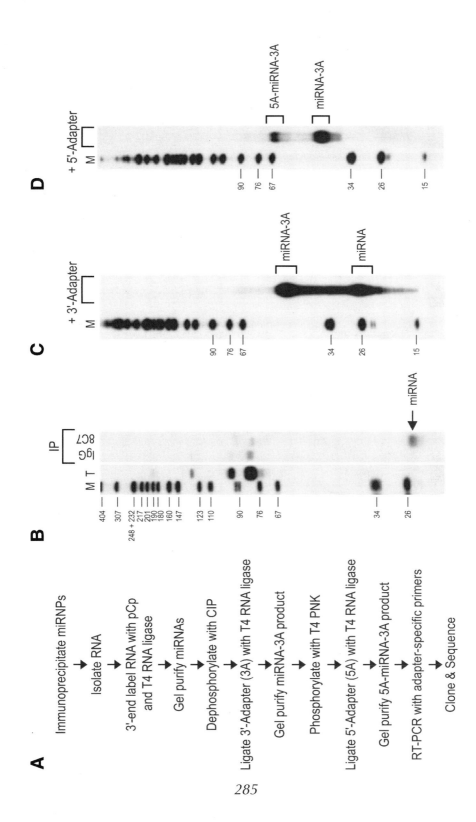

285

36. Zero Blunt PCR cloning kit (Invitrogen).
37. QIAquick PCR purification kit (Qiagen).
38. 3′, 5′-Bis [α-^{32}P] cytidine (pCp), at 3000 Ci/mmol, 10 mCi/mL (Amersham).
39. [α-^{32}P] dCTP, at 3000 Ci/mmol, 10 mCi/mL (Amersham).
40. Metaphor agarose (Cambrex).

3. Methods

The outline of the procedures and representative gels are shown in **Fig. 1**. All procedures and centrifugations are performed on ice/4°C unless otherwise indicated. Use RNase-free solutions, tubes, and pipets (*see* **Note 1**).

3.1. miRNP Immunoprecipitation

3.1.1. Binding of Antibodies to protein-G Agarose Beads

1. Bind the 8C7 monoclonal antibody or nonimmune mouse serum (IgG; serves as negative control) on protein-G agarose beads. Use 50-μL bed volume protein-G agarose beads per immunoprecipitation (IP). Wash beads three times with 1 mL lysis buffer (*see* **Notes 2** and **3**).
2. Aspirate last wash, taking care not to dry the beads, add 1 mL lysis buffer and 20 μL ascites (or 20 μL nonimmune mouse serum for the control IP).
3. Rotate for 45 min.
4. Wash antibody beads three times with 1 mL lysis buffer.

3.1.2. Preparation of Cell Lysate

1. The cell lysate can be prepared while the antibodies are binding to the protein-G agarose beads. Wash HeLa cells (approx 10^8 cells) with PBS. Scrape if adherent, with PBS, and pellet. If cells are in suspension, pellet and wash twice with PBS (*see* **Note 4**).
2. Resuspend pellet with equal volume of lysis buffer; add RNasin to a final concentration of 0.1 U/μL.

Fig. 1. (*Opposite page*) MicroRNP immunoprecipitation and directional microRNA cloning. (**A**) Outline of experimental steps. (**B**) Immunoprecipitations (IPs) were performed from HeLa cells with the anti-eiF2C2 (anti-Argonaute 2) antibody 8C7 or with nonimmune mouse serum (IgG). RNA was isolated, 3′-end labeled with pCp and analyzed on 10% urea/PAGE (polyacrylamide gel electrophoresis). Total (T) refers to RNA extracted from HeLa cells prior to IP. Marker (M) is radiolabeled pBR322 DNA–Msp I digest. (**C**) 3′-Adapter (3A) was ligated to gel-purified miRNAs with T4 RNA ligase and the ligation reaction was resolved on 15% urea/PAGE. miRNA-3A ligated products and unligated miRNAs are indicated. (**D**) 5′-Adapter (5A) was ligated to gel-purified miRNA-3A with T4 RNA ligase and the ligation reaction was resolved on 15% urea/PAGE. The final 5A-miRNA-3A ligation products and unligated miRNA-3A are indicated.

3. Sonicate briefly (three times, 8–10 s each) using 40% output (Sonics Vibra-Cell sonicator or equivalent).
4. Centrifuge cell lysate at 20,000g (in Eppendorf microfuge) for 10 min. Collect supernatant and save aliquot (as reference for total). Discard pellet.

3.1.3. Immunoprecipitation and RNA Isolation

1. Use half of the supernatant with your specific antibody beads (i.e., 8C7 beads) and use the other half with the negative control antibody. Perform IPs in microfuge tubes. If the volume of the IP is less than 1 mL, add lysis buffer to 1 mL. Rotate in the cold room for 1 h.
2. Wash beads five times with 1 mL lysis buffer.
3. Add 300 µL of lysis buffer to the washed beads and 300 µL of buffer-saturated phenol. Vortex for 30 s to 1 min.
4. Spin at top speed in microfuge for 2 min, at room temperature (RT). Collect aqueous (upper) phase. Beads should be at the bottom of the tube (*see* **Note 5**).
5. Extract with equal volume of phenol/chloroform/isoamyl alcohol, pH 7.9, at RT. Collect aqueous phase (*see* **Note 6**).
6. Add 10 U, RNase-free DNAse I. Incubate 37°C for 15 min.
7. Extract with equal volume of phenol/chloroform/isoamyl alcohol, pH 7.9, at RT. Collect aqueous phase.
8. Precipitate RNA (by adding 0.1 vol 3 M sodium acetate, pH 5.2, 3 µL glycogen, and 3 vol of 100% ethanol. Place at −80°C for 15 min or at −20°C overnight). Spin at top speed in Eppendorf for 30 min at 4°C.
9. Carefully aspirate supernatant and wash pellet with 1 mL of 70% ethanol. Spin as in **step 8** for 10 min. Aspirate supernatant and air-dry the pellet.
10. Resuspend pellet in 15 µL H_2O. Proceed with miRNA isolation or store RNA at −80°C (*see* **Note 7**).

3.2. miRNA Purification

3.2.1. 3′-End Labeling of RNA

1. For each labeling reaction combine the following (total reaction volume of 10 µL): 4 µL RNA; 1 µL of 10X T4 RNA ligase buffer; 0.5 µL of RNasin; 3 µL DMSO; 1 µL [5′-^{32}P]pCp; 0.5 µL of T4 RNA ligase.
2. Place tubes at 4°C (cold room) overnight.
3. To the 10 µL of reaction in **step 1,** add 200 µL H_2O and 200 µL phenol/chloroform/isoamyl alcohol, pH 7.9. Extract as described above.
4. Collect aqueous supernatant, add equal volumes of 5 M ammonium acetate and 3 µL glycogen and 2.5 vol 100% ethanol. Place at −80°C for 15 min or at −20°C overnight. Spin at top speed in Eppendorf for 30 min at 4°C. Aspirate supernatant.
5. To the RNA pellet, add 200 µL H_2O and repeat **step 4** (two precipitations with ammonium acetate result in removal of most free pCp).

6. Aspirate supernatant and wash RNA pellet with 70% ethanol; spin at top speed in Eppendorf for 10 min at 4°C. Aspirate supernatant and air-dry the pellet.
7. To the pellet, add 20 μL of RNA loading buffer. You can check 3 μL on a 15% Urea/PAGE gel before loading the entire sample for miRNA gel purification.

3.2.2. miRNA Gel Purification

1. Preparation of urea/PAGE solutions.
   a. For 1 L of 10% urea/PAGE, combine the following in a glass beaker: 480 g urea, 250 mL of 40% acrylamide/bis (19 : 1), 50 mL of 10X TBE, and H_2O up to 1 L. Stir until completely dissolved, filter-sterilize, and store up to a year at RT in an aluminum-covered bottle (to protect from light).
   b. For 1 L of 20% urea/PAGE, combine the following in a glass beaker: 480 g urea, 500 mL of 40% acrylamide/bis (19 : 1), 50 mL of 10X TBE, and H_2O up to 1 L. Stir until completely dissolved, filter-sterilize, and store as described in **step 1a**.
   c. To prepare 15% urea/PAGE, simply make a 1 : 1 solution of 10% and 20% urea/PAGE solutions.
2. Cast the gel apparatus using 0.75-mm combs and 18×24-cm glass plates. Dispense 40 mL of 15% urea/PAGE solution in a 50-mL Falcon tube. To polymerize, add 200 μL of 10% APS and 30 μL of TEMED, mix well, and immediately pour the gel.
3. Load most or all of the labeled RNA. Also, load labeled DNA or RNA markers (see below for preparation of radiolabeled DNA markers) and the RNA from the negative control antibody. Leave two or three lanes spacing between samples. There is no need to heat the RNA prior to loading. Run gel until the bromophenol blue dye is 1 cm from the bottom of the gel.
4. Disassemble glass plates and lift gel on a piece of old, exposed film. Cover with Saran Wrap and expose wet gel to film by placing it in a cassette, between two intensifying screens, at −80°C. Use radioactive pen or other means (e.g., preflashing) to align gel with film. The exposure time varies with amount of RNA precipitated and loaded on gel. Usually, approx 5 h is sufficient.
5. Excise gel piece corresponding to labeled miRNAs with a clean razor blade and place in microfuge tube.
6. Add 3 vol of 0.3 *M* sodium acetate, pH 5.2, and rotate in cold room for approx 4 h. Collect supernatant and add 3 vol of 0.3 *M* sodium acetate, pH 5.2, to the gel pieces; rotate in a cold room overnight. Pool the supernatants and precipitate RNA by adding 3 μL of glycogen and 2.5 vol ethanol, as described above. Spin, aspirate supernatant, and wash RNA pellet with 70% ethanol; air-dry. Dissolve RNA in 19 μL of H_2O and add 1 μL RNasin (*see* **Note 8**).

3.3. Ligation of Adapters to miRNAs

3.3.1. Dephosphorylation

1. Combine the following (total reaction volume of 30 μL): 20 μL RNA; 6 μL H_2O; 3 μL of 10X NEB3 buffer; 1 μL of CIP alkaline phosphatase.
2. Incubate 37°C for 30 min.

3. Inactivate enzyme by incubating at 65°C for 20 min.
4. Add 170 μL H_2O. Add equal volume of phenol/chloroform/isoamyl alcohol, pH 7.9, extract, and ethanol precipitate as described above. Wash pellet with 70% ethanol and redissolve the air-dried pellet with 9 μL H_2O and 1 μL RNasin (*see* **Note 9**).

3.3.2. 3′-Adapter Ligation and Gel Purification

1. Combine the following (total reaction volume of 20 μL): 2.5 μL of H_2O, 10 μL RNA, 2 μL of 10X T4 RNA ligase buffer, 3 μL DMSO, 1.5 μL (30 U) of 3 T4 RNA ligase, 1 μL (100 pmol) of 3′-adapter (3A at 100 μ*M*).
2. Ligate overnight in the cold room (*see* **Note 10**).
3. Add equal volume of 2X RNA loading buffer and load samples in 15% urea/PAGE.
4. Gel excise and elute appropriate size product as described above. Dissolve pelleted RNA in 15.6 μL H_2O and add 1 μL RNasin.

3.3.3. Phosphorylation

1. Combine the following (total reaction volume of 20 μL): 16.6 μL RNA, 2 μL of 10X PNK buffer, 0.4 μL of 100 m*M* stock (=2 m*M* final) of ATP, 1 μL (10 U) of T4 Polynucleotide kinase.
2. Incubate 37°C for 30 min.
3. Add 180 μL H_2O. Add equal volume of phenol/chloroform/isoamyl alcohol, pH 7.9, extract, and ethanol precipitate as described above. Wash pellet with 70% ethanol and redissolve the air-dried pellet with 9 μL H_2O and 1 μL RNasin.

3.3.4. 5′-Adapter Ligation and Gel Purification

Perform as described in **Subheading 3.3.2.** but use 5′-adapter (5A) instead. Gel-purify and elute appropriate ligation product as described in **Subheading 3.3.2.** Dissolve RNA pellet in 19 μL H_2O and 1 μL RNasin.

3.4. Amplification of Adapter-Ligated miRNAs and Cloning

3.4.1. Reverse Transcription (RT)

1. Combine the following: 10 μL RNA, 1 μL RT primer (20 μ*M*), 1 μL dNTPs (10 m*M*).
2. Heat above mix at 90°C for 3 min. Place on ice and spin.
3. Add the following to the above while still on ice (total reaction volume of 20 μL): 4 μL of 5X First Strand buffer, 2 μL DTT (0.1 *M*), 1 μL RNasin, 1 μL of Superscript II RT.
4. Incubate at 37°C for 30 min to 1 h.
5. Inactivate enzyme by incubating at 65°C for 20 min. Chill, spin, and store at −20°C or proceed with PCR.

3.4.2. PCR

1. For a 50-mL reaction, assemble the following in thin-wall PCR tubes: 36 μL H_2O, 5 μL of 10X buffer for Pfu Turbo polymerase (containing $MgCl_2$), 1 μL dNTPs

(10 m*M*), 5 μL template (from RT reaction), 1 μL 5P primer (10 μ*M*), 1 μL RT primer (10 μ*M*), 1 μL Pfu turbo.

2. For the negative control PCR, combine the same reagents as above but with 5 μL H$_2$O instead of template. This PCR will show the position and amount of nonspecific PCR products, such as primer-dimers.

3. PCR cycle program: 94°C for 2 min; 30 cycles of 94°C for 40 s, 52°C for 50 s, 72°C for 50 s; 72°C for 2 min; 4°C hold.

3.4.3. Cloning

Cleanup PCRs (including the "negative control PCR") with QIAquick PCR purification kit. Elute with 40 μL elution buffer (provided in the kit). Check 7-μL aliquots of PCR products and markers on 3.5% metaphor agarose. Proceed with cloning in the pCR-Blunt vector using the Zero Blunt PCR Cloning Kit (Invitrogen; kanamycin-resistant vector), followed by sequencing.

3.5. Analysis of Sequenced Clones

1. The directional cloning of RNAs allows easy identification of the strand polarity of the cloned miRNAs, as their 5′ ends always follow the 5A adapter.

2. Extract sequences that are found between the 5A and 3A adapters and analyze them by performing BLAST searches in annotated miRNA and genomic databases. First, determine whether the cloned RNAs correspond to known miRNAs by searching the miRNA registry (http://www.sanger.ac.uk/Software/Rfam/mirna/index.shtml) *(21)*.

3. If the cloned RNAs do not correspond to known miRNAs, search in annotated genomic databases (e.g., at the National Center for Biotechnology Information, [NCBI], http://www.ncbi.nlm.nih.gov/BLAST/ and http://www.ncbi.nih.gov/Genomes/) *(22)* to determine whether the cloned RNAs map to the genome. You may also search Expressed Sequence Tag (EST) databases (e.g., at http://www.ncbi.nlm.nih.gov/BLAST/) because most animal miRNAs are expressed as longer pri-miRNAs, and the sequences of many pri-miRNAs are found in EST databases. If the 3′-most nucleotide of the cloned RNAs is a cytosine and it is not found in the genomic or EST sequences, it likely represents the ligated pCp, in which case it needs to be removed. However, not all cloned RNAs will contain the ligated pCp.

4. Extract flanking genomic sequences and determine, by using RNA folding programs (e.g., mfold, http://www.bioinfo.rpi.edu/applications/mfold/old/rna/form1.cgi) *(23)*, if they have the capacity to form stem-loop structures and whether the cloned RNAs are derived from one of the stems of these putative precursors. If they do, it is very likely that the cloned RNAs correspond to novel miRNAs. As most miRNAs are conserved, determine whether the cloned, putative miRNAs and their precursors are conserved in other species. Verification that the cloned RNAs represent authentic miRNAs requires the demonstration that they are expressed as approx 22-nt transcripts (typically by Northern blot; for a detailed protocol on miRNA Northern blots, see Chapter 17) *(8)*.

5. Prior to publishing the novel miRNAs, obtain unique names from the miRNA Registry, which is the official miRNA repository *(21)* and deposit sequences with NCBI.

3.6. Sequence and Properties of Adapters and Primers

1. The adapters consist of DNA–RNA oligonucleotides and with certain modifications to allow directional ligation to the miRNAs using T4 RNA ligase. The sequences are as follows: 3′-adapter (3A): 5′-pUUU aac cgc atc ctt ctc gaa ttc tgt idT-3′; and 5′-adapter (5A): 5′-tac gaa ttc taa tac gac tca ctg AAA-3′ (p = phosphate; capital letters = RNA; small letters = DNA; idT = 3′ inverted deoxythimidine, a blocking group to prevent ligations with the 3′-end of 3A).
2. The RT-PCR primers consist of DNA oligos. RT primer (RT): 5′-aca gaa ttc gag aag gat gcg gtt aaa-3′; and 5′ primer (5P): 5′-tac gaa ttc taa tac gac tca ctg aaa-3′ (*see* **Note 11**).

3.7. Preparation of Radiolabeled pBR322 DNA–Msp I Digest Markers

1. Combine the following (total reaction volume of 20 μL): 1 μL (= 1 μg) of pBR322/Msp 1 Digest (NEB), 2 μL of 10X *Eco*PolI (Klenow) buffer, 5 μL [α-^{32}P]dCTP, 1 μL DNA polymerase I (Klenow), 11 μL H_2O.
2. Incubate at 30°C for 15 min. Add 200 μL H_2O.
3. Add 200 μL phenol/chloroform/isoamyl alcohol, pH 7.9. Vortex, extract, and ethanol-precipitate labeled marker as described above. After the final wash with 70% ETOH, dry pellet and resuspend in 100 μL RNA loading buffer.
4. Heat labeled marker at 95°C for 3 min. Cool on ice. Spin briefly. This is the stock (very hot) marker. Make a 1:100 dilution of an aliquot from stock in RNA loading buffer to make working concentration marker. Store stock and dilution at −20°C. Marker is good for at least 3 mo, but adjustments need to be made on how much to load (either from diluted marker or from stock as the marker becomes old).

4. Notes

1. The strategy outlined above can be adopted for immunopurification and cloning of RNAs from any RNP that contains small RNAs, provided that these RNAs contain 5′ and 3′ ends that are amenable to ligation.
2. The anti-Gemin4 monoclonal antibody 17D10 *(6)* can also be used for immunopurification of miRNPs and miRNA isolation. In principal, other antibodies directed against Argonaute proteins can also be used to immunoprecipitate miRNPs and RISCs.
3. This protocol is designed for preparative miRNP immunoprecipitations for the purpose of cloning miRNAs. It can be scaled down 10-fold for analytical purposes (i.e., use 2 μL of ascites and approx 10^7 HeLa cells, which corresponds to 1 confluent 100-mm tissue culture plate).
4. Other mammalian cell lines can also be used for miRNP IPs and miRNA cloning *(24)*.

5. If, instead of phenol, phenol/chloroform is used for the first extraction, the beads will be at the interphase; this could result in loss of substantial material from the aqueous phase.

6. Acidic phenol/chloroform can also be used when extracting miRNAs, as RNA partitions in the aqueous phase under both acidic and basic conditions, in contrast to DNA, which remains in the organic phase in acidic buffers. However, avoid using acidic phenol/chloroform when extracting miRNAs that have been ligated to adapters as the adapters contain DNA.

7. Purified RNA can also be used for other applications, such as Northerns.

8. Elution of RNA from the gel may be performed for 2 h using a thermomixer at 37°C (e.g., Eppendorf thermomixer at 1200 rpm). Recovery rate of RNA eluted from gel is 50–70%.

9. Treatment with alkaline phosphatase removes phosphates from the 5′-end as well as the 3′-end of pCp-labeled miRNAs. After dephosphorylation, labeled miRNAs have free-hydroxyl groups at both ends. The radioactive phosphorus in pCp is at the a position of the 5′-end of cytidine, which has formed a phosphodiester bond with the hydroxyl group found at the 3′-ends of miRNAs during ligation with T4 RNA ligase. For this reason, dephosphorylation does not remove the label from the RNA.

10. Incubation can also be performed for 1 h at 37°C. However, overnight ligations with T4 RNA ligase at 4°C result in improved ligation efficiencies.

11. We have also designed a newer generation of adapters and primers. The sequences of the new adapters are as follows: 3′-adapter (3AD): 5′-pUCC aga tct ata gtg tca cct aaa tcg tat gga idT-3′; and 5′-adapter (5AD): 5′-tcc gaa att aac cct cac taa agG GA-3′; where p = phosphate; capital letters = RNA; small letters = DNA; idT = 3′-inverted deoxythymidine). The DNA oligonucleotide primers for RT-PCR are as follows: newRT primer (containing the Sp6 minimal promoter sequence): 5′-cat acg att tag gtg aca cta tag-3′; and new 5′ primer (containing the T3 minimal promoter sequence): 5′-att aac cct cac taa agg ga-3′. These new adapters and primers offer the following advantages:

   a. If the two adapters ligate together (during initial RNA ligation steps or during PCR-[e.g., as primers-dimers]), a *Bam*HI site is created (GGATCC). Digestion of the PCR product with BamHI prior to ligation in the cloning vector helps to reduce this background.

   b. Sp6 and T3 RNA polymerases can be used for in vitro transcription (e.g., after cloning, the plasmid can be linearized with *Bgl*II (AGATCT), and T3 RNA polymerase can be used to generate RNA transcripts, which are in sense orientation to the directionally cloned product.

Acknowledgments

We are grateful to Dr. G. Dreyfuss for the 8C7 and 17D10 antibodies. This work was supported by grants from the NIH, the Department of Pathology & Laboratory Medicine, University of Pennsylvania School of Medicine and the PENN Genomic Institute (Z.M.).

References

1. Lee, R., Feinbaum, R., and Ambros, V. (2004) A short history of a short RNA. *Cell* **116,** S89–S92, and one page following S96.
2. Ruvkun, G., Wightman, B., and Ha, I. (2004) The 20 years it took to recognize the importance of tiny RNAs. *Cell* **116,** S93–S96, and two pages following S96.
3. Lagos-Quintana, M., Rauhut, R., Lendeckel, W., and Tuschl, T. (2001) Identification of novel genes coding for small expressed RNAs. *Science* **294,** 853–858.
4. Lau, N. C., Lim, L. P., Weinstein, E. G., and Bartel, D.P. (2001) An abundant class of tiny RNAs with probable regulatory roles in *Caenorhabditis elegans*. *Science* **294,** 858–862.
5. Lee, R. C. and Ambros, V. (2001) An extensive class of small RNAs in *Caenorhabditis elegans*. *Science* **294,** 862–864.
6. Mourelatos, Z., Dostie, J., Paushkin, S., et al. (2002) miRNPs: a novel class of ribonucleoproteins containing numerous microRNAs. *Genes Dev.* **16,** 720–728.
7. Carrington, J. C. and Ambros, V. (2003) Role of microRNAs in plant and animal development. *Science* **301,** 336–338.
8. Ambros, V., Bartel, B., Bartel, D. P., et al. (2003) A uniform system for microRNA annotation. *RNA* **9,** 277–279.
9. Nelson, P., Kiriakidou, M., Sharma, A., Maniataki, E., and Mourelatos, Z. (2003) The microRNA world: small is mighty. *Trends Biochem. Sci.* **28,** 534–540.
10. Murchison, E. P. and Hannon, G. J. (2004) miRNAs on the move: miRNA biogenesis and the RNAi machinery. *Curr. Opin. Cell Biol.* **16,** 223–229.
11. Bartel, D. P. (2004) MicroRNAs: genomics, biogenesis, mechanism, and function. *Cell* **116,** 281–297.
12. Lee, Y., Ahn, C., Han, J., et al. (2003) The nuclear RNase III Drosha initiates microRNA processing. *Nature* **425,** 415–419.
13. Hutvagner, G. and Zamore, P. D. (2002) A microRNA in a multiple-turnover RNAi enzyme complex. *Science* **297,** 2056–2060.
14. Nelson, P. T., Hatzigeorgiou, A. G., and Mourelatos, Z. (2004) miRNP : mRNA association in polyribosomes in a human neuronal cell line. *RNA* **10,** 387–394.
15. Elbashir, S. M., Lendeckel, W., and Tuschl, T. (2001) RNA interference is mediated by 21- and 22-nucleotide RNAs. *Genes Dev.* **15,** 188–200.
16. Hamilton, A. J. and Baulcombe, D. C. (1999) A species of small antisense RNA in posttranscriptional gene silencing in plants. *Science* **286,** 950–952.
17. Carmell, M. A., Xuan, Z., Zhang, M. Q., and Hannon, G. J. (2002) The Argonaute family: tentacles that reach into RNAi, developmental control, stem cell maintenance, and tumorigenesis. *Genes Dev.* **16,** 2733–2742.
18. Baulcombe, D. (2002) DNA events. An RNA microcosm. *Science* **297,** 2002–2003.
19. Lerner, M. R. and Steitz, J. A. (1979) Antibodies to small nuclear RNAs complexed with proteins are produced by patients with systemic lupus erythematosus. *Proc. Natl. Acad. Sci. USA* **76,** 5495–5499.
20. Choi, Y. D. and Dreyfuss, G. (1984) Isolation of the heterogeneous nuclear RNA–ribonucleoprotein complex (hnRNP): a unique supramolecular assembly. *Proc. Natl. Acad. Sci. USA* **81,** 7471–7475.

21. Griffiths-Jones, S. (2004) The microRNA Registry. *Nucleic Acids Res.* **32,** D109–D111 (database issue).

22. McGinnis, S. and Madden, T. L. (2004) BLAST: at the core of a powerful and diverse set of sequence analysis tools. *Nucleic Acids Res.* **32,** W20–W25.

23. Zuker, M. (2003) Mfold web server for nucleic acid folding and hybridization prediction. *Nucleic Acids Res.* **31,** 3406–3415.

24. Dostie, J., Mourelatos, Z., Yang, M., Sharma, A., and Dreyfuss, G. (2003) Numerous microRNPs in neuronal cells containing novel microRNAs. *RNA* **9,** 180–186.

17

Detection of MicroRNAs and Assays to Monitor MicroRNA Activities In Vivo and In Vitro

Marianthi Kiriakidou, Peter Nelson, Stella Lamprinaki, Anup Sharma, and Zissimos Mourelatos

1. Introduction

MicroRNAs (miRNAs) are approx 22-nucleotide (nt) regulatory RNAs derived from endogenous genes and processed from longer (approx 70 nt, in animals) precursor RNAs (pre-miRNAs) *(1–8)*. miRNAs bind to Argonaute (Ago) proteins, such as Ago-2 (also known as eIF2C2) *(6,9)*, and typically associate with additional proteins to form microribonucleoproteins (miRNPs) *(6)*. Another class of approx 22-nt RNAs, termed short, interfering RNAs (siRNAs), is functionally related to miRNAs *(10,11)*. siRNAs are processed from double-stranded RNA (dsRNA), bind to Argonaute proteins, and could assemble with additional proteins to form complexes termed RNA-induced silencing complexes (RISCs) *(12)*. Ago-2 is the endonuclease that cleaves the RNAs targeted by miRNAs or siRNAs *(13–16)*. miRNPs and RISCs are the effector complexes that mediate translational repression or endonucleolytic cleavage of cognate mRNAs. The function of miRNAs is largely dictated by the degree of complementarity between the miRNA and its RNA target. If the complementarity is extensive, the Ago2 found in miRNPs/RISCs, cleaves a single phosphodiester bond on the target RNA, located across from the middle of the guide si/miRNA *(11,17)*. If the complementarity is partial, the stability of the target mRNA is not affected, but its translation is repressed *(18–20)*. In both cases, near perfect complementarity of the proximal (towards the 5′ end) portion of miRNAs is required for target mRNA recognition *(21–26)*, and if the complementarity extends beyond the 10th nucleotide of the miRNA, target mRNA cleavage occurs *(27,28)*. We describe methods for

From: *Methods in Molecular Biology, vol. 309: RNA Silencing: Methods and Protocols*
Edited by: G. G. Carmichael © Humana Press Inc., Totowa, NJ

the detection of miRNAs by Northern blots, reporter-based assays that monitor miRNA-directed gene expression regulation in vivo, and an in vitro assay that recapitulates miRNA-dependent endonucleolytic cleavage of RNA targets.

2. Materials

1. HeLa cells.
2. Phosphate-buffered saline (PBS) (Fisher).
3. Lysis buffer: 20 mM Tris-HCl, pH 7.4, 200 mM sodium chloride, 2.5 mM magnesium chloride, 0.05% NP-40, 1 tablet of EDTA-free protease inhibitors per 50 mL of lysis buffer.
4. 1 M HEPES–KOH, pH 7.9. For 250 mL; dissolve 59.5 g of HEPES in 150 mL of H$_2$O. Bring pH to 7.9 with 5 M KOH. Add H$_2$O to bring volume to 250 mL. Filter-sterilize and store at 4°C.
5. Buffer D: 20 mM HEPES–KOH, pH 7.9, 100 mM KCl, 0.2 mM EDTA, 0.5 mM dithiothreitol (DTT).
6. TE buffer: 10 mM Tris-HCl, pH 8.0, 1 mM EDTA.
7. 10X Annealing buffer: 100 mM Tris-HCl, pH 7.5, 1 M NaCl, 10 mM EDTA.
8. 10X MIB buffer: 400 mM potassium acetate, pH 7.4, 20 mM MgCl$_2$, 10 mM DTT. Store in aliquots at –20°C.
9. TRIzol (Invitrogen).
10. RNasin (Promega).
11. Recombinant protein-G agarose beads (Invitrogen).
12. Sonicator (Sonics Vibra-Cell or equivalent).
13. Lipofectamine 2000 (Invitrogen).
14. Chroma-10 spin columns (Stratagene).
15. Millipore water.
16. Ethidium bromide (10 mg/mL, Invitrogen).
17. ULTRAHyb-Oligo (Ambion).
18. 20X SSC buffer: 0.3 M sodium acetate, 3 M sodium chloride, pH 7.0.
19. 10% Sodium dodecyl sulfate (SDS).
20. T4 Polynuclotide kinase (T4 PNK; New England Biolabs [NEB]).
21. T4 RNA ligase (NEB).
22. T4 DNA ligase (NEB).
23. pBR322 DNA–Msp I digest (DNA markers; NEB)
24. pRL-TK vector (Promega).
25. pGL3 vector (Promega).
26. Protease inhibitor tablets, EDTA free (Roche).
27. Nonimmune mouse serum or mouse IgG.
28. Anti-miRNP antibodies.
29. Phenol, buffer saturated (pH 7.5–7.8; Invitrogen).
30. Phenol/chloroform/isoamyl alcohol (25 : 24 : 1), pH 7.9 (Fisher).
31. Chloroform.

32. RNA Loading buffer 1–2X: 95% formamide, 18 m*M* EDTA, xylene cyanol, bromophenol blue (Ambion).
33. Glycogen (5 mg/mL; Ambion).
34. Ethanol, 100%.
35. Isopropanol.
36. 3 *M* Sodium acetate, pH 5.2.
37. SE 400 Sturdier Gel electrophoresis apparatus with 18×16-cm glass plates (Amersham).
38. Urea (Ambion).
39. Acrylamide/bis 19 : 1 40% (w/v) solution (Ambion)
40. 10X TBE (Ambion).
41. 10% (w/v) Ammonium persulfate (APS; dissolved in H_2O).
42. TEMED (Bio-Rad).
43. Dulbecco's modified Eagle's medium (DMEM), high glucose (4.5 g/L; Invitrogen).
44. Trypsin–EDTA, 10X (Invitrogen).
45. L-Glutamine, 200 m*M* (Invitrogen).
46. Fetal bovine serum (FBS) (Hyclone).
47. DMEM-complete medium (90% DMEM, high glucose, 10% fetal bovine serum (FBS), 2 m*M* L-glutamine).
48. Dual luciferase assay (DLR) kit (Promega).
49. Luminometer (e.g., Turner TD-20/20; Promega).
50. (γ-^{32}P)-ATP; at 3000 Ci/mmol, 10 mCi/mL (Amersham).
51. 3′, 5′-Bis (α-^{32}P) cytidine (pCp), at 3000Ci/mmol, 10 mCi/ml (Amersham).
52. 30-mL Centrifuge tubes (Sarstedt).
53. Stratalinker, ultraviolet (UV) crosslinker (Stratagene).
54. Hybond N+ nylon membrane (Amersham).

3. Methods

Representative experiments demonstrating (1) Northern blot of miRNAs, (2) in vitro miRNP-mediated, target RNA cleavage and, (3) reporter-based assays for monitoring of miRNA activity, in vivo, are shown in **Figs. 1**, **2**, and **3**, respectively.

3.1. Northern Blots for the Detection of miRNAs

3.1.1. Isolation of Total RNA Using TRIzol

The protocol describes methods to isolate total RNA from 1 confluent, 100-mm plate or from 1 well of a 24-well-plate of HeLa cells. Listed in parentheses are the volumes or conditions that apply to the RNA isolation from the 24-well plate (*see* **Note 1**).

1. Wash cells with 10 mL (1 mL) of ice-cold PBS.
2. Aspirate. Add 6 mL (1 mL) of TRIzol and pipet vigorously. Transfer to a 30-mL Sarstedt push-cap tube (microfuge tube). Cap and let stand at room temperature (RT) for 5 min.

3. Add 1.2 mL (0.2 mL) chloroform. Shake tube vigorously for 15 s. Incubate at RT for 5 min.
4. Centrifuge at 12,000g for 15 min at 4°C using SS-34 Sorvall or equivalent (Eppendorf or equivalent tabletop centrifuge).
5. Collect aqueous (upper) layer to a new tube. Add 3 mL (0.5 mL) of isopropanol. Incubate at RT for 10 min.
6. Centrifuge at 12,000g for 10 min at 4°C.
7. Aspirate supernatant and wash pellet with 4 mL (0.8 mL) ice-cold 70% ETOH.
8. Centrifuge at 12,000g for 15 min at 4°C.
9. Dry pellet and resuspend in RNAse-free water. Store at –80°C.

3.1.2. 5′-End Labeling of DNA Oligonucleotide Probe

1. Synthesize a DNA oligonucleotide that is fully complementary to the miRNA of interest.
2. Assemble the following in a 20 µL reaction: 1 µL (= 4 pmol) of DNA Oligonucleotide (4 µM), 2 µL of 10X buffer for T4 polynucleotide kinase (T4 PNK), 3 µL of (γ-^{32}P)ATP; at 3000 Ci/mmol, 10 µCi/µL (1 µL = 3.3 pmol), 1 µL T4 PNK, 13 µL H2O.
3. Incubate at 37°C for 30 min.
4. Add 2 µL of 0.5 M EDTA (pH 8.0) to terminate reaction.
5. Add 1 µL of glycogen (at 2 mg/mL).
6. Pass through a Chroma-10 spin column (or equivalent) to remove unincorporated nucleotides. Final elution volume is approx 23 µL. Store probe at –20°C.

3.1.3. Denaturing Gel Electrophoresis and Blotting

1. Preparation of urea/PAGE (polyacrylamide gel electrophoresis) solutions.
   a. For 1 L of 10% urea/PAGE, combine the following in a glass beaker: 480 g urea, 250 mL of 40% acrylamide/bis (19 : 1), 50 mL of 10X TBE, and H_2O up to 1 L. Stir until completely dissolved, filter-sterilize, and store up to a year at RT in an aluminum-covered bottle (to protect from light).
   b. For 1 L of 20% urea/PAGE, combine the following in a glass beaker: 480 g urea, 500 mL of 40% acrylamide/bis (19 : 1), 50 mL of 10X TBE, and H_2O up to 1 L. Stir until completely dissolved, filter-sterilize, and store as described in **step 1a**.
   c. To prepare 15% urea/PAGE, simply make a 1 : 1 solution of 10% and 20% urea/PAGE solutions.
2. Cast the gel apparatus using 1.5-mm combs and 18 × 16-cm glass plates. Dispense 40 mL of 15% urea/PAGE solution in a 50-mL Falcon tube. To polymerize, add 200 µL of 10% APS and 30 µL of TEMED, mix well, and immediately pour the gel.
3. Wash wells with running buffer, using a syringe fitted with an 18-guage, or thinner, needle.
4. Prerun gel for 20 min at 400 V in running buffer.
5. Wash wells again with running buffer.

6. As a marker, use radiolabeled DNA or RNA (e.g., pBR322 DNA–Msp I digest; for preparation, *see* Chapter 16).
7. Load marker and samples and run gel until the bromophenol blue is 1 cm from the bottom of the gel. Leave at least one empty lane between marker and samples. (The marker will transfer on the membrane during blotting). Load total RNA (1 µg up to 50 µg) per lane to detect miRNAs. Many miRNAs are abundant and we find that 5 µg of total RNA is sufficient.
8. Place gel in a tray with running buffer containing 0.5 µg/mL of ethidium bromide.
9. Agitate gently for 30 min. (This step washes excess urea and stains the gel).
10. Photograph gel. You can judge quality of RNA and loading by comparing the intensity of tRNAs and 5.8S and 5S RNAs.
11. Electoblot for 45 min at 150 mA using Hoeffer Semidry-transfer (or similar) apparatus. Soak membrane (Hybond-N+) and blot papers in running buffer.
12. Rinse membrane with Millipore H_2O, blot excess liquid, and UV crosslink using Stratalinker (or equivalent) apparatus while the membrane is still damp.
13. The Membrane is now ready for hybridization. The membrane can be stripped and reprobed up to four times.

3.1.4. Hybridization

1. Prehybridize membrane at 37°C for 1 h in ULTRAHyb-Oligo (Ambion). Use 10 mL per 100 cm² of membrane. Use constant rotation or agitation/rocking for this and all subsequent steps.
2. Add probe in a small amount of ULTRAHyb-Oligo, mix and, add this mixture into the bag that contains the membrane in the prehybridization solution (there is no need to change the solution).
3. Hybridize overnight at 37°C.
4. Remove hybridization buffer and wash membrane for 10 min at RT with 2X SSC and 0.1%SDS.
5. Repeat wash as in **step 4**.
6. Do a final wash for 10 min at 37°C with 0.1X SSC and 0.1% SDS.
7. Drain excess fluid but keep membrane damp (if membrane is dried, it is impossible to strip it for another round of hybridizations). Place in heat-sealable pouch and expose on a phosphorimager screen (or film; *see* **Fig. 1**).

3.2. In vitro, miRNA-Mediated, Target RNA Cleavage

We describe a method for in vitro miRNA-directed target RNA cleavage using immunoprecipitated miRNPs as a source for the miRNP endonuclease and let-7a as the miRNA guiding cleavage of synthetic RNA targets. A major advantage of using immunopurified miRNPs over cell extracts is the elimination of contaminating nucleases. This allows the use of short, synthetic, end-labeled, oligonucleotides as RNA targets and bypasses the need for in vitro transcription and capping of RNA targets from DNA templates (uncapped RNAs are quickly degraded by nonspecific nucleases in cell extracts).

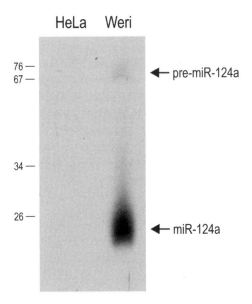

Fig. 1. Detection of a neuronal specific miRNA (miR-124a) by Northern blot. Total RNA (5 μg) was isolated from Weri cells (a human retinoblastoma cell line) or HeLa cells, fractionated on a 15% urea/PAGE (polyacrylamide gel electrophoresis) gel, transferred to a membrane and hybridized with a 5′-end radiolabeled oligonucleotide probe complementary to miR-124a (sequence of DNA oligo probe: 5′-TGGCATTCAC-CGCGTGCCTTAA-3′). Mature miR-124a and pre-miR-124a are only expressed in the neuronal-derived Weri cells but not in the epithelial-derived HeLa cells. Nucleotide sizes (from pBR322 DNA–Msp I marker) are indicated on the left.

Quantification of specific miRNAs in immunopurified miRNPs can be performed by quantitative Northern blots using synthetic RNAs (whose sequence is the same with the miRNA of interest) as standards. See **Fig. 2** for a representative gel of let-7a-directed target RNA cleavage using immunopurified miRNPs from HeLa cells.

3.2.1. miRNP Immunoprecipitation

3.2.1.1. BINDING OF ANTIBODIES TO PROTEIN-G AGAROSE BEADS

1. Bind the 8C7 (anti-human eIF2C2 monoclonal antibody) or nonimmune mouse serum (IgG; serves as negative control) on protein-G agarose beads. Use 30-μL bed volume protein G Agarose beads per immunoprecipitation (IP). Wash beads three times with 1 mL lysis buffer. (*see* **Note 2**)
2. Aspirate last wash, taking care not to dry the beads; add 1 mL lysis buffer and 4 μL ascites (or 4 μL nonimmune mouse serum for the control IP).
3. Rotate for 45 min.
4. Wash antibody beads three times with 1 mL lysis buffer.

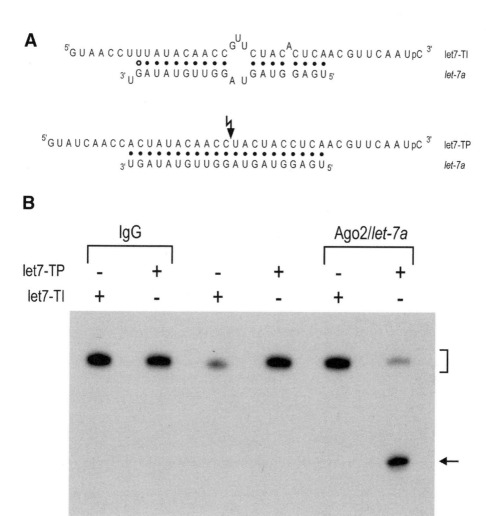

Fig. 2. miRNA-directed, target RNA cleavage using immunopurified Ago2/*let7a*-miRNA containing miRNPs as a source for the Ago2 endonuclease. (**A**) Schematic of the RNA targets used and the potential base pairing with *let-7a*; the (5′-^{32}P) of pCp is shown. Cleavage site is indicated with a lightning bolt. (**B**) The indicated, 3′-end radio-labeled RNA targets were incubated with Ago2/*let-7*-containing immunopurified miRNPs or with IgG (negative control). The RNA targets are indicated by bracket and the 3′-cleavage product by arrow. Only let7-TP, the RNA target that bears extensive complementarity to *let-7* miRNA is cleaved at a position across from the middle of the *let-7* miRNA (i.e., between the 10th and 11th nucleotide of *let-7*).

3.2.1.2. Preparation of Cell Lysate

1. The cell lysate can be prepared while the antibodies are binding to the protein-G agarose beads. Wash HeLa cells (approx 2×10^7 cells, corresponding to two 100-mm confluent plates) with PBS. Scrape if adherent, with PBS, and pellet. If cells are in suspension, pellet and wash two times with PBS.
2. Resuspend pellet with 2.5 mL of lysis buffer; add RNasin to a final concentration of 0.1 U/μL.
3. Sonicate briefly (three times, 8–10 s each) using 40% output (Sonics Vibra-Cell sonicator or equivalent).
4. Centrifuge cell lysate at 20,000*g* (in Eppendorf microfuge) for 10 min. Collect supernatant and discard pellet.

3.2.1.3. Immunoprecipitation

1. Use half of the supernatant with your specific antibody beads (i.e., 8C7 beads) and use the other half with the negative control antibody. Perform IPs in microfuge tubes. If the volume of the IP is less than 1 mL, add lysis buffer to bring to 1 mL. Rotate in the cold room for 1 h.
2. Wash beads five times with 1 mL of lysis buffer.
3. Wash beads two times with 1 mL of buffer D.
4. Aspirate supernatant (but do not dry beads) and estimate bed volume of bead slurry.
5. Add equal volume of glycerol to the bead slurry (to make a 50% glycerol stock). Mix well and store at –20°C (*see* **Note 3**).

3.2.2. RNA Probe Synthesis Labeling and Gel Purification

Synthesize RNA oligonucleotide probe complementary to the miRNA of interest (e.g., let-7a) and either 5′-end label with T4 polynucleotide kinase and (γ-^{32}P)-ATP (as described above for 5′-end labeling of DNA oligos) or 3′-end label with T4 RNA ligase and 3′, 5′-bis(α-^{32}P)-cytidine (pCp), as described below.

3.2.2.1. 3′-End Labeling

1. For each labeling reaction combine the following (total reaction volume 20 μL).
2. Place tubes at 4°C (cold room) overnight. 1 μL (=4 pmol) of Synthetic target RNA (4 μ*M*), 2 μL of 10X T4 RNA ligase buffer, 0.5 μL RNasin, 6 μL DMSO, 3 μL of [5′-^{32}P]pCp at 3000Ci/mmol, 10 mCi/mL (1 mL = 3.3 pmol), 6.5 μL H$_2$O, 1 μL of T4 RNA ligase.

3.2.2.2. Gel Purification of Labeled RNA

1. Upon completion of 3′-end labeling (or 5′-end labeling) of RNA, add to the reaction an equal volume of RNA loading buffer (there is no need to extract the reaction with phenol or ethanol precipitate the RNA).
2. Heat sample at 95°C for 1 or 2 min. Cool on ice. Spin briefly in a microfuge tube and load on the gel. Run gel until the bromophenol blue dye is 1 cm from the bottom of the gel.

3. Disassemble glass plates and lift gel on a piece of old, exposed film. Cover with Saran Wrap and expose wet gel to film for a few seconds; align gel with film (e.g., by preflashing).
4. Excise gel piece corresponding to the labeled RNA with a clean razor blade, place in microfuge tube, and count radioactivity (to calculate how many counts correspond to each picomole of RNA probe).
5. Add 3 vol of 0.3 *M* sodium acetate, pH 5.2, and rotate in a cold room for approx 4 h. Collect supernatant and add 3 vol of 0.3 *M* sodium acetate, pH 5.2, to the gel pieces; rotate in a cold room overnight. Pool the supernatants and precipitate RNA by adding 3 mL of glycogen and 2.5 vol of ethanol, as described above. Spin, aspirate supernatant, and wash RNA pellet with 70% ethanol; air-dry. Dissolve RNA with H_2O to make a stock concentration of 100,000 counts per minute (cpm)/μL. Make a working concentration of 20,000 cpm/μL.

3.2.3. In Vitro Cleavage Assay

1. For a 10-μL reaction, assemble the following on ice: 5 μL of miRNPs in glycerol, 1 μL of 10X MIB (miRNA buffer), 3 μL H_2O, 1 μL (approx 20,000 cpm) of end-labeled RNA target.
2. Incubate for 60 min at 37°C (or perform time-course).

3.2.3.1. ISOLATION OF RNA FROM IN VITRO CLEAVAGE REACTIONS

1. To the 10-μL reaction, add 190 μL H_2O to bring volume to 200 μL.
2. Add equal volume (200 μL) phenol (buffer saturated, pH 7.5–7.9). Vortex for 1 min.
3. Spin in Eppendorf at top speed for 2 min at RT.
4. Collect aqueous (upper) phase in clean Eppendorf tube. Add equal volume of phenol/chloroform/isoamyl alcohol (pH 7.9 or acidic). Vortex for 1 min.
5. Spin in Eppendorf at top speed for 2 min at RT.
6. Collect aqueous phase in clean microfuge tube and add 0.1 vol 3 *M* sodium acetate pH 5.2, 3 mL glycogen, and 3 vol 100% ETOH. Mix well.
7. Precipitate RNA at –80°C for 15 min or at –20°C overnight.
8. Spin in Eppendorf at top speed for 30 min at 4°C.
9. Discard supernatant and wash pellet with 70% ice-cold ETOH.
10. Spin in Eppendorf at top speed for 10 min at 4°C.
11. Discard the supernatant and dry the pellet. Count and resuspend in RNA loading buffer.
12. Analyze with 15% (or 20%) denaturing gel electrophoresis.

3.3. Renilla Luciferase-Based Assays for Monitoring of miRNA Activity, In Vivo

MicroRNAs interact with miRNA recognition elements (MREs)—short sequences found in their mRNA targets. MREs can be placed in the 3′-untranslated region (3′-UTR) of a reporter construct to make the reporter responsive to the miRNA of interest. If the complementarity between the MRE and the miRNA of interest is extensive, target mRNA cleavage occurs, whereas if it is partial, the translation of the mRNA target is repressed. In both cases, near perfect complementarity of the proximal (toward the 5′ end) portion of miRNAs (encompassing the first 7 to 8 nucleotides [nt] of the miRNA) is required for target mRNA recognition. In general, placing multiple MRE copies in the 3′-UTR of the reporter, with partial complementarity to the miRNA of interest, augments the inhibitory effect on translation. However, other sequences flanking the MREs can be important in some cases. The extent of miRNA-mediated translational inhibition of specific mRNA targets may be dictated by many factors, including the strength of the miRNA : MRE binding, number of MREs, accessibility of MREs, and concentration of the miRNA and its mRNA target(s).

A single MRE with optimal binding characteristics to the miRNA of interest can be used to achieve robust translational inhibition. This approach has been used to test various binding configurations between MREs placed in the 3′-UTR of a Renilla Luciferase-expressing mRNA, and the let-7b miRNA, in cells that express endogenous let-7b *(22)* (*see* **Fig. 3**).

3.3.1. Construct Design and Cloning

MicroRNA recognition elements are synthesized as sense and antisense DNA oligonucleotides, with flanking sequences containing restriction enzymes (*see* **Fig. 3**). As the negative control, the sequence of the MRE can be scrambled or point mutations that disrupt binding of the 5′ end of the miRNA with the MRE can be engineered (*see* **Fig. 3**). MREs (and scrambled or mutant MREs) are cloned in the 3′-UTR of the Renilla Luciferase mRNA in the pRL-TK vector, between the *Xba*I and *Not*I sites.

3.3.1.1. ANNEALING OF DNA OLIGONUCLEOTIDES

1. Dissolve DNA oligonucleotides in TE buffer at a stock concentration of 100 μ*M*.
2. Anneal sense and antisense DNA oligos (100-μL reaction): 1 μL of sense DNA oligo (100 μ*M*), 1 μL of antisense DNA oligo (100 μ*M*), 10 μL of 10X Annealing buffer, 88 μL H_2O.
3. Place tube in a 65°C water bath for 20 min.
4. Remove from water bath and let stand at RT for 1–2 h.
5. Store at –20°C.

3.3.1.2. VECTOR PREPARATION

1. Restriction digest of pRL-TK vector (30-μL reaction): 3 μL of pRL-TK vector (at 1 μg/mL), 23 μL H_2O, 3 μL of 10X NEBuffer 3, 0.5 μL *Xba*I, 0.5 μL *Not*I.

Fig. 3. (*Opposite page*) Renilla Luciferase assays to monitor miRNA activity in cultured cells. (**A**) Strategy for cloning of MREs (microRNA recognition elements) in 3′-UTR of Renilla Luciferase (pRL-TK vector). MREs are synthesized as DNA

A

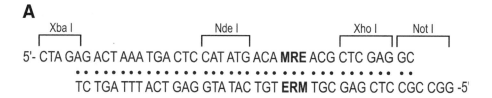

MRE of pRL-TK, LIN28-wt   5'-GCA CAG CCT ATT GAA CTA CCT CAT

MRE of pRL-TK, LIN28-M1:   5'-GCA CAG CCT ATT GAA CTA CCC TCA T

B

C

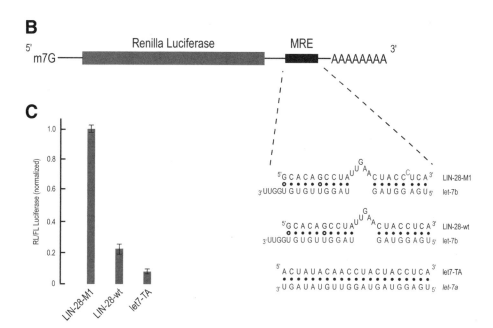

Fig. 3. (*Continued*) oligonucleotides, embedded in the backbone shown (top strand = sense). The DNA oligos are annealed and cloned in pRL-TK digested with *Xba*I and *Not*I. The *Nde*I and *Xho*I restriction sites are unique (i.e., not found in the pRL-TK vector). (**B**) Schematic representation of the reporter construct (red: coding region) and potential base-pairing between MREs and the let-7b or let-7a miRNAs (highly homologous miRNAs, normally expressed in HeLa cells). The MRE sequence designated as LIN-28-wt is found in the 3'-UTR of the human and mouse lin-28 mRNA and is responsive to let-7b. In contrast, a point mutation that is predicted to disrupt interaction with the let-7b miRNA (LIN-28-M1) abolishes let-7b-mediated translational repression. (**C**) HeLa cells were cotransfected with Renilla Luciferase (RL) constructs bearing the indicated MREs in the 3'-UTR, along with Firefly Luciferase (FL). Results shown are average values (with standard deviations) of normalized RL/FL activities obtained from three separate experiments.

2. Incubate at 37°C overnight.

3. Load entire reaction on a 1% agarose gel (containing ethidium bromide [EB]) along with markers and a small aliquot of undigested vector.

4. Gel-purify linearized (digested) vector using QIAGEN gel purification kit (or equivalent). Elute gel-purified vector in 40 µL EB (provided in kit; or 40 µL TE buffer). Adjust concentration to approx 50 ng/µL.

3.3.1.3. CLONING

1. Ligate annealed MREs (sense and antisense DNA oligos) to vector (10-µL reaction): 1 µL of pRL-TK/*Xba*I–*Not*I vector (approx 50 ng/µL), 1 µL of annealed DNA oligos, 1 µL of 10X T4 DNA ligase buffer, 6.5 µL H_2O, 0.5 µL of T4 DNA ligase.

2. Ligate overnight at 16°C.

3. Transform competent bacterial cells (e.g., TOP10) using standard techniques and plate transformants of Luria–Bertani (LB)–ampicillin plates.

4. Grow colonies, isolate plasmid, and screen for the presence of insert by digesting plasmids with *Xho*I, which should linearize only the plasmids that contain inserts (pRL-TK vector does not contain any *Xho*I sites). Confirm by sequencing.

3.3.2. Transfection

This protocol is designed for double transfections of HeLa cells grown in 24-well plates, using Firefly Luciferase (pGL3 plasmid; for transfection normalization) and Renilla Luciferase (pRP-TK) constructs bearing MREs for miRNA of interest in the 3′-UTR of the Renilla Luciferase mRNA. Each well is transfected with 0.5 µg of each plasmid (total amount of DNA is 1 µg) lipoplexed in 2 µL of Lipofectamine 2000 (LF2000). Perform experiments in triplicates. Use sterile plasticware and aseptic techniques.

3.3.2.1. PLATING CELLS IN 24-WELL PLATES

One confluent, 100-mm dish of HeLa cells can provide enough cells for two 24-well plates (48 wells) that will be ready for transfection the next day.

1. Grow HeLa cells (e.g., one 100-mm dish) to confluence.

2. Aspirate medium and wash cells with 10-mL PBS.

3. Aspirate PBS and apply 1 mL of trypsin/EDTA (10X) solution. Shake gently. Incubate at 37°C for 1 min.

4. Shake gently to detach all cells and add 9 mL of complete DMEM medium to the plate. Pipet up and down gently to resuspend cells and dilute the trypsin.

5. Aliquot half of the cells (5 mL) to a 50-mL Falcon tube that already contains 20 mL of complete DMEM medium. Mix well and dispense 1 mL of this solution to each well of a 24-well plate.

6. Place 24-well plate in incubator. Cells will be confluent and ready for transfection the next day (*see* **Note 4**).

3.3.2.2. TRANSFECTION OF CELLS IN 24-WELL PLATES

Volumes and Procedures for a Single Well

1. Aspirate medium. Replace with 0.5 mL of complete DMEM without antibiotics, per well.
2. Place cells back in the incubator
3. For each well:
   i. Combine 0.5 µg of each plasmid in 50 µL of DMEM in one Eppendorf sterile tube
   ii. In a separate tube, add 2 µL of LF2000 in 50 µL of DMEM
4. Incubate at RT for 3–5 min.
5. Mix the contents of the two tubes together and let sit at RT for 15–20 min (but no more that 30 min). During this incubation, the LF2000 forms lipoplexes with the DNA. The final volume in the tube is 100 µL.
6. Apply the above mixture (100 µL) to one well using a sterile pipet tip and dispense gently at the side of the well (do not squirt directly onto cells).
7. Place plate in the incubator and perform dual luciferase assay, 16–24 h later.

Volumes and Procedures for an Entire 24-Well Plate with Transfections Performed in Triplicate

Master mixes for transfection of a 24-well plate using pGL3 as the normalization construct are described (prepare enough mix for 28 wells to compensate losses from pipetting) for transfections performed in triplicates.

1. For every 3 wells; dispense 1.5 µg of reporter construct (Renilla Luciferase) in 1 microfuge tube (8 tubes for the 24 wells).
2. In a separate microfuge tube, add 1.4 mL of DMEM. Add 14 µg of pGL3 plasmid and mix well.
3. Aliquot 150 µL of above mix to each tube from **step 1.**
4. In a separate microfuge tube, add 1.4 mL of DMEM. Add 56 µL of LF2000 and mix well.
5. Let tubes stand at RT for 3 min.
6. Apply 150 µL of LF2000–DMEM (from **step 4**) to each of the eight tubes from **step 3** (these tubes contain the Firefly and Renilla Luciferase constructs dissolved in 150 µL of DMEM). Mix. The final volume in each tube is now 300 µL. Let stand at RT for 15 min.
7. Apply 100 µL of the mix from **step 6** to each well. Place plate in the incubator and perform a dual Luciferase Assay, 16–24 h later.

3.3.3. Dual Luciferase Assay

This protocol is for double transfections using Firefly luciferase (pGL3 plasmid) and Renilla Luciferase constructs with the dual luciferase (DLR) kit assay from Promega and the Turner TD-20/20 (or similar) luminometer.

1. Turn on the luminometer (Turner TD-20/20 or equivalent, equipped with program for DLR readings and microfuge adapter) and printer.

2. Prepare a 1X working concentration of PLB (passive lysis buffer; stock solution is 5X). Keep at RT.
3. Remove cells from incubator and aspirate medium. Wash each well with 1 mL PBS.
4. Aspirate PBS and add 100 µL of 1X PLB to each well. Shake plate for at least 20 min at RT.
5. Label 24 Eppendorf tubes with the sample number. Thaw the LAR-II reagent (prepared as per Promega's instructions) and dispense 100 µL in each tube. You will need 2.4 mL for one 24-well plate, but thaw 3 mL to compensate for pipet losses.
6. Thaw the Stop-n-Glow reagent (prepared as per Promega's instructions) and place on ice. You will need 2.4 mL for one 24-well plate, but thaw 3 mL.
7. Remove 15 mL of lysed cells from the first well and add them to the first Eppendorf tube containing 100 µL LAR-II reagent. Mix well and read luminescence. This reading corresponds to the activity of the Firefly Luciferase.
8. When the screen of the luminometer displays "GO," remove the tube from the luminometer and add 100 µL of Stop-n-Glow reagent. Mix, place tube back in the luminometer, and press "Go." The second reading is the activity of Renilla Luciferase. The values of the two luciferases and the ratio are automatically printed and shown on the screen (*see* **Note 5**).
9. Repeat **steps 7** and **8** with the other samples.
10. Cover the 24-well plate and seal with parafilm and store at –20°C in case you need to repeat the measurements. The samples are still usable for several months.

4. Notes

1. The same protocol can be used to isolate total RNA from any cell line. To isolate RNA from cell lines grown in suspension, simply pellet cells, wash with PBS, pellet again, and use 1 mL of TRIzol per 5×10^6 cells.
2. The anti-Gemin4 monoclonal antibody 17D10 *(6)* can also be used for immunopurification of miRNPs that contain the AgO_2 endonuclease.
3. The endonuclease is still active in miRNPs after a year of storage at –20°C, with 50% glycerol.
4. By adjusting the numbers of cells plated, the wells can be plated more that 1 d ahead of the transfection day. For example, to prepare 24-well plates to be transfected 3 d after plating, add 1 mL (= 1/10 cells) from **step 6** of **Subheading 3.3.2.1.**) to a 50-mL Falcon tube that contains 24 mL of complete DMEM medium. Mix well and apply 1 mL of this solution to each well of a 24-well plate.
5. In addition to determining protein levels by the DLR assay, the amount of Renilla Luciferase mRNA (and the Firefly Luciferase mRNA as normalization control) can be quantified by Northern blots or RNA protection assays to determine the levels of the Renilla mRNA bearing the MRE. For this, either transfect additional wells and isolate total RNA with TRIzol, or use techniques that allow simultaneous isolation of protein and total RNA from the same well. (e.g., by using the PARIS kit from Ambion).

Acknowledgments

We are grateful to Dr. G. Dreyfuss for the 8C7 and 17D10 antibodies. This work was supported by grants from the NIH (M.K., P.N., Z.M.), a Pfizer Fellowship for Rheumatology and Immunology (M.K.), the Department of Pathology & Laboratory Medicine, University of Pennsylvania School of Medicine, and the PENN Genomic Institute (Z.M.).

References

1. Lee, R., Feinbaum, R., and Ambros, V. (2004) A short history of a short RNA. *Cell* **116,** S89–S92, and one page following S96.
2. Ruvkun, G., Wightman, B., and Ha, I. (2004) The 20 years it took to recognize the importance of tiny RNAs. *Cell* **116,** S93–S96, and two pages following S96.
3. Lagos-Quintana, M., Rauhut, R., Lendeckel, W., and Tuschl, T. (2001) Identification of novel genes coding for small expressed RNAs. *Science* **294,** 853–858.
4. Lau, N. C., Lim, L. P., Weinstein, E. G., and Bartel, D. P. (2001) An abundant class of tiny RNAs with probable regulatory roles in *Caenorhabditis elegans*. *Science* **294,** 858–862.
5. Lee, R. C., and Ambros, V. (2001) An extensive class of small RNAs in *Caenorhabditis elegans*. *Science* **294,** 862–864.
6. Mourelatos, Z., Dostie, J., Paushkin, S., et al. (2002) miRNPs: a novel class of ribonucleoproteins containing numerous microRNAs. *Genes Dev.* **16,** 720–728.
7. Carrington, J. C., and Ambros, V. (2003) Role of microRNAs in plant and animal development. *Science* **301,** 336–338.
8. Ambros, V., Bartel, B., Bartel, D. P., et al. (2003) A uniform system for microRNA annotation. *RNA* **9,** 277–279.
9. Carmell, M. A., Xuan, Z., Zhang, M. Q., and Hannon, G. J. (2002) The Argonaute family: tentacles that reach into RNAi, developmental control, stem cell maintenance, and tumorigenesis. *Genes Dev.* **16,** 2733–2742.
10. Hamilton, A. J. and Baulcombe, D. C. (1999) A species of small antisense RNA in posttranscriptional gene silencing in plants. *Science* **286,** 950–952.
11. Elbashir, S. M., Lendeckel, W., and Tuschl, T. (2001) RNA interference is mediated by 21- and 22-nucleotide RNAs. *Genes Dev.* **15,** 188–200.
12. Murchison, E. P. and Hannon, G. J. (2004) miRNAs on the move: miRNA biogenesis and the RNAi machinery. *Curr. Opin. Cell Biol.* **16,** 223–229.
13. Liu, J., Carmell, M. A., Rivas, F. V., et al. (2004) Argonaute2 is the catalytic engine of mammalian RNAi. *Science* **305,** 1437–1441.
14. Song, J. J., Smith, S. K., Hannon, G. J., and Joshua-Tor, L. (2004) Crystal structure of argonaute and its implications for RISC slicer activity. *Science* **305,** 1434–1437.
15. Meister, G., Landthaler, M., Patkaniowska, A., Dorsett, Y. Teng, G., and Tuschl, T. (2004) Human argonaute2 mediates RNA cleavage targeted by miRNAs and siRNAs. *Mol. Cell.* **15,** 185–197.

16. Rand, T. A., Ginalski, K., Grishin, N. V., and Wang, X. (2004) Biochemical identification of argonaute2 as the sole protein required for RNA-induced silencing complex activity. *Proc. Natl. Acad. Sci. USA* **101,** 14,385–14,389.

17. Hutvagner, G. and Zamore, P. D., (2002) A microRNA in a multiple-turnover RNAi enzyme complex. *Science* **297,** 2056–2060.

18. Olsen, P. H., and Ambros, V. (1999) The lin-4 regulatory RNA controls developmental timing in *Caenorhabditis elegans* by blocking LIN-14 protein synthesis after the initiation of translation. *Dev. Biol.* **216,** 671–680.

19. Seggerson, K., Tang, L., and Moss, E. G. (2002) Two genetic circuits repress the *Caenorhabditis elegans* heterochronic gene lin-28 after translation initiation. *Dev. Biol.* **243,** 215–225.

20. Nelson, P. T., Hatzigeorgiou, A. G., and Mourelatos, Z. (2004) miRNP : mRNA association in polyribosomes in a human neuronal cell line. *RNA* **10,** 387–394.

21. Lai, E. C. (2002) Micro RNAs are complementary to 3′ UTR sequence motifs that mediate negative post-transcriptional regulation. *Nat. Genet.* **30,** 363–364.

22. Stark, A., Brennecke, J., Russell, R. B., and Cohen, S. M. (2003) Identification of *Drosophila* microRNA targets. *PLoS Biol.* **1,** 1–13.

23. Enright, A. J., John, B., Gaul, U., Tuschl, T., Sander, C., and Marks, D. S. (2003) MicroRNA targets in *Drosophila*. *Genome Biol.* **5,** R1.

24. Lewis, B. P., Shih, I. H., Jones-Rhoades, M. W., Bartel, D. P., and Burge, C. B. (2003) Prediction of mammalian microRNA targets. *Cell* **115,** 787–798.

25. Doench, J. G. and Sharp, P. A. (2004) Specificity of microRNA target selection in translational repression. *Genes Dev.* **18,** 504–511.

26. Kiriakidou, M., Nelson, P. T., Kouranov, A., et al. (2004) A combined computational-experimental approach predicts human microRNA targets. *Genes Dev.* **18,** 1165–1178.

27. Martinez, J. and Tuschl, T. (2004) RISC is a 5′ phosphomonoester-producing RNA endonuclease. *Genes Dev.* **18,** 975–980.

28. Haley, B. and Zamore, P. D. (2004) Kinetic analysis of the RNAi enzyme complex. *Nat. Struct. Mol. Biol.* **11,** 599–606.

18

High-Throughput Analysis of MicroRNA Gene Expression Using Sensitive Probes

Mouldy Sioud and Øystein Røsok

1. Introduction

MicroRNAs (miRNAs) represent a new class of noncoding RNAs whose functions are, in most cases, unknown, but are believed to play important biological roles *(1)*. These tiny RNAs are genome encoded as primary transcripts, referred to as pri-miRNAs, which are processed in the nucleus by Dorsha, a ribonuclease, into pre-miRNAs approx 60–80 nucleotides (nt) long that form stem-loop hairpin structures *(2)*. Subsequent to transport to the cytoplasm, pre-miRNAs are processed by Dicer to approx 22-nt mature miRNAs, which are incorporated into the ribonucleoprotein (RNP) complex that can direct either mRNA cleavage or translation arrest (*see* **Fig. 1**). The first identified miRNAs in *Caenorhabditis elegans* are lineage-4 (*lin-4*) and lethal-7 (*let-7*) *(3,4)*. Recent data supported a role for *lin-4* and *let-7* as posttranscriptional negative regulators for several genes, including *lin-14, lin-28,* and *lin-41* genes *(5)*. In general miRNAs bind to the 3′ untranslated regions (UTRs) of their target mRNAs with imperfect homology *(6)*.

So far, the biological function of miRNAs has been elucidated in only a few examples *(7,8)*. To uncover the biological function of miRNAs, it is necessary to identify their targets. In contrast to plant miRNAs, their human counterparts exhibited only partial homology to their potential target mRNAs *(1)*. Thus, computational approaches to find the target genes might not be helpful without experimental validation. To enable the identification of target genes, one should identify which miRNAs are expressed in particular cell types and under which conditions. In principle, such analysis would guide the search of gene targets. Alternatively, transfection of human cells with either synthetic micro RNAs or

From: *Methods in Molecular Biology, vol. 309: RNA Silencing: Methods and Protocols*
Edited by: G. G. Carmichael © Humana Press Inc., Totowa, NJ

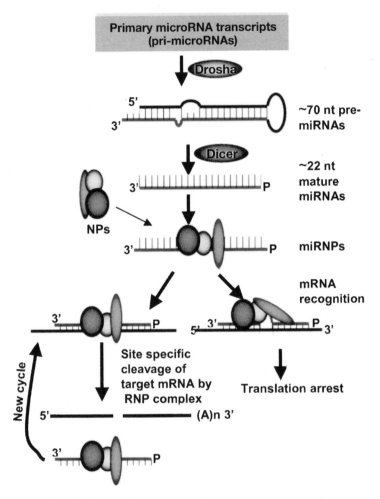

Fig.1. Schematic representation of miRNA processing.

plasmid-expressing micro RNAs and subsequent gene expression profiling using microarry technology should identify micro RNA targets genes.

High-throughput gene expression profiling is a powerful tool for investigating transcriptional activity in various cell types *(9)*. Identical array copies usually hybridized with labeled cDNA probes prepared from polyA+ using standard methods for making cDNA libraries. The resulting images captured, for example, on a phosphorimager are analyzed using software that quantifies the signal of each spot corresponding to individual clones. The intensity of each spot is expected to be proportional to the amount of mRNA present within the analyzed samples. In contrast to mRNAs, mature miRNAs are approx 22 single-stranded RNA species and the preparation of labeled probes for gene profiling can be critical. In contrast to end labeling using either T4 polynucleotide kinase or RNA ligase, cDNA labeling incorporates multiple radioactive bases, and thus the specific activity of such probes is expected to be high. This chapter focuses on the preparation of sensitive cDNA probes from short RNAs and their use in the quantification of miRNA gene expression. In addition purified micro RNAs can be 3′ end labeled by poly (A) polymerase, which adds several labeled ATP molecules.

2. Materials

2.1. RNA and DNA Oligonucleotides

1. Short RNA templates were chemically made by Eurogentec (Belgium).
2. Dimer DNA oligonucleotides corresponding to mature miRNAs were chemically made by Invitrogen (USA) (*see* **Note 1**).

2.2. Buffers and Solutions

1. 5X PowerScripts reverse transcription buffer: 250 mM Tris-HCl (pH 8.3), 375 mM KCl, and 15 mM MgCl$_2$.
2. TE buffer: 10 mM Tris-HCl, pH 8, and 1 mM EDTA.
3. 10X TBE buffer: 890 mM Tris-base, 890 mM boric acid, and 20 mM EDTA, pH 8.0.
4. 20X SSC solution: 3 M NaCl, 0.3 M Na$_3$ citrate.
5. Washing solution: 2X SSC and 1% sodium dodecyl sulphate (SDS).
6. 20X SSPE: 3.6 M NaCl, 0.2 M sodium phosphate, and 0.02 M EDTA, pH 7.7.
7. 100X Denhardt's solution: 2% (w/v) BSA (bovine serum albumin), 2% (w/v) Ficoll, and 2% (w/v) PVP (polyvinylpyrrolidone).
8. Hybridization solution: 5X SSPE, 10% dextran sulfate (w/v), 1% Denhardt's solution (w/v), 1% SDS (w/v), and 1 µg/mL sheared salmon testes DNA.

2.3. Enzymes and Other Reagents

1. PowerScripts reverse transcriptase (Clonetec).
2. Ribonuclease inhibitor (RNasin).
3. 10X dNTP mixture (5 mM each dCTP, dGTP, dTTP).
4. 95% Ethanol.
5. 70% Ethanol (diluted with diethyl pyrocarbonate [DEPC]-treated water).
6. 3 M Sodium acetate (pH 4.5).
7. [α-^{32}P] dATP (10 µCi/µL; 3000 Ci/mmol).
8. Phenol : chloroform : isoamyl alcohol (25 : 24 : 1, v/v/v) TE saturated.
9. 30% Acrylamide/bis acrylamide solution (19 : 1).
10. $N,N′$-Tetramethylethylenediamine (TEMED).
11. 10% Ammonium persulfate (APS).
12. Diethyl pyrocarbonate (DEPC). This reagent is a suspected carcinogen and should be handled with great care.
13. Dithiothreitol (DTT), 100 mM.
14. Sheared salmon testes DNA.

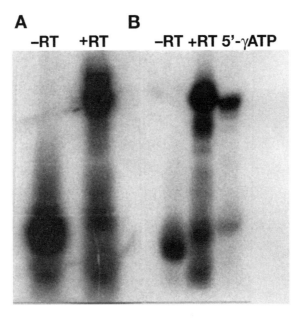

Fig. 2. Reverse transcription of small RNAs using a random hexamer primer. (**A**) An arbitrary single-stranded siRNA (21 nt) was in vitro reverse transcribed in the presence of [α-^{32}P]ATP as described in **Subheadings 2.**, and **3.** Aliquots (2 µL) were analyzed by a 15% polyacrylamide gel with of 7 *M* urea. –RT = reverse transcriptase not included, +RT = reverse transcriptase included. (**B**) Using the same conditions, chemically synthesized human *let*-7d miRNA was reverse transcribed. As marker, a 5'- labeled *let*-7d RNA was included.

3. Methods

3.1. Direct Random-Primed cDNA Synthesis on Small RNAs (21 nt)

Given the short nature of mature miRNAs (approx 22 nt), as a first step for testing the idea that specific cDNA probes might be prepared from these tiny miRNAs, we have investigated whether small chemically synthesised 21-nt RNAs can be reverse transcribed using a random hexamer oligonucleotide.

1. Use the following reverse transcription conditions: 10 µL of 5X Reverse transcription buffer, 1 µL (0.1 µg) of chemically synthesized RNA template, 1 µL (0.1 µg) of random hexamer primer (0.1 µg) in 5 µL 100 m*M* DTT, 10 U RNase inhibitor, 2 µL of 5m*M* dNTPs (without ATP), 3 µL of [α-^{32}P]dATP (2500 Ci/mmol), 10 U PowerScripts reverse transcriptase. Complete to 20 µL with DEPC-treated water and then incubate at 37°C for 1–2 h.
2. Subsequent to incubation, analyze an aliquot (1 µL) using a 15% denaturing polyacrylamide gel to verify the quality of the cDNA probe and estimate the percentage incorporation of labeled nucleotides using a phorphorimager system. **Figure 2** shows a representative example of in vitro cDNA synthesis using

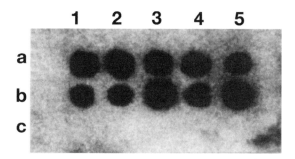

Fig. 3. RNA oligonucleotide arrays indicate specific cDNA synthesis. Ten chemically synthesized single-stranded siRNA (21 nt) were spotted onto Hybond-N+ nylon membrane (lanes a and b) and then hybridized with a cDNA probe prepared from the 10 templates. As specificity controls, additional five chemically synthesized single-stranded RNAs (21 nt) were spotted onto the Hybond-N+ nylon membrane (lane c).

chemically made small RNAs (e.g., *let-7*b miRNA) as templates. Despite using a random hexamer primer, the reverse-transcribed RNA species exhibited similar length as the 22-nt RNA template. This observation would indicate that the priming is preferentially occurring at the 3′ end of the RNA template. Similar results were obtained with all tested small RNA templates (21 nt).

3.2. RNA and Oligonucleotide Arrays

To confirm the specificity of the prepared cDNA probes, dot blot experiments were performed. In these experiments, 10 small RNA molecules (21 nt), including the let-7d were reverse transcriptase as described in **Subheading 3.1.** and then used as a probe to hybridize a nitrocellulose array containing the corresponding RNA templates. As controls for specificity, five arbitrary small RNA molecules were included in the array. The arrays were prepared and hybridized using the following conditions:

1. Spot RNA templates (0.2 μg) onto Hybond-N+ nylon membrane (Amersham Life Science), prewetted in 10X SSC, in 1 to 2 μL aliquots. Alternatively, a commercial dot blotting apparatus can be used.
2. Allow the membranes to air-dry. Wrap the membranes in Saran Wrap and place RNA side down on a transilluminator for 1–2 min (ultraviolet [UV] crosslinking).
3. Prehybrize the array at 60°C for at least 4 h in hybridization buffer containing sheared salmon testes DNA (1 μg/mL).
4. Add the cDNA probes to the prehybridization solution and incubate overnight with continuous agitation at 37°C.
5. The next day, wash the filters by incubating them in 2X SSC buffer, 0.2 SDS at room temperature for 10 min.
6. Replace the solution and incubate for 5 min with continuous agitation at 37°C.
7. Remove filter, wrap in Saran Wrap, and expose it to a PhosphorImaging screen. **Figure 3** shows the hybridization signals. The data indicate that the in vitro

synthesized cDNA probes do hybridize to their RNA templates (lanes a and b). In contrast, no detectable hybridization signals were obtained with the control molecules (lane c), confirming the specificity of the designed probes.

3.3. Detection of miRNA expression in Human Cells

As mentioned earlier, Dicer processes pre-miRNAs to the approx 22-nt mature miRNAs. Despite the use of a random hexamer primer, we have found that small RNA can be successfully reverse transcribed. In the next experiments, we have investigated whether this type of cDNA probes can be useful in monitoring miRNA gene expression in human cells. These experiments involve several steps, including the preparation of total RNA, small RNAs, cDNA synthesis, and hybridization.

3.3.1. Preparation of Low-Molecular-Weight RNAs

Small RNA species were isolated mainly as described by Hamilton and Baulcombe *(7)*. Total RNA was isolated using the guanididium thiocyanate phenol–chloroform method *(10)*. High-molecular-weight RNA was precipitated with 10% polyethylene glycol (800) and 0.5 *M* NaCl at 4°C for 30 min. Low-molecular-weight RNA was precipitated from the supernatant by the addition of 3 vol of ethanol for 24 h at –70°C. After precipitation, samples were washed with 70% ethanol, air-dried, and then dissolved in 50 µL DEPC-treated water. Other methods for preparing small RNAs such as the QIAGEN Resin Kit can be used (*see* **Note 2**).

3.3.2. Preparation of Small RNAs

1. Once you have prepared low-molecular-weight RNA and checked the quality of your preparation (*see* **Note 2**), apply the sample (approx 50–100 µg) on 15% denaturing polyacrylamide gels. As size markers, include a small RNA (21 nt).
2. After electrophoresis, stain the gel with ethidium bromide, cut out the portion of the gel that contains small cellular RNAs (16–26 nt in length) and then cut the gel portion into small pieces.
3. Add DEPC-treated water (0.5–1 mL), 50 U RNase inhibitor, and incubate the sample for 4 h with continuous agitation at 37°C.
4. After incubation, extract the sample with phenol/chloroform, being careful not to pipet any chloroform.
5. Add sodium acetate (0.3 *M* final) and precipitate with ethanol (3 vol) at –70°C for at least 2 h.
6. Spin in a microcentrifuge at 14,000 rpm for 30 min at 4°C and carefully remove the supernatant.
7. Gently wash the pellet with 200 µL of 70% ethanol and air-dry the pellet for approx 10 min or until pellet is dry.
8. Dissolve the pellet RNA into 20 µL of DEPC-treated water and RNasin (10 U). This RNA can be used for the preparation of the [32]P- or ^{33}P-labeled cDNA probes. RNA can be stored at –70°C until use.

3.3.3. Preparation of cDNA Probes

1. To reverse transcribe the prepared small RNAs, use the same conditions as chemically synthesized RNA templates (*see* **Subheading 3.1.**). However, use approx 0.5–1 µg of small RNA and 1 µg of random hexamer primer and increase the final reaction volume to 50 µL.
2. Before using the labeled probes, you should analyze a small aliquot (2 µL) using a 15% polyacrylamide gel. This will verify that the cDNA synthesis works in your hands and will help you to estimate the quality of the cDNA. **Figure 4A** shows an example of direct random-primed cDNA synthesis on gel-purified miRNAs from untreated HL60 cells, TPA-treated HL60 cells, and SKBR3 cells (*see* **Note 3**).

3.4. Profiling miRNA Expression in HL60 Cells

To illustrate the utility of the prepared probes, the cDNA probes shown in **Fig. 4A** were hybridized to an array containing 30 random microRNA sequences (*see* **Note 1**). The oligonucleotide arrays were prepared as described in **Subheading 3.2.**, except that samples were heated 90°C for 1 min prior spotting onto Hybond-N+ nylon membrane. Hybridization with cDNAs to oligonucleotide array and washing were performed as described in **Subheading 3.2.** As shown in **Fig. 4B,** hybridization with the prepared cDNA probes detected the expression of miRNAs in HL60 cells. Notably, a differential miRNA expression between untreated and TPA-treated HL60 cells was found (*11*).

In this chapter, we have examined the specificity and sensitivity of using short cDNA probes to analyze the expression of miRNAs. The designed probes should provide significantly better sensitivity than those derived from 5′ labeling because several radiolabeled bases can be incorporated (*see* **Note 4**). Similar to cDNA probes derived from mRNAs, cDNA probes derived from mature miRNAs and/or pre-miRNAs can also be useful in comparative expression analysis between samples (*see* **Note 5**). For instance, tumors vs normal tissues and stem cells vs their differentiated counterparts. cDNA synthesis and dC-tailing by reverse transcriptase offers the possibility of linear amplification of microRNA probes (*see* **Note 6**). In addition to cDNA, purified microRNAs can be 3′ end labeled using poly(A) polymerase, which incorporates several ATP molecules, thus providing high sensitivity as well (*see* **Note 7**).

4. Notes

1. DNA oligonucleotides corresponding to the mature miRNAs were designed based on the published sequences (http://www.sanger.ac.uk/Software/Rfam/mirna).
2. Before embarking on your experiments, you should always check the quality of the prepared small RNAs by running a small aliquot on 15% polyacrylamide gel. The quality of the RNA used to make your cDNA probe is the single most important factor for making good probes and the use microarray technology. Commercially available kits for the isolation of low-molecular-weight RNA can

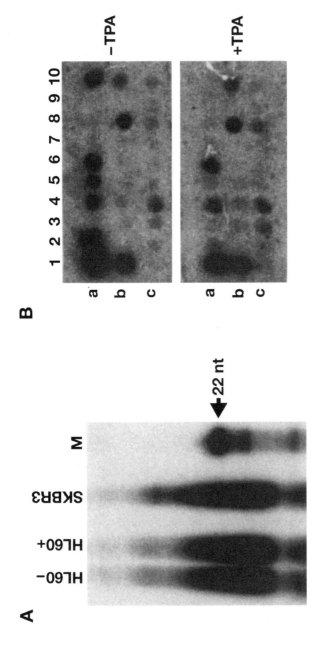

Fig. 4. High-throughput analysis of miRNA expression in human cells. (A) cDNA synthesis. PAGE-purified small RNAs (approx 16–26 nt) prepared from untreated HL60 cells, TPA-treated HL60 cells, or SKBR3 cell line were in vitro reverse transcribed. Subsequently, 2-μL samples were analyzed using a 15% polyacrylamide gel with 7 M urea. As a marker, a 5'-labeled RNA oligonucleotide (22 nt) was included (lane M). (B) Profiling miRNA gene expression by oligonucleotide arrays. Dimer oligonucleotides representing 30 arbitrary miRNA sequences were spotted onto Hybond-N+ nylon membranes and then hybridized with cDNA probes made from small cellular RNAs prepared from either untreated or TPA-treated HL60 cells.

318

be used (e.g., QIAGEN Resin Ambion). Prepared low-molecular-weight RNAs population contain both mature and miRNA precursors. Both RNA can be reverse transcribed using a random hexamer primer.

3. To remove unincorporated ^{32}P-labeled nucleotides from labeled cDNA probes, use spin columns or ethanol precipitation. Note that in the experiments shown in **Figs. 3** and **4**, we did not remove unincorporated [α-^{32}P]dATP.

4. Purified miRNA can be 5′ labeled and used as the probe. However, we found that the sensitivity of such probes is less than those prepared by reverse transcription.

5. To detect rare miRNAs, very sensitive cDNA probes are required. This can be done by using two labeled nucleotides (e.g., ATP and CTP). Each array can be stripped and reprobed. For effective stripping of the probes, store the membranes at –20°C when not in use.

6. Notably, first-stand synthesis and dC tailing by reverse transcriptase would facilitate the addition of oligonucleotide containing the T7 RNA polymerase promoter. After primer extension with DNA polymerase to generate double-stranded cDNA, the template can be transcribed using T7 RNA polymerase. This would result in linear amplification of microRNAs. This method has been used for mRNA amplification for the preparation of microarray probes.

7. Purified microRNAs can be 3′ end-labeled using either RNA ligase or poly(A) polymerase. In contrast to RNA ligase, however, poly(A) polymerase can incorporate several labeled ATP molecules, thus providing high sensitivity comparable to that obtained with the cDNA-labeling method. In the case of 3′ end-labeling the filter arrays must contain the microRNA antisense strands.

Acknowledgments

This work was supported in part by the Norwegian Cancer Society. We thank Dr. Anne Dybwad for critical reading of the manuscript.

References

1. Bartel, D. P. (2004) MicroRNAs: genomics, biogenesis, mechanism, and function. *Cell* **116**, 281–297.

2. Lee, Y., Jeon, K., Lee, J. T., Kim, S., and Kim, V. N. (2002) MicroRNA maturation: stepwise processing and subcellular localization. *EMBO J.* **21**, 4663–4670.

3. Lee, R. C., Feinbaum, R. L., and Ambros, V. (1993) The *C. elegans* heterochronic gene *lin*-4 encodes small RNAs with antisense complementarity to *lin*-14. *Cell* **75**, 843–854.

4. Reinhart, B. J., Slack, F. J., Basson, M., et al. (2000) The 21 nucleotide let-7 RNA regulates developmental timing in *Caenorhabditis elegans*. *Nature* **403**, 901–906.

5. Johnson, S. M., Lin, S. Y., and Slack, F. J. (2003) The time of appearance of the *C. elegans* let-7 microRNA is transcriptionally controlled utilizing a temporal regulatory element in its promoter. *Dev. Biol.* **259**, 364–379.

6. Lai, E. C. (2002) MicroRNAs are complementary to 3′UTR motifs that mediate negative post-transcriptional regulation. *Nat. Genet.* **30**, 363–364.

7. Hamilton, A. J. and Baulcombe, D. C. (2000) A species of small antisense RNA in posttranscriptional gene silencing in *Drosophila* cells. *Nature* **404,** 293–296.

8. Carrington, J. and Ambros, V. (2003) Role of microRNAs in plants and animal development. *Science* **301,** 336–338.

9. King, H. C. and Sinha A. A. (2001) Gene expression profile analysis by DNA microarrays. *JAMA* **286,** 2280–2288.

10. Chomczymski, P. and Sacchi, N. (1987) Single step method of RNA isolation by acid guanidium thiocyanate phenol chloroform extraction. *Anal. Biochem.* **162,** 156–160.

11. Sioud, M. and Røsok, Ø. (2004) Profiling microRNA expression using sensitive cDNA probes and fitter arrays. *Biotechniques* **37,** 578–580.

19

Modification of Human U1 snRNA for Inhibition of Gene Expression at the Level of Pre-mRNA

Peng Liu, Mary Louise Stover, Alexander Lichtler, and David W. Rowe

1. Introduction

U1 snRNA is a component of the U1 snRNP complex, which contains seven common snRNP proteins and three specific U1 snRNP proteins *(1)*. It initiates spliceosome association with pre-mRNA by defining the 3′ boundary of exons *(2)*. As the splicing reaction proceeds, U1 snRNP and the other spliceosome components are sequentially released from the transcript *(3)*.

Previously, it has been observed that in the human immunodeficiency virus (HIV) genome, the polyadenylation (pA) signal within the 5′ long terminal repeat is located immediately downstream of the transcription start site and upstream of the major 5′ splice site (5′ss) *(4)*. In this orientation, U1 snRNA binds to the 5′ss and suppresses the upstream pA, allowing formation of the full-length transcript. However, placing this 5′ss site further from this pA signal reduces expression of the full-length transcript because it is truncated at the now-activated upstream cleavage-pA site *(5)*. Persistent U1 snRNA binding to a site in proximity to the pA signal could account for the observation that a cryptic or an unpaired 5′ss within the terminal exon of an mRNA also prevents cytoplasmic accumulation of that mRNA, such as within mouse polyomavirus, bovine papillomavirus, and the U1A gene *(6–8)*.

Based on these observations, we have developed a U1 snRNA anti-RNA strategy for inhibiting gene expression. We have demonstrated that the modified U1 snRNA is able to suppress expression of a number of transgenes by targeting unique sequences located within the terminal exon in both transient and stable transfection protocols *(9–11)* with constructs which include the SV40 promoter driving β-galactosidase, CAT, GFP, and luciferase, as well as endogenous genes such as type I collagen and osteocalcin *(12)*. We have also found

From: *Methods in Molecular Biology, vol. 309: RNA Silencing: Methods and Protocols*
Edited by: G. G. Carmichael © Humana Press Inc., Totowa, NJ

that the modified U1 snRNA-mediated gene inhibition is dependent on hybrid formation between a specific targeted sequence and the 5′ end of the modified U1 snRNA *(10,12)*. It is able to reduce gene expression by limiting the polyadenylation of the targeted pre-mRNA transcript.

2. Materials

1. Human U1snRNA express vectors. These vectors will be provided by Rowe's laboratory if requested.
2. TOP 10 *Escherichia coli* competent cells (Invitrogen).
3. Oligonucleotide primers (5′ mutagenic primer and 3′ primer with deletion of *Pst*I site).
4. Restriction enzymes, T4 DNA polymerase (5 U/μL), and T4 ligase.
5. dNTP mix: 10 mM each of dATP, dGTP, dCTP, and dTTP.
6. Calf intestinal alkaline phosphatase (CIP)/bacterial alkaline phosphatase (BAP).
7. Agarose and DNA sequencing gel equipment.
8. Ampicillin.
9. 5X T4 DNA polymerase buffer: 165 mM Tris-acetate (pH 7.9), 330 mM sodium acetate, 50 mM magnesium acetate, 5 mM dithiothreitol (DTT).
10. 2.5 M $CaCl_2$.
11. TE buffer: 10 mM Tris-HCl and 1 mM EDTA.
12. 2X HEPES-buffered saline solution.
13. SDS-PAGE (sodium dodecyl sulfate–polyacrylamide gel electrophoresis) equipment.
14. Complete medium:F12 medium containing 5% fetal calf serum (FCS) with 2 mM glutamine, 1% nonessential amino acids, 100 U of penicillin/mL, and 100 μg of streptomycin/mL.
15. Autoclaved, distilled water.
16. GTC: 50 g guanidine thiocyanate, 0.5 g sodium lauroyl sarcosinate, 2.5 mL of 1 M Na-citrate, pH 7. Add autoclaved H_2O to 100 mL and filter. Before use, add 0.7 mL of 2-mercaptoethanol and 0.33 mL of Antifoam A (Sigma).
17. 10X SSC: 1.5 M NaCl, 0.15 M Na-citrate, pH 7.0.

3. Methods

The methods describe in this section outline (1) the construction of U1 targeting constructs, (2) the introduction of the U1 antitargeting constructs into cells, and (3) the assessment of target gene activity.

3.1. Construction of U1 Targeting Constructs

The parental recombinant U1 snRNA gene consists of the five snRNA-specific enhancer elements in the 315-bp promoter, the U1 coding sequence, and a unique 3′ termination sequence *(13,14)* (*see* **Fig. 1**). The wild-type U1 snRNA will be referred to herein as U1 snRNA and is subcloned into the plasmid pUC18 (*see* **Fig. 2**). The modified construct is identified as the U1 antitarget

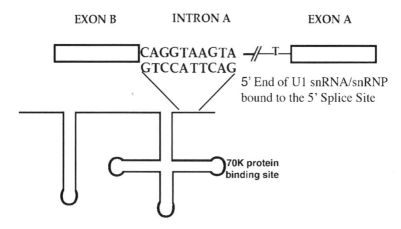

Fig. 1. A depiction of U1 snRNA, which recognizes the 5' splice donor site of pre-mRNA (A) and a schematic structure of the U1 snRNA gene.

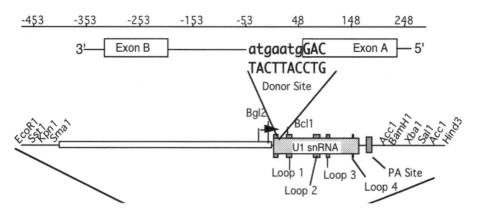

Fig. 2. Schematic of U1 snRNA gene in pUC18 plasmid.

gene followed by the first base number of the targeted sequence (e.g., U1 anti-target xxx, in which the xxx indicates the 5'-most base of the complementary RNA transcript sequence). The target numbering begins from the AUG translation start codon.

3.1.1. Design of 5' Mutagenic Primer

The following steps are described to design a 5' primer for modification of U1 snRNA:

1. Identification of a targeted sequence. A candidate 10-bp target must be selected within the terminal exon or 3'-UTR (untranslated region) of the targeted gene. This selected sequence should not be homologous to other RNA expressed in a

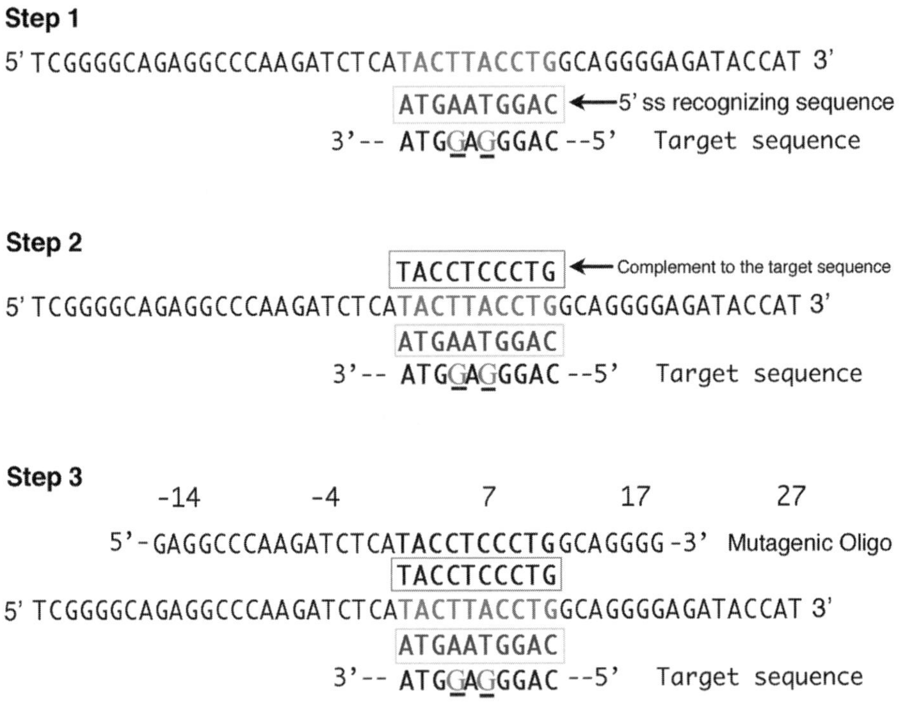

Step 1

5' TCGGGGCAGAGGCCCAAGATCTCATACTTACCTGGCAGGGGAGATACCAT 3'

                         | ATGAATGGAC | ◄— 5' ss recognizing sequence

                  3'-- ATGGAGGGAC --5'   Target sequence

Step 2

                        | TACCTCCCTG | ◄— Complement to the target sequence

5' TCGGGGCAGAGGCCCAAGATCTCATACTTACCTGGCAGGGGAGATACCAT 3'

                        | ATGAATGGAC |

                  3'-- ATGGAGGGAC --5'   Target sequence

Step 3

    -14   -4   7   17   27

   5'-GAGGCCCAAGATCTCATACCTCCCTGGCAGGGG-3' Mutagenic Oligo

                    | TACCTCCCTG |

5' TCGGGGCAGAGGCCCAAGATCTCATACTTACCTGGCAGGGGAGATACCAT 3'

                        | ATGAATGGAC |

                  3'-- ATGGAGGGAC --5'   Target sequence

Fig. 3. A schematic description of designing a 5' mutagenic primer. Step 1: position the targeted sequence under the consensus sequence from 5' to 3'; step 2: make a complement of this sequence; step 3: make a mutagenic oligo as described in the text.

    particular cell type, as determined by carrying out a BLAST search of the mouse, rat, and human expressed sequence tag (EST) databases. Selection of at least three targeted sites for testing optimal inhibitory action is recommended (*see* **Note 1**).
2. Find the consensus 5'ss recognizing sequence (CAG/GTAAGTA) of U1 snRNA, then position the targeted sequence under the consensus sequence from 5' to 3', and make a complement of this sequence as shown in steps 1 and 2 in **Fig. 3**. In **Fig. 3**, a 10-bp targeted sequence is chosen from the rat osteocalcin gene as an example *(12)*.
3. Replace the consensus U1 5'ss recognizing sequence with the complement of the target sequence to make the 5' oligo primer from the *Bgl*II site to the *Nla* site as shown in step 3 in **Fig. 3**.
4. Order the above 5' (mutagenic) primer and the 3' (selection) primer (5' AGTGC-CAAGCTTGCATGCCAGCAGGTC 3').

3.2. PCR-Directed Mutagenesis of U1 snRNA

    U1 antitarget vectors are created by PCR mutagenesis of the 5' sequence, between bases +1 and +10, which normally complements the 5' splice donor (*see* **Fig. 3**). The 5' (mutagenic) primer as described in **Subheading 3.1.1.** contains a

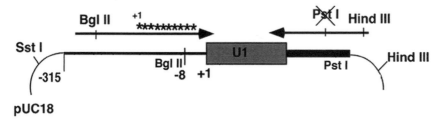

Fig. 4. A PCR-directed strategy for modification of U1 snRNA.

*Bgl*II restriction site for insertion into position 8 bp in the U1 promoter. The 3′ (selection) primer (5′ AGTGCCAAGCTTGCATGCCAGCAGGTC 3′) extends through the U1 termination sequence into the pUC18 polylinker, terminating with a *Hin*dIII site. A base change (underlined) is made to destroy a *Pst*I site proximal to the *Hin*dIII site to allow selection against plasmids containing the original gene (*see* **Fig. 3**).

1. Reaction procedure: In a PCR tube, add the following and gently mix: 1 µL of U1 snRNA (10 ng/µL), 10 µL of 5X PCR buffer, 8 µL of dNTP mix, 1 µL of 5′ primer (0.25 ng/µL), 1 µL of 3′ primer (0.25 ng/µL), 0.5 µL of T4 DNA polymerase (5 U/µL), 35 µL H₂O, for a total of 50 µL.
2. Polymerase chain reaction is performed using the following cycling conditions: 94°C for 5 min, followed by 30 cycles of 94°C for 30 s, 45°C for 30 s and 72°C for 2 min, terminating at 72°C for 7 min, cooled and held at 4°C.
3. Add 10 µL of 5 *M* NaCl and 30 µL of H₂O into the PCR tube.
4. Extract with 50 µL phenol/50 µL chloroform, precipitate with 2 vol of absolute ethanol at –20°C for 30 min, and centrifuge in an Eppendorf centrifuge at maximum speed for 10 min.
5. Dissolve the PCR pellet in 20 µL of H₂O and run 5 µL of dissolved PCR product on a mini (PAGE) gel. A single band is visualized at + about the 260-bp size, as shown in **Fig. 4**.

3.3. Restriction Enzyme Digestion of PCR Production and U1 snRNA

1. Separately digest the PCR product and U1 snRNA with a combination of *Bgl*II and *Hin*dIII in a 1.5-mL microcentrifuge tube.
2. Add 20 µL of PCR products, 1.5 µL of *Hin*dIII, 3 µL of its enzyme buffer, and 5.5 µL of H₂O to a tube and incubate at 37°C for 1 h.
3. Add 1.5 µL of 1 m*M* NaCl and 1.5 µL of *Bgl*II and incubate for an additional 1 h.
4. Add 2 µL of tRNA (10 mg/mL), 5 µL of 5 *M* NaCl, and 62 µL of H₂O to the tube and extract with 100 µL phenol/100 µL chloroform, followed by precipitation with 2 vol of 100% ethanol. Redissolve the pellet in 10 µL of TE buffer.
5. For U1 snRNA, add 10 µL of U1 snRNA (0.5 µg/µL), 1.5 µL of *Hin*dIII, 3 µL of its enzyme buffer, and 5.5 µL of H₂O to a tube and incubate at 37°C for 1 h.
6. Add 1.5 µL of 1 m*M* NaCl and 1.5 µL of *Bgl*II and incubate for another 1 h.
7. Add 10.5 µL of H₂O and 1.5 µL of CIP/BAP and incubate at 37°C for 30 min.

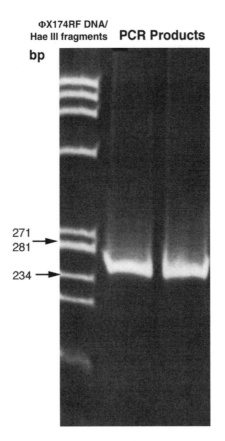

Fig. 5. An example of visualized PCR products on mini-PAGE gel.

8. Add 2 μL of tRNA, 5 μL of 5 *M* NaCl, and 52 μL of H_2O to the tube, extract with 100 μL phenol/100 μL chloroform, and precipitate again with 2 vol of 100% ethanol. Redissolve the pellet in 20 μL of TE buffer.

3.4. Ligation

Add the digested PCR product and U1 snRNA fragment to a 1.5-mL microcentrifuge tube for the ligation reaction. This step allows insertion of the PCR fragment into U1 snRNA between *Bgl*II and *Hind*III sites: 9 μL of PCR products, 3 μL of U1 snRNA, 4 μL of T4 ligase buffer (5X), 2 μL of T4 ligase, 2 μL of H_2O, for a total of 20 μL.

The reaction is performed at room temperature for 2 h.

3.5. Transformation

1. When the ligation reaction is completed, carry out transformation in a ice-cooled 1.5-mL microcentrifuge tube by adding 4 μL of ligation solution and 100 μL of

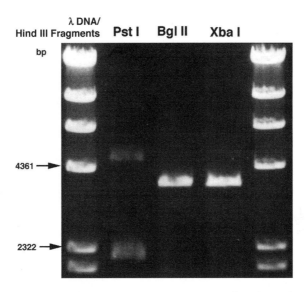

Fig. 6. An example of diagnostic restriction-enzyme digestion of a positive clone with *Pst*I, *Bgl*II, and *Xba*I, showing uncut for *Pst*I and a single band for *Bgl*II and *Xba*I.

TOP 10 *E. coli* competent cells (Invitrogen), incubating on ice for 1 min and heat shocking at 37°C for 1 min.

2. Plate 100 μL of transformed cells on an antibiotic-resistant Luria–Bertani (LB) plate and incubate at 37°C overnight.

3.6. Screening of Positive Clones by Restriction-Enzyme Digestion

1. From the transformation plate, pick 10–20 colonies and incubate for 8–12 h in LB medium with antibiotics.
2. Prepare the plasmid DNA with either the QIAprep Spin Miniprep kit or alkaline lysis mini-preparation and dissolve in 20 μL of TE buffer.
3. Digest 2 μL of harvested plasmid DNA in a 1.5-mL microcentrifuge tube, with 1 μL of either *Hin*dIII or *Pst*I in the presence of 2 μL of enzyme buffer and 15 μL H_2O at 37°C for 2 h.
4. Visualize digested and undigested DNA fragments on a 1% agarose gel. The resulting clones are selected by the absence of a *Pst*I site (shown in **Fig. 5**) and confirmed by sequencing.
5. Prepare a large-scale preparation of plasmid DNA from the selected clone using either a QIAGEN Plasmid Maxi Kit or CsCl/ethidium bromide centrifugation (*see* **Note 2**) (**Fig. 6**).

3.7. Cell Culture and Transfection

1. NIH 3T3 fibroblasts are used as an example for transfection. Grow cells and passage every 3 d in F12 medium containing 5% FCS with 2 m*M* glutamine, 1% nonessential

amino acids, 100 U of penicillin/mL, and 100 µg of streptomycin/mL. Confluent adherent cells are passaged in a 100-mm culture disk and grown to 70–80% confluence on the day of transfection.

2. Transfect using a calcium phosphate precipitate protocol, with 10 µg of U1 DNA, 2 µg target DNA, and 1 µg of TK-luciferase DNA per 100-mm dish. Three 100-mm-diameter plates or three 35-mm-diameter wells of six-well plates (Falcon) can be used for each experimental group, with at least three transfections per data point. Analysis is performed on the cell extract harvested 48 h following the transfection.

3. Feed the cells with 9 mL of DMEM with 10% FCS prior to transfection.

4. Prepare Calcium Phosphate/DNA mixture. For each 100-mm plate, mix 50 µL of 2.5 M CaCl$_2$, 500 µL of 2X HBS, pH 7.5, and 5–30 µg DNA in a final volume of 500 µL, adjusted with H$_2$O.

5. Place 500 µL of 2X HBS, pH 7.5, in a sterile 15-mL conical microcentrifuge tube. Use a mechanical pipettor attached to a plugged 1-mL pipet to bubble this solution and add the above Calcium Phosphate/DNA mixture dropwise with a Pasteur pipet. Immediately vortex the solution for 5 s.

6. Allow a precipitate to form for 15 min in a laminar-flow hood at room temperature. At the end of this time, the solution should appear slightly cloudy.

7. Mix the precipitate by pipetting up and down once and then add the precipitate in a slow dropwise fashion to the cells.

8. Incubate the cells with the Calcium Phosphate/DNA for 4–6 h in a humidified CO$_2$ incubator at 37°C.

9. Remove the medium and perform a glycerol shock with a sterile 10% glycerol solution for 3 min.

10. Add 5 mL of 1X PBS to the glycerol solution of the cells, agitate to mix, and remove the solution. Wash twice with 5 mL of 1X PBS. Feed the cells with the culture medium.

11. Stable cotransfection experiments can be performed utilizing 10 µg of the U1 construct, 2.0 µg of target gene, and 0.5 µg of a RSV hygromycin resistance or SV2neo selection vector. The transfected cultures are selected with 100 µg/mL hygromycin or 200–400 µg/mL G418 for at least 3 wk, and individual colonies are randomly picked and expanded. The remainder of the colonies are pooled and expanded.

3.8. RNA Extraction and Assessment of Target Gene Activity

Analysis is performed on the cell extract harvested 48 h following the transfection for transient cotransfection and several weeks after stable selection.

1. Add 700 µL or 2 mL of TRIzol (BRL) directly to 35-mm or 100-mm confluent plates.

2. Collect the cell lysate from individual plates and incubated on ice for 15 min.

3. Add 1/5 vol chloroform to the lysate, mix thoroughly, and incubate on ice for 15 min.

4. Precipitate the aqueous phase with 0.5 vol isopropanol (original TRIzol reagent volume), resuspend the pellet in GTC solution, and precipitate with 1 vol of isopropanol and wash in 80% alcohol.

5. Resuspend the final pellet in 25–200 µL of DEPC-treated H$_2$O (*6*).

U1snRNA   **U1anti-OC**

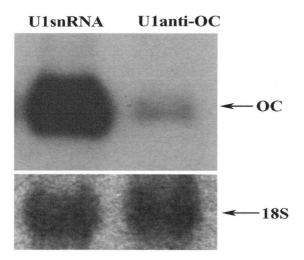

Fig. 7. An example of reduced expression of endogenous gene by U1 antitarget. U1 antiosteocalcin (OC) remarkably reduces the osteocalcin expression in ROS17/2.8 cells.

3.9. Northern Analysis

1. Electrophorese 5–10 µg of RNA in a formaldehyde/MOPS denaturing 1–1.5% agarose gel for 3 h at 85 V.
2. Transfer the RNA to an S&S nylon-reinforced nitrocellulose membrane overnight by capillary action in 10X SSC.
3. Wash the blot for 10 min in 10X SSC to remove any formaldehyde and gel particles and then UV crosslink two times at 1200 µJ using a shortwave UV illuminator. Dry the blot completely prior to hybridization.
4. Use 3×10^6 to 5×10^6 counts per milliliter of probe in hybridizations for 12 h at 42°C. Wash the blots and expose to X-ray film according to the manufacturer's instruction (*see* **Fig. 7**).

4. Notes

1. Although any 10-bp segment can be chosen for the targeted sequence, we prefer to use a sequence similar to the consensus 5′ss recognizing sequence but having at least 1- to 2-bp mismatches. The mismatches should be placed in the middle of the segment (*see* **Subheading 3.1.1.**).
2. Alternatively, the human U1 gene can be cloned into pBC-SK(+/–) (Stratagene, La Jolla, CA) (*see* **Fig. 8**). To avoid the variability of stable cotransfection, the hygromycin-resistance gene was inserted into the *Acc*I site downstream of the U1 snRNA after ablation of an *Xba*I site in the RSV gene. To clarify the selection of a clone containing the modified U1 snRNA insert, a 2.6-kb *Bgl*II–*Xba*I stuffer fragment was substituted for the 10-bp splice donor site to create the U1 snRNA stuffer vector.

   In this case, the 5′ primer (mutagenic primer) contains the *Bgl*II site that is present at –8 bp in the U1 promoter. The 3′ primer extends beyond the *Xba*I site,

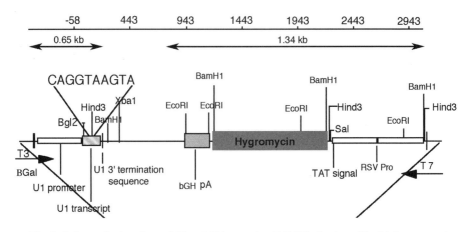

Fig. 8. Schematic drawing of U1 snRNA gene in pBC SK+/– plasmid with hygromycin-selection gene.

which is downstream of the U1 termination sequence. PCR fragments are cut with *Bgl*II/*Xba*I and replace the 2.3-kb stuffer fragment to produce a set of modified U1 snRNAs, designated U1 anti-target(xxx). DNA sequencing was performed to confirm that the mutations were successfully introduced into the U1 recognition sequence. This version of U1 snRNA contains the *Hin*dIII site in loop 3, which can be used to distinguish the modified U1snRNA from endogenous one by RNase protection assay (*see* **Subheading 3.6.**).

Acknowledgments

This work was supported by a grant from PHS, NIH, R01-AR30426, and from the Osteogenesis Imperfecta Foundation.

References

1. Will, C. L. and Luhrmann, R. (2001) Spliceosomal UsnRNP biogenesis, structure and function. *Curr. Opin. Cell. Biol.* **13,** 290–301.
2. Robberson, B. L., Cote, G. J., and Berget, S. M. (1990) Exon definition may facilitate splice site selection in RNAs with multiple exons. *Mol. Cell. Biol.* **10,** 84–94.
3. Konforti, B. B., Koziolkiewicz, M. J., and Konarska, M. M. (1993) Disruption of base pairing between the 5′ splice site and the 5′ end of U1 snRNA is required for spliceosome assembly. *Cell* **75,** 863–873.
4. Ashe, M. P., Griffin, P., James, W., and Proudfoot, N. J. (1995) Poly(A) site selection in the HIV-1 provirus: inhibition of promoter-proximal polyadenylation by the downstream major splice donor site. *Genes Dev.* **9,** 3008–3025.
5. Ashe, M. P., Pearson, L. H., and Proudfoot, N. J. (1997) The HIV-1 5' LTR poly(A) site is inactivated by U1 snRNP interaction with the downstream major splice donor site. *EMBO J.* **16,** 5752–5763.

6. Chomczynski, P. and Sacchi, N. (1987) Single-step method of RNA isolation by acid guanidinium thiocyanate-phenol-chloroform extraction. *Anal. Biochem.* **162,** 156–159.
7. Furth, P. A., Choe, W. T., Rex, J. H., Byrne, J. C., and Baker, C. C. (1994) Sequences homologous to 5′ splice sites are required for the inhibitory activity of papillomavirus late 3′ untranslated regions. *Mol. Cell. Biol.* **14,** 5278–5289.
8. Gunderson, S. I., Polycarpou-Schwarz, M., and Mattaj, I. W. (1998) U1 snRNP inhibits pre-mRNA polyadenylation through a direct interaction between U1 70K and poly(A) polymerase. *Mol. Cell.* **1,** 255–264.
9. Fortes, P., Cuevas, Y., Guan, F., et al. (2003) Inhibiting expression of specific genes in mammalian cells with 5′ end-mutated U1 small nuclear RNAs targeted to terminal exons of pre-mRNA. *Proc. Natl. Acad. Sci. USA* **100,** 8264–8269.
10. Beckley, S. A., Liu, P., Stover, M. L., Gunderson, S. I., Lichtler, A. C., and Rowe, D. W. (2001) Reduction of target gene expression by a modified U1 snRNA. *Mol. Cell. Biol.* **21,** 2815–2825.
11. Liu, P., Gucwa, A., Stover, M. L., Buck, E., Lichtler, A., and Rowe, D. (2002) Analysis of inhibitory action of modified U1 snRNAs on target gene expression: discrimination of two RNA targets differing by a 1 bp mismatch. *Nucleic Acids Res.* **30,** 2329–2339.
12. Liu, P., Kronenberg, M., Jiang, X., and Rowe, D. (2004) Modified U1 snRNA suppresses expression of a targeted endogenous RNA by inhibiting polyadenylation of the transcript. *Nucleic Acids Res.* **32,** 1512–1517.
13. Murphy, J. T., Burgess, R. R., Dahlberg, J. E., and Lund, E. (1982) Transcription of a gene for human U1 small nuclear RNA. *Cell* **29,** 265–274.
14. Murphy, J. T., Skuzeski, J. T., Lund, E., Steinberg, T. H., Burgess, R. R., and Dahlberg, J. E. (1987) Functional elements of the human U1 RNA promoter. Identification of five separate regions required for efficient transcription and template competition. *J. Biol. Chem.* **262,** 1795–1803.

Index